PRACTICAL STATISTICS

PRACTICAL STATISTICS
For Non-Mathematical People

DR RUSSELL LANGLEY

DAVID & CHARLES   NEWTON ABBOT

ISBN 0 7153 5039 0

First published 1968
by PAN BOOKS LTD

2nd (*Revised*) *Printing, 1970*

First hardback edition 1971
DAVID & CHARLES (PUBLISHERS) LIMITED

Printed in Great Britain
by Redwood Press Limited Trowbridge Wilts
for David & Charles (Publishers) Limited
Newton Abbot Devon

PREFACE

THIS IS A book on Statistical Inference. It deals with methods of extracting the truth from numerical data. It should suit anyone who wants a straightforward account of this most important subject.

Knowledge in this field has grown tremendously in the past few decades, and the time is ripe for a 'translation for laymen'. It has therefore been my aim to present, in plain English, the basic concepts and a carefully selected set of significance tests, particularly for the benefit of those who are not mathematically minded. Technical terms have been avoided as much as possible, and the choice of material has been governed by three considerations – usefulness, simplicity, and safety.

All procedures are illustrated by examples which are worked out in full, and plenty of questions (with answers) are provided to enable readers to become proficient with the techniques described.

I am very grateful to Professor E. J. Williams and Miss Alison Doig of the Statistics Department, Melbourne University, and to Mr H. J. Halstead of the Mathematics Department, Royal Melbourne Institute of Technology, for their advice on a number of problems, and to those authors and publishers who have kindly allowed me to make use of their examples and Tables, as acknowledged in the text.

I am indebted to the Literary Executor of the late Sir Ronald A. Fisher, FRS, Cambridge, to Dr Frank Yates, FRS, Rothamsted, and to Messrs. Oliver & Boyd Ltd, Edinburgh, for permission to adapt Tables 27 and 28 from their book *Statistical Tables for Biological, Agricultural, and Medical Research*.

Special thanks are also due to my secretary, Miss Barbara Mills, for her typing, checking of all Tables, and for helpful criticisms of the manuscript from a non-mathematician's point of view.

RUSSELL LANGLEY
M.B., B.S., D.D.M., M.A.C.D.

82 Collins St.,
Melbourne

NOTE

The word 'data' is a Latin plural noun, but nowadays it is commonly used in English as a group term in the singular, and will be used as such throughout this book.

CONTENTS

CHAPTER 1

INTRODUCTION

On Being Misled by Numbers

WHO HASN'T BEEN fooled by numbers at one time or another? For numbers are peculiar things. On the one hand, they are undoubtedly essential for the precise description of many observations ('rockets go terribly fast' is rather vague, isn't it?), and yet on the other hand, we all know that numbers can be very misleading at times.

Some people even get to the stage of mistrusting all numerical observations. You will hear them say, 'Go on, you can prove almost anything you want to with figures.' Which implies, of course, that you can prove almost nothing with figures. I have even heard it said that with Statistics you can prove that a man is perfectly comfortable when he is standing with one foot in a pail of iced water, and the other foot in a pail of boiling water! Such jibes are due to ignorance, born out of the unhappy experience of being misled by figures in the past. But surely the answer to this is to learn enough about figures to make sure that you won't be duped again.

This book deals with this particular problem. It might even have been subtitled, 'How to Avoid Being Misled by Numbers'.

There are 8 ways in which numerical data is likely to be misleading, viz. –

(1) Arithmetical errors.
(2) False percentages.
(3) Fictitious precision.
(4) Misleading presentation.
(5) Incomplete data.
(6) Faulty comparisons.
(7) Improper sampling.
(8) Failure to allow for the effect of chance.

Let us begin by looking at the first 6 traps; the last 2 on this list will be dealt with in detail in subsequent chapters.

Arithmetical Errors

It is well known that we tend to accept things in print as being true, chiefly on the strength of the fact that they have been printed. This applies both to numbers and to ideas. Yet, in spite of all due care, arithmetical mistakes do occasionally creep into print, so if the subject is one which matters to you, it is best to check the author's calculations before accepting them.

In a critical article on the first Kinsey Report, Professor W. A. Wallis pointed out that there were so many arithmetical mistakes that it was not even clear how many men had actually been studied. On one page of the Report it is stated that the observations were made on a total of 12,214 men, while on another page is a map showing 427 dots, each of which is said to represent 50 men; if so, there were $50 \times 427 = 21,350$ altogether. Or again, one table shows the number of men 30 years of age or less as being 11,467, while in the very next table the same group total is shown as 11,985. When two such figures differ, it stands to reason that at least one of them must be wrong. (*Journ. Amer. Statist. Assoc.*, 1949, pp. 463–84.)

False Percentages

We all learnt something about percentages when we were young. But sometimes we forget little details, which in the present case would leave us wide open for swallowing a heap of false figures. For instance –

(a) *Beware of adding percentages.* 'The price of men's haircuts must be increased. In the past 2 years, wages have risen 10%, combs, brushes, and other materials have gone up 8%, shop rentals have gone up 10%, and electric light bills have gone up 5% – a total rise of exactly 33%.' But this total is wrong. If each of the items making up the cost of each haircut had risen 10%, the total cost would only rise by 10%.

(b) *Beware of decreasing percentages.* 'Apples are 100% cheaper than last year.' Does this really mean they're giving apples away free? For 100% less than any quantity is zero.

How about this one: 'Because of the shocking weather conditions, this year's wheat crop is 120% less than last year's.' This is quite impossible, for this year's crop can't be less than zero.

All percentage changes must be based on the original level. So, a rise in wages from £20 to £30 is a 50% increase; if wages now fall to £20 again, the downgrading is a fall of £10 from £30, which is a 33% decrease.

(c) *Beware of huge percentages.* 'J. B. earns 1,000% more than Smithy.' Sounds a colossal difference, doesn't it? Yet it is exactly equivalent to saying '11 times'. (Not '10 times', as might be thought, for 100% more = twice, 200% more = 3 times, and so on.) You can take it that, as a general rule, people using huge percentages are doing so to exaggerate their claim. In which case, they are apt to be biased, anyway.

(d) *Beware of percentages unaccompanied by the actual numbers.* 'In a special experiment, we found that 83·3% people got relief from Dumpties within 60 seconds.' They conveniently forgot to mention that the experiment concerned 6 people, 5 of whom got the stated relief. And if you test enough small groups, sooner or later you're almost certain to get one group to suit your purpose, purely by chance.

Fictitious Precision

Surely no one would be fooled by the apparent accuracy of a figure given in the World Almanac of 1950, that there were 8,001,112 people in the world who spoke Hungarian. I like that final 12. It suggests that when this count was made, exactly 12 toddlers had just learned to say 'Pa-Pa' (which is 'Dad-Dad' in Hungarian).

However, this same kind of fault can appear in much more sophisticated forms, as illustrated in the following excerpt from *How to Lie with Statistics* by Darrell Huff (Gollancz, 1962) –

Ask a hundred citizens how many hours they slept last night. Come out with a total of, say, 783·1. Any such data is far from precise to begin with. Most people will miss their guess by fifteen minutes or more, and there is no assurance that the errors will balance out. We all know someone who will recall five sleepless minutes as half a night of tossing insomnia. But go ahead, do your arithmetic, and announce that people sleep an average of 7·831 hours a night. You will sound as if you knew precisely what you were talking about.

So don't be too impressed by a result simply because it is quoted to 10 or so decimal places. Make sure that the degree of precision claimed is warranted by the evidence.

You will often be able to detect fictitious precision by asking: *How could anyone have found that out?*

While a healthy scepticism is desirable, you must be prepared to sometimes come across results which at first seem to be incredible, and yet which are true. For example, if someone tells you that there are 3,300 fish in a certain lake, you are entitled to wonder how anyone could possibly know such a thing. Yet it could be quite a reliable figure. It is found as follows. Catch 100 fish in the lake, tag them with special markers, and put them back in the lake. Return a couple of months later, catch another sample of 100 fish, and see how many of this sample are tagged fish from the first catch. Suppose there are 3. Then it can be induced that if 100 tagged fish represent $\frac{3}{100}$ of the total fish population of the lake (i.e. 100 = $\frac{3}{100}$ total), this total must be $\frac{100}{3} \times 100 = 3,333$, or in round figures, 3,300.

Misleading Presentation

One of the tricks about numbers is that there is often a variety of ways of presenting the same numerical fact, and some of these ways seem to suggest a different conclusion from others. As Darrell Huff says –

You can, for instance, express exactly the same fact by calling it a 1% return on sales, a 15% return on investment, a $10,000,000 profit, an increase in profits of 40% (compared with 1935–39 average), or a decrease of 60% from last year. The method to choose is the one that sounds best for the purpose at hand!

Even diagrams, charts, and graphs, which are excellent for presenting a numerical message so that it can be noted at a single glance, are not immune from malpresentation. Sometimes it seems that the man who prepared the chart was really hoping you'll only take a single glance, for if you look twice you may notice that units are omitted, or that the values shown do not agree with those in the body of the text (you can always call it a printing error if you're caught at this one!), or most extraordinary of all, the neat conjuring trick which Huff calls the 'gee-whiz graph' (see Fig. 1).

Incomplete Data

The little figures that aren't mentioned can result in an awful lot of numerical distortion.

Time Magazine (June 12, 1964) quoted 'some sober statistics compiled in recent months by various state authorities' concerning the safety of driving in large cars versus small cars. Three independent reports showed the risk of being killed in an accident was up to 5½ times greater for persons in small cars as compared with large cars. This would make a good advertizing point if you were selling large cars,

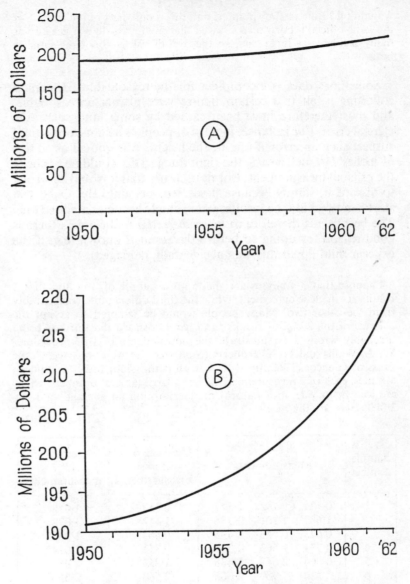

Fig. 1. Graph A shows a gentle increase of 15% over 12 years. Exactly the same data is presented as a 'gee-whiz graph' in Graph B, simply by expanding the vertical scale. Beware of graphs which lack zero points.

wouldn't it? But, as *Time* pointed out, this is only part of the story. For the same official reports also showed that small cars do not get into as many accidents as large cars, so the overall risk is about the same in both.

Sometimes data is incomplete for the reason that the person collecting it felt that certain figures were plainly unreasonable, and must therefore have been caused by some fault such as a clerical error. For instance, in a list of people's heights, one would suspect such an error if one of the heights was quoted as '8 feet 6½ inches'. In such a case, the right thing to do would be to check the original measurement. But note clearly that it would be wrong to discard it, simply because it seemed very unlikely. Once you start hand-picking the results, your sample becomes a biased one. The best rule is therefore to *never discard a result* unless there is good reason for doing so before the result is known (e.g. if the experimental apparatus was accidentally damaged).

Suppose that 3 analyses are made on a sample of ore, and that 2 results are in close agreement, while the third differs quite considerably from the other two. Many people would be tempted to accept the average of the 2 closest results, and would discard the third as being 'probably wrong'. To illustrate the unsoundness of this procedure, W. A. Wallis and H. V. Roberts (*Statistics – A New Approach*, Free Press of Glencoe, 1960, pp. 140–1, with permission) took 10 random samples, each of 3 measurements, from a large table of numbers which are known to vary in a natural manner around an average value of 2·000. Here are the results –

Sample number	Measurements			Average of 2 closest measurements	Average of all 3 measurements
1	0·724	0·782	1·547	0·753	1·018*
2	1·682	1·201	0·336	1·442*	1·073
3	0·623	0·743	2·495	0·683	1·287*
4	4·334	1·663	0·045	0·854	2·014*
5	0·864	2·642	5·436	1·753*	2·981
6	2·414	1·989	2·666	2·540	2·356*
7	1·506	2·364	0·763	1·134	1·544*
8	3·048	2·037	2·759	2·904	2·615*
9	2·347	4·816	1·536	1·942*	2·900
10	2·637	2·563	1·893	2·600	2·364*

In each case, the average which is closest to the true average of 2·000 is marked with an asterisk. The Table shows that averaging the 2 closest measurements gives the better result in 3 cases, whereas averaging all 3 measurements gives the better result in 7 cases. And this is so, in spite of the fact that in 5 of the samples there is a distinct temptation to discard a 'wild' measurement (as in Sample #3).

As Wallis and Roberts point out, the ultimate folly of rejecting an extreme observation was demonstrated when 'shortly after 7 o'clock on the morning of December 7, 1941, the officer in charge of a Hawaiian radar station ignored data solely because it seemed so incredible'. For those of you too young to remember this incident, it refers to the surprise attack on Pearl Harbour by Japanese bombers, by which they declared war on the USA.

Faulty Comparisons

Apart from being used to describe things, numbers are often used for comparing things. Whenever this is done, care must be taken to ensure that the things being compared are genuinely fit to be compared. Darrell Huff (*loc. cit.*) gives a couple of good examples of illogical comparisons –

The death rate in the American Navy during the Spanish–American War was 9 per 1,000. For civilians in New York City during the same period it was 16 per 1,000. This suggests that it was safer to be in the Navy than out of it. But the groups are not really comparable. The Navy is made up mostly of young healthy men, whereas the civilian population includes infants, the old, and the ill, all of whom have a higher death rate wherever they are.

Hearing that it cost $8 a day to maintain each prisoner in Alcatraz, a U.S. senator exclaimed, 'It would be cheaper to board them in the Waldorf-Astoria!' Well, it wouldn't really, because it's not fair to compare the total maintenance cost per prisoner at Alcatraz with the rent of a hotel room; after all, guarding and feeding prisoners must cost something.

The trick in these examples is to compare 2 things which sound as if they are fit to be compared when, in fact, they are not. The very preciseness of the numbers themselves helps to carry the illusion. How about this newspaper report –

The figures just released by the National Safety Council show that the most reckless age for car drivers is 20 to 29 years. This age group accounted for 31·6% of the accidents on our roads last year, compared with 23·3% for the 30 to 39 years group, 16·2% for the 40 to 49 years group, 9·4% for the 50 to 59 years group, 11·0% for the 60 and over group, and an exemplary 8·5% for the under 20 year-olds.

Looks like those kids were full of caution, while their older brothers were full of beer. But did anyone say there were equal numbers of drivers in each of these age groups? Because other-wise these figures may indicate nothing more than the relative number of drivers in each age group.

However, the usual cause for being misled by comparisons is either that the things being compared are biased samples, or that the effect of chance has not been properly assessed. Which brings us to the main subject matter of this book.

CHAPTER 2

NATURE OF PROBABILITY

Absolute and Probable Truth

THE ONLY KIND of conclusion which can be absolutely true is one which is *implied* by the premisses on which it rests. The conclusion in such a case is really contained within the meaning of the premisses. A simple example of such a deduction is that if $A > B$, and $B > C$, then A must be larger than C. (The sign '$>$' means 'is larger than'.)

A nice instance of absolute truth is the conclusion reached by M. Cohen and E. Nagel (in *An Introduction to Logic*, Routledge & Kegan Paul Ltd., 1963) that there are at least two persons in New York City who have the same number of hairs on their heads. This piece of absolute truth was not discovered by counting the hairs on the eight million inhabitants of that city, but by studies which revealed that (1) the maximum number of hairs on human scalps could never be as many as 5,000 per square centimetre, and (2) the maximum area of the human scalp could never reach 1,000 square centimetres; from these premisses one can correctly infer that no human being could ever have $5,000 \times 1,000 = 5,000,000$ hairs on his head. As this number is less than the population of New York City, it follows by *implication* that at least two New Yorkers must have the same number of scalp hairs.

Of course, you can always question the truth of the underlying premisses, although in the present instance your chances of finding them wrong would be comparable with the likelihood of finding a man 35 feet tall. At a practical level, this means no chance at all.

Nevertheless, in the vast majority of cases we must be satisfied with conclusions based on incomplete evidence. For example, atoms can't be seen, so our belief in the atomic structure of matter rests on indirect evidence. Yet this belief is almost certainly true, because a multitude of observations all point to the same conclusion, and a host of predictions based on the atomic hypothesis has also come true to act as additional confirmation.

As time goes by and more evidence accumulates, the closer our conclusions will approach absolute truth. If contrary evidence crops up, our conclusions must be revised.

The degree of probability of a thing being true is the subject we shall now investigate. It is a fascinating study, and though its origins go back over 300 years, it is only now beginning to make its impact felt on our everyday lives. It goes by the fancy name of Statistical Inference, but we shall be avoiding technical names in the interests of simplicity and clarity. Of one thing you can be sure – you'll be hearing a lot more about this subject in the future, and the people who count will be ones with a working knowledge of it. H. G. Wells (d. 1946) foresaw this when he said –

Statistical thinking will one day be as necessary for efficient citizenship as the ability to read and write.

The Laws of Chance

The study of probability began during the Italian Renaissance when men sought to develop systems for winning at dice. Gamblers consulted scholars for assistance, and the first clear exposition on the subject was written by none other than the famous mathematician–astronomer, Galileo Galilei, in about 1620. The problem posed was this. If 3 dice are thrown at the same time, what is the total score which will occur most frequently? Experienced dice players believed 10 and 11 were the commonest totals, but was this really so?

Galileo's answer is contained in a 4-page essay entitled *Sopra le Scoperte dei Dadi*, which appears to have been written for his patron, the Grand Duke of Tuscany. A translation of this essay is given in Professor F. N. David's *Games, Gods and Gambling* (Griffin & Co., 1962, Appendix 2), and it is discussed by the same author in *Biometrika Journal*, 1955, pp. 11–12. Galileo begins by pointing out that a die (=the singular of 'dice') has 6 faces, and when thrown it can equally well fall on any face; in other words, there are 6 possible outcomes with any throw, and all are equally likely if the die is honest. The probability of showing any specified face is therefore 1 chance in 6 throws. But, he goes on, if a second die is also thrown, each face of the first die can be combined with each face of the second one, to form a total of $6 \times 6 = 36$ possible combinations, which can be tabulated like this –

I	II	I	II	I	II	I	II	I	II	I	II
1	1	2	1	3	1	4	1	5	1	6	1
1	2	2	2	3	2	4	2	5	2	6	2
1	3	2	3	3	3	4	3	5	3	6	3
1	4	2	4	3	4	4	4	5	4	6	4
1	5	2	5	3	5	4	5	5	5	6	5
1	6	2	6	3	6	4	6	5	6	6	6

Galileo explained that each one of these combinations (1 and 1, 1 and 2, etc.) could be expected to appear once in every 36 throws. Now it is obvious that if the scores of the two dice are added together, the possible totals are not all equally likely. For instance, to make a total of 2, you must throw a 1 and 1, and this combination will only be likely to occur once in 36 throws, whereas a total score of 4 can be made with a 1 and 3, a 2 and 2, or a 3 and 1, so this total could be expected on an average of 3 times in every 36 throws. Therefore if you bet on a total of 4, you will win three times oftener than on 2.

He was then able to say that when 3 dice are thrown together, each of the above 36 pairs could combine with any of the 6 faces of the third die, so that there will be $6 \times 6 \times 6 = 216$ possible outcomes. By adding up the scores of each possible combination (the totals range from 3 to 18), he showed that totals of 10 and 11 would in fact be the equally commonest outcomes, but they were only going to beat their nearest challengers (9 and 12) by a tiny margin – once in 108 times. We can only marvel at those gamblers who had detected this difference in frequency (a difference of less than 1%) purely from practical experience!

This short essay was the only thing that Galileo wrote on the subject of probability, but in it he exposed, with great clarity of reasoning, the 4 fundamental Laws of Chance. The whole science of Statistics has been built on these foundations, so let us have a closer look at them.

First Law – The Proportionate Law

Whenever something (such as throwing a die) can have more than one result, if all the possible results have an equal chance of occurring, the probability of any one of them occurring in a single trial will be the proportion which that particular result bears to all the possible results.

Galileo said this in so many words when he explained that a die has 6 faces, and each has the same chance of being on top when the die stops rolling, so the probability that any specified face will show with a single throw of the die will be 1 chance in 6.

Actually, this Law really amounts to a technical definition of *probability* as *relative frequency*, or proportion of times a certain result is to be expected. It applies not only to potential outcomes such as the result of die or coin tossing, but also and with equal force to the outcome of unbiased sampling from existing groups of things. With 1 ace of hearts in a pack (=deck) of 52 playing cards, the probability of picking this ace blindly from the pack is thus 1 chance in 52; this can also be described as a probability of $\frac{1}{52}$ or 0·0192, or 1·92%.

In simple cases like unbiased coin tossing, the probability of a specified result can be foretold from the nature of the procedure, but there are many practical situations (comparable with tossing a bent coin) in which the probabilities can only be found by observing the results of actual trials.

Second Law – The Law of Averages

Whenever something (such as throwing a die) can have more than one result, if all the possible results have an equal chance of occurring, the results that will be observed in a number of trials (throws) will generally vary to some extent from the inherent proportions, but the extent of this variation will become progressively less as the number of trials is increased.

This Law was not specifically mentioned by Galileo, but he must have understood it intuitively to have arrived at his conclusions. Anyway, surely everyone knows that if you throw a die 6 times, you will rarely get a 1, 2, 3, 4, 5, and 6 (in any order) in those 6 throws.

Look at it this way. The Proportionate Law states that if you throw an unbiased die, the probability of getting any one score (say, 2) is 1 chance in 6 throws. Now note carefully that this does *not* mean that in 6 throws you must always get exactly one 2. Try it, and you will find that in some groups of 6 throws there will be no 2s at all, sometimes there will be one 2, sometimes two 2s, sometimes three 2s, and occasionally four or more 2s in the batch. But, in accordance with the Proportionate Law, the commonest number of 2s in batches of 6 throws will be one; the average will thus be $\frac{1}{6}$.

Likewise with coin tossing. It is not justifiable to say that a coin or the method of tossing is biased simply because you don't get exactly 2 heads in a group of 4 tosses. True, the probability of getting a head is $\frac{1}{2}$, but even with 1,000 tosses you may not get exactly 500 heads. But let us see what happens to the proportion of heads when the sample size is increased. I tossed a coin 10 times and got 4 heads; this is a proportion of $\frac{4}{10} = 0\cdot40$. I then did a second trial, tossing the coin 200 times, which yielded 106 heads; this is a proportion of $\frac{106}{200} = 0\cdot53$. Here you see the Law of Averages at work, for in the smaller sample the observed proportion differed from the expected 0·50 by 0·10, whereas the proportion of heads in the larger sample differed from 0·50 by only 0·03.

Notice that it is the observed *proportions* that approach the theoretical expectation. The difference between the observed and expected *numbers* actually increases as the number of trials increases. Thus in the above example, the small sample showed 4 heads when the expected number was 5, a difference of 1, while in the large sample there were 106 heads instead of the expected 100, which is a difference of 6.

This tendency to vary from the exact proportions also applies to sampling from existing groups of things. Thus if there is an equal number of red- and black-faced cards in a pack of playing cards, the probability of drawing a red card blindly from the pack will be $\frac{1}{2}$. But a draw of 4 cards will not always produce 2 reds and 2 blacks. However, if you repeated this experiment 1,000 times you would certainly find that this 2-and-2 result was the commonest outcome, and the proportions of total red and black cards would slowly but surely get closer and closer to the 50% mark as the experiment proceeded.

This tendency for any particular set of observations to vary from the exact proportions is attributed to Chance. It is very important, and we shall have a lot more to say about it later on.

Third Law – The Addition Law

Whenever something (such as throwing a die) can have more than one result, the probability of alternative results occurring in a single trial will be the sum of their individual probabilities.

This can be illustrated by calculating the probability of getting *either* a 2 *or* a 3 in a single throw of a die. Each number has a chance of $\frac{1}{6}$, so this Law tells us that the chance of getting one or other score will be –

$$\frac{1}{6} + \frac{1}{6} = \frac{2}{6}, \quad \text{i.e. 1 chance in 3.}$$

Galileo demonstrated this Law when he showed that the chance of making a total score of 4 with 2 dice is the chance of getting a 1 and a 3, plus the chance of getting a 2 and 2, plus the chance of getting a 3 and 1, which is altogether 3 chances in 36.

This Law also applies to groups of existing things. Thus the chance of drawing an ace of hearts *or* the ace of spades blindly from a normal pack of playing cards will be –

$$\tfrac{1}{52} + \tfrac{1}{52} = \tfrac{2}{52}, \text{ i.e. 1 chance in 26.}$$

Fourth Law – The Multiplication Law

Whenever something (such as throwing a die) can have more than one result, the probability of getting any particular combination of results in 2 or more independent trials (whether consecutively or simultaneously) will be the product of their individual probabilities.

The first thing to appreciate about this Law is that it makes no difference to the probabilities whether the events occur one after the other or at the same time (so long as they are independent, i.e. the result of each event exerts no influence on the outcome of the others). Thus if the probability of getting two 1s by throwing a die twice works out at 1 chance in 36, the probability of getting two 1s by throwing a pair of dice together will also be 1 chance in 36.

Galileo demonstrated this Law clearly when he tabulated all the possible combinations that can occur when dice are thrown simultaneously, in the manner shown on page 21. The Multiplication Law simply puts into words what is obvious from that tabulation, namely that as the probability of getting any particular number on each die is 1 chance in 6 (i.e. $\tfrac{1}{6}$), the probability of getting any particular combination of numbers with the 2 dice (say a 3 with the first die, and a 1 with the second die) will be $\tfrac{1}{6} \times \tfrac{1}{6} = \tfrac{1}{36}$, i.e. 1 chance in 36 throws.

Notice that the probability of getting a score of 3 and 1 with 2 dice *if the order is not specified* will be 1 chance in 18, for this score can be made in 2 ways (a 3 with the first die and a 1 with the second die, or a 1 with the first die and a 3 with the second die); each of these ways has a probability of $\tfrac{1}{36}$, so these probabilities must be added in accordance with the Addition Law –

$$\tfrac{1}{36} + \tfrac{1}{36} = \tfrac{1}{18}$$

As with the other Laws of Chance, the Multiplication Law applies not only to the outcome of such things as die or coin tossing, but also to the outcome of taking samples from existing groups of things. Thus the probability of drawing any specified card (say, the jack of diamonds) from a shuffled pack of 52 playing cards is $\frac{1}{52}$, so the probability of getting a specified card from each of 2 such packs (say, the jack of diamonds from one pack, and the queen of diamonds from the other one) will be the product of their individual probabilities (which is the same as their proportions), which is –

$$\frac{1}{52} \times \frac{1}{52} = \frac{1}{2,704}, \quad \text{i.e. 1 chance in 2,704.}$$

An amusing illustration of the Multiplication Law is given by M. J. Moroney in *Facts From Figures* (Penguin, 1962). A young man told his girl-friend: 'Statistics show, my dear, that you are one in a billion.' This compliment was perfectly true, for she had the 4 qualities that he had been searching for, namely –

> Grecian nose   0·01
> Platinum blonde hair   0·01
> Eyes of different colours   0·001
> First-class knowledge of Statistics   0·00001

Beside each feature is shown its incidence in the female population. The chance of finding such a combination will therefore be the product of these proportions –

$$0·01 \times 0·01 \times 0·001 \times 0·00001 = 0·000000000001$$

– which is precisely 1 chance in an English billion.

On a more serious plane, *Time Magazine* (January 8, 1965) reported that a young prosecutor made legal history by using the above technique to obtain a conviction against a Californian couple charged with robbery. A witness saw a blonde white woman with a pony-tail hairdo running from the scene of the crime and departing in a yellow car driven by a bearded Negro. Police arrested a married couple who fitted the above description and who owned a yellow car. The prosecutor explained the Multiplication Law to the jury, and proceeded to estimate that, in the city concerned, the probability of a white woman being blonde was 1 in 4; this was multiplied by the probability of a white woman wearing her hair in a pony-tail, and in turn by the probability of a man being a Negro, and of having a beard, and of being in possession of a yellow car, and finally by the 1 in 1,000 probability that a Negro man would be accompanied by a white woman. This calculation reduced the chance of finding another such couple in that city down to 1 in 12 million, which was accepted by the jury as circumstantial evidence of proof of identity beyond any reasonable doubt, and jail sentences were imposed.

When taking samples from existing groups of things, the group is reduced in size as sampling proceeds. Allowance must be made for this if the sample reduces the parent group to any appreciable extent (say, 10% or more). For example, suppose a bag contains 2 black marbles and 4 white ones; the marbles are identical in all respects except colour. If 2 marbles are drawn blindly from the bag, what is the probability that they will both be black ones? In a case like this, we must allow for the fact that the first marble drawn is not put back into the bag. We start with 6 marbles in the bag, 2 of which are black, so by the Proportionate Law, the probability of drawing a black marble in the first draw will be $\frac{2}{6}$. If the first draw gives a black marble, the bag will then contain 1 black and 4 white marbles. This means that there will be 5 possible outcomes of the second draw, of which drawing a black marble will be 1. The probability of drawing a black marble in the second draw will therefore be $\frac{1}{5}$. The Multiplication Law then tells us that the probability of both of these events happening (i.e. of both drawn marbles being black) will be the product of $\frac{2}{6}$ and $\frac{1}{5}$, which is –

$$\frac{2}{6} \times \frac{1}{5} = \frac{2}{30} = \frac{1}{15}, \quad \text{i.e. 1 chance in 15.}$$

These basic Laws of Chance can be extended to even more complex problems. For example –

For many years doctors have been seeking a test which will detect any cancer in a person. So far, the search has proved as fruitless as the old alchemist's search for the Philosopher's Stone, but suppose that one day such a test is discovered and proves to be 95% reliable. By this figure is meant that the test will be positive in 95% of persons who have cancer, and negative in 95% of persons who do not have cancer. If this test is applied to a large group of patients of whom 0·5% have cancer, what is the probability that a patient with a positive test has really got cancer?

We have said that the probability that a person in the group has cancer is 0·5%, which is a proportion of $\frac{5}{1000}$ or 0·005. We know, too, that the probability that a person with cancer will have a positive test is 95%, which is 0·95. Now the Multiplication Law tells us that the probability of both of these happening together is the product of their individual proportions (or probabilities), thus –

$$0·005 \times 0·95 = 0·00475$$

But we must also allow for the possibility that the person has not got cancer (100 − 0·5 = 99·5% = 0·995 of the test group), and yet gives a positive test, which we are told can happen in 5% = 0·05 of the non-cancer patients. The probability of this combination will be –

$$0·995 \times 0·05 = 0·04975$$

The probability that any person (with or without cancer) will give a positive test will therefore be the sum of these, in accordance with the Addition Law, thus –

$$0.00475 + 0.04975 = 0.05450$$

The probability that a positive test will be due to cancer will therefore be the proportion –

$$\frac{\text{Probability of cancer and a positive test}}{\text{Probability of a positive test}} = \frac{0.00475}{0.05450} = 0.09$$

This means that only 9 out of every 100 positive tests will actually be due to cancer. Such a test would not be of much use for the mass screening of a population. The reliability would therefore have to be much greater than 95% to be of practical importance. (Example from Dr J. Dunn and S. Greenhouse, *Cancer Diagnostic Tests*, USA Public Health Service Publication #9, 1950.)

One further thing needs to be mentioned before we leave this section. We are now in a position to realize that since the word 'probability' means 'relative frequency', the idea of the outcome of a single event having a certain probability has only got a general meaning. When we say that the probability of getting a head in one particular toss of a coin is $\frac{1}{2}$, we mean that in *the long run* such tosses give heads $\frac{1}{2}$ of the time. Think about this for a moment, for if you don't get this point, you may well fall into 'the gambler's fallacy', as follows.

An old story tells of a gambler living in one of the French provinces who was about to become a father. As the exciting event grew near, he happened to notice that a whole succession of girls was being born in his village. By the time his wife was due, there had been 19 girls born in a row. Now the gambler knew from the Multiplication Law that the chance of 20 girls being born in succession, like getting 20 heads in 20 coin tosses, is $(\frac{1}{2})^{20}$, which is 1 chance in a million. What would you bet that the gambler's wife would have, a boy or a girl?

The 'gambler's fallacy' was to bet heavily in favour of the baby being a boy, on the ground that the chance of it being a girl was now only 1 in a million. His probability calculation was wrong. Each birth is an independent event, and what has gone before has no effect on what shall follow. Every individual birth has a probability of being a boy of $\frac{1}{2}$. It is only when the group of births are considered as a whole that the chances are so high. After all, the gambler was betting on the outcome of a single birth, not on the outcome of a whole group of births. The gambler would only have been justified if he had bet heavily in advance that there would not be 20 girls in a group of 20 successive births.

Questions

From now on, groups of questions will be found at the end of each section. Do not proceed until you have answered them correctly, for they will exemplify and expand statements in the text. By keeping all your calculations together in an exercise book, you will compile a very useful supplement to this book. There should be no snags, for the arithmetic has purposely been kept as simple as possible, and nothing will be asked which is not dealt with in the text. Answers to all questions (often with comments) are given on page 360 onwards. If these questions are being used for class work, any objection to the provision of these answers can easily be remedied by changing the numerical data within the framework of each question.

Now see if you have properly understood the basic Laws of Chance by testing yourself with the following questions.

Q1. If 9·1% of people at a party have false teeth, what is the probability of picking a person with false teeth in a single go at a game of Blind Man's Bluff?

Q2. What is the probability of getting 4 heads in 4 tosses of a fair coin?

Q3. A man rolls a pair of dice and gets two 6s. He then rolls them again, and again gets two 6s. How frequently would this be likely to happen if the dice are true (i.e. not loaded)?

Q4. Make a systematic list of all the possible outcomes of tossing a coin 3 times (HHH, HHT, etc.). Then say what the probability is of getting 2 heads and 1 tail, in any order, in 3 tosses.

Q5. If 2 men toss a fair coin repeatedly, one always betting on heads, the other always on tails, would they tend to end up more or less even, or would one of them eventually go broke?

Q6. If you draw 4 cards blindly from an ordinary pack of 52 playing cards, what is the chance of drawing the 4 aces, if drawn cards are not replaced in the pack?

Q7. If you toss a fair coin twice, what is the probability of getting either 2 heads or 2 tails?

Q8. A, B, and C are 3 men who go to a party wearing similar hats. They put their hats on a hall table, and then proceed to get very drunk. When the party ends, each man grabs one of the 3 hats without noticing whether it is his own or not.

(a) Write down all the different ways in which the 3 hats could be distributed among the 3 men. Then say what is the probability that they will have picked their correct hats.

(*b*) Show how you can arrive at the same answer using the Multiplication Law.

Q9. A rapid screening test for detecting skin cancer by fluorescence under ultra-violet rays was described in the A.M.A. *Archives of Dermatology* in May 1963. The test was found to be positive in 68 out of 75 proven skin cancers, and in 5 out of 191 non-cancerous lumps such as moles and warts. If this test becomes popular, and if $\frac{1}{3}$ of the patients to whom it is applied have skin cancer, what is the probability of missing a skin cancer with this test (i.e. what is the probability that a patient will have a skin cancer and yet give a negative test)?

Q10. If 2 marbles are drawn blindly from a bag containing 100 red marbles well mixed with 400 white marbles, what is the probability that the drawn marbles will be –

(*a*) Both red?

(*b*) Both white?

(*c*) If the drawn marbles are neither both red nor both white, the only other possibility is that 1 will be red and 1 will be white; what will be the probability of this combination?

*   *   *

We have seen how the probability of certain events can be calculated by applying the 4 basic Laws of Chance. In situations encountered in practice, however, the calculations required are often much more complicated. To make things easier, men have therefore sought (and found) the general formulae which underlie almost every conceivable situation. These formulae are the mathematical keys to these problems. Three of them are especially important, and will be dealt with now; do not take fright if they are mathematically beyond you, for later on we shall see that their practical application can be made about as simple as looking up someone's telephone number.

Binomial Formula

The word 'binomial' means 'consisting of 2 names or classes'. It is a convenient descriptive term for any situation in which there are only 2 possible outcomes. Thus coin tossing is binomial, because the outcome must be either heads or tails; likewise childbirth is binomial, because the baby can only be male or female. The term is also applied to groups of things which can be divided into 2 distinct classes. For instance, a bag of marbles is a binomial group if it contains only black and white marbles; a group of people is binomial if it consists of men and women, or if it is divisible into married and single people. In each case the 2 classes are distinguished from each other by one class having some feature or characteristic and the other class lacking it: positive or negative, success or failure, present or absent, perfect or damaged, and so on.

We have seen how to calculate the probability of certain outcomes of a binomial situation by means of the basic Laws of Chance. Thus to determine the probability of getting 3 heads in 3 tosses of a coin, we apply the Multiplication Law and find that the answer is –

$$(\tfrac{1}{2})^3 = \tfrac{1}{2} \times \tfrac{1}{2} \times \tfrac{1}{2} = \tfrac{1}{8}, \text{ i.e. 1 chance in 8.}$$

To find the probability of getting 2 heads and 1 tail in a set of 3 tosses (as in Question 4, p. 28), you made a list of all the possible outcomes, and found the answer by applying the Proportionate Law. The only trouble with this method is that as the number of tosses increases, the amount of work piles up rapidly. You would need to list more than a thousand possible outcomes for 10 tosses, and over a million outcomes for 20 tosses. Obviously this is not very practical.

Fortunately, the general formula underlying such probabilities is known. If p is the probability of an event having a certain outcome (say, heads), and q is the probability of the only alternative outcome (i.e. tails), then the probabilities of various outcomes occurring in n trials are given by the terms of a formula which was first worked out by a Swiss mathematician, Jacques Bernoulli, and published in 1713 (eight years after his death). It is known as the *Binomial Formula* –

$$p^n + \frac{n}{1} p^{n-1} q + \frac{n(n-1)}{2 \times 1} p^{n-2} q^2 +$$
$$\frac{n(n-1)(n-2)}{3 \times 2 \times 1} p^{n-3} q^3 + \ldots + q^n$$

For a simple illustration of the application of this formula, consider again the question of getting 2 heads and 1 tail in 3 tosses of a coin. Here $p = \tfrac{1}{2}$, $q = \tfrac{1}{2}$, and $n = 3$, so the Binomial Formula becomes –

$$\left(\frac{1}{2}\right)^3 + \left[\frac{3}{1} \times \left(\frac{1}{2}\right)^2 \times \frac{1}{2}\right] + \left[\frac{3 \times 2}{2 \times 1} \times \left(\frac{1}{2}\right)^1 \times \left(\frac{1}{2}\right)^2\right] + \left(\frac{1}{2}\right)^3$$

The first term of this formula is the probability of getting 3 heads, the second term is the probability of getting 2 heads and

1 tail, the third term is the probability of getting 1 head and 2 tails, and the fourth term is the probability of getting 3 tails. We want the second term –

$$\tfrac{3}{1} \times (\tfrac{1}{2})^2 \times \tfrac{1}{2} = \tfrac{3}{1} \times \tfrac{1}{2} \times \tfrac{1}{2} \times \tfrac{1}{2} = \tfrac{3}{8}$$

This means the chance of getting 2 heads and 1 tail in 3 tosses will be 3 chances in 8; on the average it will occur 3 times in every 8 trials of 3 tosses each.

This formula can be applied to any binomial situation. However, we shall soon see that there are shortcuts which greatly simplify the job. But read on to get the hang of the underlying principles.

Poisson's Probability Formula

In a binomial situation we can name the proportions of the 2 constituent classes, p and q. Consider now what happens when one of these proportions becomes very, very small. Imagine a few blades of grass in a square mile of desert. The proportion of grass to bare ground is so small that it is virtually impossible to express it as a fraction; in such a case it is more convenient, and also more accurate, to work in terms of the *incidence* of the feature, so we count the blades of grass per square mile. Other examples of this kind would include the number of flaws in the insulation of a length of electric cable, the number of misprints per page of a book, the number of germs per pint of milk, the number of raisins per ounce of raisin bread, the number of fish in a pond, and so on. Situations like these, characterized by having some feature in very small proportions, should always be considered in terms of their incidence rather than their proportions, for working out their relative areas, volumes, or weights will always be more or less inaccurate. The situation is even more extreme when it comes to things which happen in a period of time, such as the arrival of people at a telephone booth, the number of accidents or breakdowns per week in a factory, the number of goals kicked in a football match, or the number of storms each year in Timbuktu. It's not hard to keep an accurate count of the number of accidents, storms, etc., but it is obviously not possible to say, except by a wild guess, how many such incidents did not occur during the observation period.

All such cases are described as being *isolated occurrences* in time or space. The number of occurrences can be stated, but not the number of non-occurrences. Yet without this last bit of information we cannot apply the Binomial Formula, for it calls for both p and q.

In 1837 a French mathematician called Siméon Poisson discovered how to modify the Binomial Formula to suit the situation of isolated occurrences, and found, rather remarkably, that when isolated occurrences happen as a result of chance (randomly, and independently of one another), if you know the average number (m) of such occurrences in the past, the probability of 0, 1, 2, 3, 4, etc., occurrences happening in the future can be predicted quite accurately by the successive terms of what is known as *Poisson's Probability Formula* –

$$\text{e}^{-m} + \frac{m}{1}\text{e}^{-m} + \frac{m^2}{2 \times 1}\text{e}^{-m} +$$
$$\frac{m^3}{3 \times 2 \times 1}\text{e}^{-m} + \frac{m^4}{4 \times 3 \times 2 \times 1}\text{e}^{-m}, \quad \text{etc.}$$

– where e $= 2\cdot7183$ (the base of Napierian logarithms).

The first term in this formula gives the probability of 0 occurrences, the second term gives the probability of 1 occurrence, the third term the probability of 2 occurrences, and so on.

Once again you will probably be relieved to know that the results obtained from this formula can be presented in a Table, which enables us to determine these probabilities quite simply. Who would want to work out such a formula when a Table can give the answer in 10 seconds!

Normal Probability Formula

So far we have been dealing with situations in which the outcomes are necessarily whole numbers. We count the number of heads or tails. We count the number of accidents on a road. We count the number of people who die from a certain disease. There are no in-between gradations in cases like these.

By contrast, there are many observations which involve a process of measurement. It may be of height, weight, speed, temperature, time, viscosity, voltage, loudness, brightness, and so on – in all measurements we are dealing with quantities which have an infinite number of gradations. They have a continuous scale stretching from zero to infinity. For convenience we put units on these scales, like mileposts along a road. But our accuracy is always limited by our measuring equipment, and by our ability to handle that equipment.

Hence it comes about that we always get some degree of variation whenever we measure anything a number of times. No matter how careful we are, this element of chance always intrudes. So that no blame will rest on us, we call such variations 'experimental error'.

To provide a simple example of this kind of variation, I have taken the trouble to measure the width of my desk 50 times, using a foot ruler. The results, in order of occurrence, were as follows –

54·14″ ·13 ·12 ·15 ·16 ·14 ·15 ·14 ·15 ·12
 ·13 ·15 ·15 ·13 ·13 ·15 ·15 ·17 ·14 ·16
 ·19 ·17 ·16 ·17 ·19 ·13 ·12 ·14 ·16 ·15
 ·15 ·17 ·16 ·17 ·20 ·16 ·15 ·15 ·15 ·14
 ·11 ·14 ·18 ·15 ·15 ·16 ·15 ·14 ·15 ·18

Now if you look through these results you will find that 54·11″ occurred once, 54·12″ occurred three times, 54·13″ occurred five times, and so on. Let us arrange the whole set of results in this way, to see the pattern of the variations.

Measurement	Frequency
54·10″	0
54·11″	1
54·12″	3
54·13″	5
54·14″	8
54·15″	16
54·16″	7
54·17″	5
54·18″	2
54·19″	2
54·20″	1
54·21″	0

This pattern, with the most frequently occurring measurement in the centre and a progressive falling away of frequencies on either side of it, is perfectly typical. You may care to try a similar experiment yourself – it will drive this important point home to you more than a thousand words.

A good way to picture the distribution of these frequencies is to present them as a frequency chart. This has been done in Fig. 2 on the next page.

If the number of measurements had been increased from 50 to

a few hundred, this frequency chart would have become completely symmetrical. The curve which fits this shape is shown in Fig. 3. It is called the *Normal Curve*. The frequency with which any particular value of x occurs is given by the height (y) of the curve at that point.

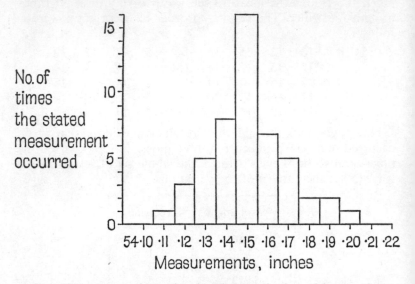

No. of times the stated measurement occurred

Measurements, inches

Fig. 2. Frequency chart, showing the distribution of frequencies of 50 measurements of the width of my desk.

The Normal Curve is of great importance in Statistics. It was discovered in 1733 by Abraham de Moivre, a refugee French mathematician living in London, in the course of solving problems for wealthy gamblers (again!), which he was obliged to do to supplement his meagre income. He found that the curve was described by this formula –

$$y = \frac{1}{S\sqrt{2\pi}}\,e^{-\frac{1}{2}(x-M)^2/S^2}$$

– where y is the height of the curve corresponding to values of x, while M and S are the mean (or average) and the standard deviation of the data, and e is again the base of Napierian logarithms. The standard deviation is a measure of the bunching of the data about the mean; and we shall deal with it in more detail in Chapter 4.

Before long, it was realized that this formula was of far more use than its originator imagined, for it was found to apply also to errors of observation in astronomy, and in fact to all kinds of physical measurements (e.g. my desk measurements). It was also found to describe the pattern of natural variations that exist between various living things of the same species (height, weight, intelligence, and so on).

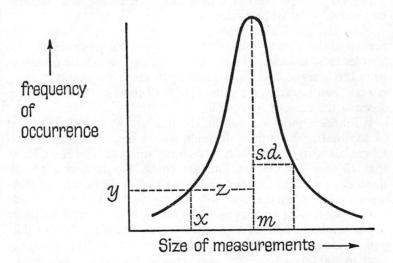

Fig. 3. The Normal Curve. The peak of the curve is directly over the arithmetic mean (*m*) of the measurements. The spread of the curve is specified by its standard deviation (*s.d.*); the greater the standard deviation, the broader the curve will be. The frequency of any particular measurement (*x*) is given by the height of the curve (*y*) at that point; its distance (*z*) from the mean, expressed in standard deviations, enables the probability of its chance occurrence to be calculated.

Even more remarkable is the fact that it also describes, with fair accuracy, the pattern of variations from average in many binomial situations. If a coin is tossed 10 times, the average number of heads will be 5; this will be the result which occurs most often, with 4 heads and 6 heads occurring next most frequently, 3 heads and 7 heads being less frequent again, and so on in the pattern of the Normal Curve. Try it for yourself. It was this, in fact, which de Moivre discovered and described in mathematical terms.

As if this was not enough, the Normal Curve also gives a close approximation to those chance variations from average which follow Poisson's Formula (provided that the average is at least 30 occurrences). For instance, if a city has an average of 30 road accidents a week, in the long run 30 will prove to be the commonest number; 29 and 31 accidents will occur somewhat less frequently, 28 and 32 still less often, and so on, following the pattern of the Normal Curve.

No wonder the Normal Curve has been called 'the veritable boy-scout knife' of Statistics.

Now the great usefulness of the Normal Curve is not only related to the wide variety of situations which it describes. It also enables us to calculate the *probability* of any particular deviation from the average, and the arithmetic to do this contains nothing harder than squares and square roots (Tables of which are provided at the back of this book).

It sounds almost too good to be true, doesn't it? But the fact of the matter is that the formula which describes the Normal Curve can also be used as a probability formula. For if we know that 10% of the observations are to be expected at a certain distance from the midpoint of all the observations (and the formula of the Normal Curve will tell us just that), it follows that the probability of making an observation of that value will be 1 in 10. Another observation, farther still from the midpoint value, will be expected to occur by chance with a lesser frequency, say once in 100 times.

Later on we shall see exactly how to do these calculations in practice. For the moment it will suffice if you have grasped the principles concerned, and seen how the Normal Probability Formula can be used as a shortcut for dealing with many binomial situations and isolated occurrences.

Notice as we proceed how an investigation which began in the service of gamblers some 350 years ago has gradually developed into an indispensable tool in the service of every branch of science, industry, and even commerce.

CHAPTER 3

SAMPLING

Basic Ideas

I HOPE YOU noticed that the following phrase was present in the first 2 basic Laws of Chance – 'if all the possible results have an equal chance of occurring'. This phrase is of such importance in all calculations of probabilities that it has been called 'the heart of Statistics'. Let us therefore look into the matter in some detail.

Given a bag of well-mixed marbles of which 10% are black and 90% are white, the probability that a blindfolded person will draw a black marble from the bag is 1 chance in 10. But in real life, when collecting a sample of people or things from a large 'parent' group, it is not as easy as you might imagine to be sure that all the individuals in the parent group do have an equal chance of being included in the sample. And yet the Laws of Chance apply only when this requisite has been satisfied.

In mobilizing USA troops in 1940, 10,000 numbers were written on slips of paper, the slips were put in capsules, and the capsules were put in a bowl and mixed. The capsules were then drawn by various blindfolded dignitaries in a public ceremony. The results showed marked departures from what would be expected if all the numbers had an equal chance of being chosen. The amount of mixing needed to mix up 10,000 capsules thoroughly had obviously been underestimated. (From W. A. Wallis and H. V. Roberts, *Statistics – A New Approach*, Free Press, 1960.)

In 1954, two scientists working for the American Cancer Society published an impressive set of figures which showed a definite association between cigarette smoking and lung cancer. They had questioned a sample group of 187,766 men about their smoking habits, and noted that over the ensuing 20 months, the number of deaths from lung cancer were as follows –

	No. of men	No. of deaths from lung cancer	Death rates from lung cancer
Non-smokers and occasional smokers	79,944	24	0·03%
Regular smokers	107,822	143	0·13%

(From E. Hammond and D. Horn, *Journ. Amer. Med. Assoc.*, 1954, Vol. 155, p. 1324.)

These observations show that the regular cigarette smokers in the sample group died from lung cancer more than 4 times as often as the non-smokers and occasional smokers. However, the chief medical statistician of the Mayo Clinic, Dr Joseph Berkson, criticized the design of this survey on the ground that the sample group had not been chosen properly. In articles in the *Proceedings of the Staff Meetings of the Mayo Clinic*, 1955, pp. 319–48, and 1959, pp. 206–24, he pointed out that the death rates from all causes in both smokers and non-smokers in the sample group were considerably lower than that of the general USA male population, which indicated that a number of people were not included in the sample group because they were very ill, and furthermore he claimed that the proportion of smokers in the sample group appeared to be less than in the general population (as evidenced by independent surveys), which suggested that a number of smokers had refused to take part in the study. He then demonstrated how such a combination of selective factors can produce an apparent association in the sample group even when there is no association whatever in the parent population. Such errors have, in fact, occurred in the past in medical research, so at least we should be warned that even very large samples cannot be trusted merely because of their size; even they have to be selected properly before it is safe to draw general conclusions from them.

A classical instance of fallacious sampling occurred in the USA in 1936, when a magazine called the *Literary Digest* sent its readers and all telephone subscribers – 10 million people in all – a questionnaire asking how they intended voting at the forthcoming presidential election. There were 2,300,000 replies, on the strength of which it was confidently predicted that the Republican candidate, Landon, would be elected. As it turned out, it was the Democratic candidate, Franklin D. Roosevelt, who won – and by a large margin. This erroneous prediction occurred because readers of this literary magazine and people who had telephones did not, in 1936, form a fair sample of the American voters.

Furthermore, don't you think it was risky to ignore such a large pro-portion of non-replies?

These 3 examples should suffice to show that statistical samp-ling can be a tricky business. Yet the bald fact remains – unless a sample is properly selected, it will very likely lead to false conclusions.

Well, then, why use samples at all? Because they *are* capable of giving accurate results, at a fraction of the cost of a complete count of the parent group, and in some cases they provide the only way of getting the desired information at all. Let us illustrate these points.

In essence, sampling consists of examining a small portion of a large 'parent' group in order to draw conclusions about that parent group. One such use of sampling is called *estimation*. A basic example of this would be to draw a handful of marbles blindly from a large bag containing a lot of well-mixed black and white marbles. Then, having counted the number of marbles of each colour in your sample, you would not be far out in reckon-ing that the rest of the marbles in the bag would have the same proportion of blacks and whites as in your sample. As a practical application of this, consider the following.

When a person travels by air, it happens not infrequently that he has to fly part of the way with one airline, and complete his journey with other airlines. For instance, if a New York executive wants to visit his branch offices in Chicago and San Francisco, he may ask a travel agent to arrange the trip for him; flight times to suit him may require flying from New York to Chicago with Trans World Airlines, from Chicago to San Francisco with United Air Lines, and finally returning from San Francisco to New York on Pan American. In such cases, the passenger pays the entire fare to the airline with which he starts his journey. This airline then pays the other airlines their due portions of the fare. Until the mid-1950s, airline companies in the USA used to work out the price of each and every such transaction, and this was a tedious clerical job which used to cost each large airline about $120,000 yearly. In an endeavour to reduce this overhead, studies were made which led to the trial of a scheme for estimating the total inter-company accounts, the estimates being based on scientifically selected samples consisting of only 12% of the inter-company dockets. This sampling scheme was tested concurrently with the complete pricing method for a period of 4 months, and proved accurate to within $700 per $1,000,000. The samp-ling method has been used exclusively ever since. (Example from W. C. Dalleck, quoted in M. Slonim, *Sampling in a Nutshell*, Simon & Schuster, 1960, pp. 113–16.)

A remarkable example of estimating from a sample occurred in the Second World War, when German industrial output was estimated by British and American statisticians from the serial numbers on captured equipment. It was like taking a random sample of marbles from a bag containing consecutively numbered marbles from 1 onwards, and then, with the help of a simple formula utilizing the size of the sample and the highest number observed in the sample, an estimate can be made of the total number of marbles in the bag. According to checks made after the war, many of these estimates were quite as accurate as those made by the Germans themselves. (From ·W. A. Wallis and H. V. Roberts, *Statistics – A New Approach*, Free Press, 1960.)

Another situation in which sampling provides the only way of getting certain information is when the articles concerned are destroyed in the testing process. Think of flash bulbs, firecrackers, or bullets. To test their efficiency or reliability, you will have to be content with testing a sample of them. Likewise with testing the length of life of electric lamps or car tyres, or testing the breaking points of articles.

Again, there are plenty of cases in which the 'parent group' is of infinite size, so can never be counted in full. When a doctor reports on such a thing as the pulse rate of malaria patients, his report can never cover more than a sample selected from all the possible past, present, and future cases of malaria. Similarly, if a scientist reports on the efficacy of a new insecticide, his conclusions are necessarily based on a sample of insects – they could never be based on the entire insect population of the world. But such samples can be sufficient. Like the case of the little boy who pulled the wings off 3 butterflies, and after due experiment, came to the conclusion that: 'Pulling the wings off butterflies makes them stone deaf. No matter how loudly I shout at them, they don't fly away.' His sample of 3 was sufficient.

Finally, let us dispel the myth that a complete count is necessarily more accurate than a properly selected sample. Try counting a couple of thousand peanuts twice and see if you get the same total on each occasion. It's not as easy as you might think. If at the same time you had to examine each peanut for grubs, your attention could be easily lulled by the monotony of the job, with consequent decreasing accuracy with increasing numbers.

Wallis and Roberts quote the case of a census that was taken in China for military and taxation purposes, which revealed a total population of 28 million. A few years later another census of the same territory for the purpose of famine relief showed 105 million. The interval between the two surveys was much too brief for this increase in population to have actually occurred; rather, the increase pointed to a very substantial evasion of the first census. Slonim, too, comments on the 'understatement of the total population in certain mountainous regions where the haze makes it difficult to distinguish a census enumerator from an income-tax investigator!'

Random Sampling

The use of samples has proved invaluable in literally thousands of different fields. But they can also lead to mistakes. So the question arises: When can a sample be trusted?

It all depends on the way the sample has been selected. A sample can be trusted (within limits that can be calculated) provided that every single individual in the parent group has an equal chance of being chosen in the sample. A sample which has been chosen in this way is called a *random sample*.

Now, the word 'random' has a specific meaning in statistical work. It does not simply mean haphazard or aimless. It does not even necessarily imply that such a sample is a typical cross-section of its parent group. It refers only to a particular way of selecting individuals from a parent group, in which care is taken to see that *every individual has the same chance of being included in the sample group*. The Laws of Chance will therefore only apply to random samples.

In practice, to get a random sample is not easy. It's not good enough to get an expert to choose what he considers to be a representative selection; after all, different experts may vary as to which individuals are 'representative'. If you are after a random sample of people, it's not good enough to stand on a street corner and select every fifth person who passes; this would exclude habitual motorists from your sample. Call on 50 homes in different areas, and you may end up with only housewives' opinions, their husbands being away at work. Pick a set of names with a pin from a telephone directory, and you automatically exclude those people who don't have a telephone. Think up a set of unrelated numbers to select your sample, and you'll find that the mind unconsciously tends to form patterns which destroy the randomness. And after the fiasco of the troop mobilization incident (p. 37), a statistician won't even trust names or numbers picked blindly from a hat.

There is only one way to be absolutely sure of getting a true random sample. It is to assign a number to each individual in the parent group, and then select the requisite number of individuals to make up the sample group by using a Table of Random Numbers. A selection of random numbers is given opposite. These numbers have been found by a kind of electronic roulette wheel and have been checked in many different ways to ensure that they are as random as possible. They are chaotic, in that there is no rule connecting any digit with its neighbours, and yet there is a kind of overall regularity, in that each digit tends to occur with equal frequency when the Table is viewed as a whole.

When using a Table of Random Numbers, the starting point should be selected by chance; the numbers can then be read to the right, left, up, or down; they can be read as single numbers if the parent group contains 2 to 9 individuals, in pairs if the parent group contains 10 to 99 individuals, or in triplets if you are choosing from a group of 100 to 999 individuals. Any number which has been encountered once is simply ignored if it occurs again. You can also ignore any numbers which exceed the number in your parent group; however, this means that if the parent group contains 100 to 200 individuals, a lot of the numbers in the Table (e.g. 251, 630, 970, etc.) will be wasted. In such a case it is quite legitimate to increase the yield of random numbers in the Table by reducing the first number in each triplet to 0 if that first number is an even number (so that 251 becomes 051), or to 1 if it is an odd number of 3 or more. Thus the top line of the Table on page 43 becomes converted to 051, 030, 188, 170, 014, etc. Note that 06 and 006 mean plain 6; likewise 051 = 51.

The way to use a Table of Random Numbers will be made clear by an example. Suppose that the Association for the Advancement of People with Double-Jointed Thumbs has 504 members, and they want to send 10 members selected at random to the National Conference. Each member has a membership number, so all we have to do is to select 10 numbers from the Table of Random Numbers on page 43.

A suitable way to choose a starting point is as follows. Put the Table of Random Numbers on a flat surface and drop a pin on it. Start from the number nearest the point of the pin. The question of which direction to read the Table is easily resolved by looking at the minute hand of your watch. Read the numbers upwards if the time is between $7\frac{1}{2}$ minutes to the hour and $7\frac{1}{2}$ minutes past the hour; read the numbers to the right if the minute hand is pointing to the right, anywhere in the quadrant between $7\frac{1}{2}$ minutes past the hour and $22\frac{1}{2}$ minutes past the hour;

TABLE OF RANDOM NUMBERS

251630	188970	014150	214129	067312	718571
595768	971114	036525	107629	372393	329505
870011	199278	426340	184776	562236	815436
251863	737509	824449	900504	921737	011470
793997	643971	169205	327821	622024	781759
451972	533283	745225	670451	525624	950966
794648	460855	581519	118782	169303	336183
761608	734325	384145	608332	598301	291413
492036	807126	143870	634580	854092	794352
906318	383847	476141	196374	805132	192246
800887	707488	722567	366616	494316	691931
678724	720000	880890	180029	481331	900541
558343	635352	541310	546655	306931	281846
474457	905617	284811	834799	826841	639528
724848	247425	593485	453524	718619	136741
861119	241816	161871	153342	442767	512211
724746	277370	758319	159970	758616	120826
412282	092904	131413	239219	763611	996794
090767	042351	357417	200699	026374	278464
225011	862790	872401	882819	329593	886279
497211	598620	953678	700441	589973	441482
157868	875508	719152	000231	230280	783326
246869	163439	753634	498916	822360	240086
776378	416056	596173	488670	490907	093399
455479	445874	284050	414980	720288	340607
949504	146521	629028	654158	342434	697835
482593	629593	891704	305520	473721	031750
519302	947665	643829	978293	478940	154687
432444	482775	982096	163646	542584	341145
428207	410888	222885	707401	525704	910350
175514	750489	683860	362880	878739	516059
221223	049031	472877	173343	928304	149117
481210	280580	418671	771059	621065	540785
073950	795521	794405	600604	780334	325852
589055	721639	618498	569932	666067	427929
504352	680467	805648	709971	594813	700622
186542	449103	045547	708185	984996	169614
592164	787417	171983	094557	589317	325726
047670	076544	637625	366947	464958	069170
374038	699590	307943	047180	326825	055111
245991	314083	458897	535841	857710	810559
987871	122147	614713	711812	776743	584853

Taken from The Rand Corporation, *A Million Random Digits* (Free Press of Glencoe, 1955, with permission).

likewise read downwards or to the left if the minute hand is in
the lower or left-hand quadrant respectively.

Suppose it turns out that we are due to start at the left-hand
end of the 4th row, and will work to the right. The first number is
251, so that member is selected. The next number is 863; there is
no member with such a number, so we ignore it and pass on to
the next one. 737, 509, and 824 are also too large (we have only
got 504 members), but the next number in the Table, 449, selects
the 2nd member for the Conference. And so we proceed until the
quota of 10 members has been filled.

Have a go at this yourself. You will then get the hang of it, and
it will only take a minute. As usual, answers are given at the end
of this book.

Questions

Q11. Divide 12 patients into 2 random groups of 6, so that one
group can be treated with one treatment, and the other group
with another treatment, for the purpose of comparing the efficacy
of the 2 treatments. The patients have been allotted numbers
from 1 to 12. Suppose that, using the Table of Random Numbers
on page 43, your pin has indicated to start at the left margin of
the 5th row from the top, and the time is ten past nine so you will
work across the row to the right. Which patients shall be put in
the first group, and which in the second?

Q12. Suppose there are 50 houses in your street, and you want
to select a random sample of 5 for some purpose or other. The
houses are numbered from 1 to 50. Suppose that your pin has
determined that you shall start at 98, at the bottom left-hand
corner of the Table of Random Numbers, and the time is 3
minutes past 5. Which houses will you select for your sample?

Q13. There are 9 racing cars in the Lakeside Grand Prix. They
are numbered 1 to 9. Select a starting order for them using a
random process. Suppose that your pin and watch tell you to
start at the top right-hand corner of the Table, and work to the
left.

*   *   *

Statistics doesn't have to be a dull subject. In his book, *Sampling
in a Nutshell* (Simon & Schuster, 1960), Morris Slonim tells of
how he applied random sampling to a study of horse racing.

A sample survey was conducted by the author (purely in the interest of science) to see how the expert handicappers fared in comparison with the 'ladies' hatpin' or random method of selection of bangtails. . . . The test covered the winter meet at Bowie in 1958. It was during this meet that a number of patrons were snow-bound after a heavy blizzard in February. It is reported that many suffered from excessive exposure, having lost their shirts before the storm began.

Horses were selected at random by use of a table of random numbers and the results were compared with the consensus of handicappers. . . .

The results of this scientific inquiry indicated that the only way to make money on horses was to seek employment in a glue factory. However, one was likely to lose considerably less money by following the consensus of the handicappers than by picking the horses at random. So if you happen to be the type who would bet on Nellie's Nag because your mother-in-law's name happens to be Nellie, we suggest you switch to the handicappers' choice.

Sample Size

The purpose of taking a sample is to find out something about its parent group. We have seen that for a sample to be statistically representative of its parent group, it must be chosen by a random process. But the sample size must also be considered. A random sample of 2 men could hardly be expected to give an accurate estimate of the average weight of men.

With random samples, the larger the sample the more accurately will it reflect the characteristics of its parent group. This accuracy increases with the square root of the sample size, so that a sample must be increased 100-fold to get a 10-fold increase in accuracy.

In practice, the main factor which determines the size of the sample needed is therefore the degree of accuracy required. That sample of 2 men would be ample if you only wanted to know the average weight of men to the nearest quarter ton, but if you wanted the answer correct to the nearest ounce, you may need a random sample of 5,000 men.

The amount of variability inherent in the parent group can also affect the size of the sample needed. The same degree of precision can be attained with a small sample taken from a relatively uniform parent group as with a large sample from a widely varying parent group. Unskilled labourers would be a fairly uniform group as regards their income, but if we were investigating their age, we would have to contend with a variation of 40 or more years.

A preliminary investigation, called a *pilot study*, is often made to gain some idea of the sample size needed.

The importance of using a sample of the right size is illustrated pointedly by the following case. When poliomyelitis vaccine was first announced some years ago, a number of clinical trials were carried out to discover whether it was effective or not. One of these trials was carried out in a town in which there were over 1,000 children. Half of the children were given the vaccine, and the other half were left untreated for comparison. After a while, an epidemic of poliomyelitis passed through the town. Everyone was pleased to find that not one of the inoculated children developed the disease. But what of the kiddies who had not been given the vaccine? None of them developed poliomyelitis either! The net result was that nothing was proved in this particular case, and the reason was simply that the sample was not big enough. In a sample of this size, only 2 cases could be expected in an average polio epidemic, so even the complete absence of cases proves nothing. It would have needed a much bigger sample to yield a significant answer. (Such proof came later.)

The reliability of information provided by a sample depends, then, on the care taken to ensure its randomness, and on the size of the sample. Note that reliability is a function of the sample *size* itself, not of the *proportion* that the sample bears to its parent group (unless the sample consists of 20% or more of the parent group). This means that a random sample of 1,000 people taken from a town of 6,000 people will not prove appreciably more accurate than a similar sample of 1,000 people taken from a city of 2,000,000 people. Strange, isn't it? It is the sample size itself which counts.

Proof of this remarkable fact can be demonstrated by an example from industrial sampling. Suppose a manufacturer of electric light globes examines samples of his product to see that the quality is up to a certain standard. The usual way is to test a small number of globes selected at random from each batch. For the purposes of our example, suppose that 20% ($=\frac{1}{5}$) of the globes that he makes are defective; the other 80% ($=\frac{4}{5}$) are perfect. Suppose furthermore that he is going to test 2 globes from each batch. Here we have a binomial situation like that of Question 10, p. 29.

If each batch consisted of 5 electric light globes, of which $\frac{1}{5}$ were defective, there would be (on the average) 1 defective globe and 4 perfect globes in each batch. What is the probability that a sample of 2 will yield 2 perfect globes (so that the batch will pass as OK)? By the Proportionate Law, the probability of the first globe being OK will be $\frac{4}{5}$. This will leave 1 defective and 3 OK globes, so the probability that the second globe will be OK will be $\frac{3}{4}$. Then, by the Multiplication Law, the probability of both of these happening will be –

$$\tfrac{4}{5} \times \tfrac{3}{4} = 0.6$$

This means that out of every 1,000 batches tested, 600 would be passed as OK.

Next, consider what is the probability of passing the batch as OK if the batch size had been larger, say 20. Such a batch would ordinarily contain $\frac{1}{5}$ of 20 = 4 faulty globes. By the same reasoning as above, the probability of picking 2 OK globes out of such a batch would be –

$$\tfrac{16}{20} \times \tfrac{15}{19} = 0.632$$

Increase the batch size to 100. It would now contain 20 faulty globes and 80 perfect ones. The probability of picking 2 perfect ones from such a batch would be –

$$\tfrac{80}{100} \times \tfrac{79}{99} = 0.638$$

If the batch size was still larger, say 10,000, with $\frac{1}{5}$ defectives and $\frac{4}{5}$ OK, the probability of choosing 2 OK globes in a sample of 2 would be –

$$\frac{8,000}{10,000} \times \frac{7,999}{9,999} = 0.640 \text{ (i.e. 640 times in each 1,000)}$$

So you see that even in a case as extreme as this, the reliability of the sample as regards the information it reveals about its parent source is hardly affected at all by the proportion the sample bears to its parent group – it is simply a function of the sample size.

Stratified Random Sampling

Having convinced you that a random sample is the ideal kind of sample for drawing any conclusions about the parent group, it must now be stated that such samples often suffer from an important drawback – their cost.

This problem is likely to arise whenever there is much variation between the individual members of a parent group. In such a case there is an appreciable risk that a *small* random sample will contain, by chance, a disproportionate number of some minority group (either over-representing them or under-representing them). To minimize this risk, a random sample in such a case needs to be relatively large. And the cost of obtaining a large random sample is often prohibitive.

What's to be done? Well, consider the following.

Suppose that a merchant, in order to reduce the cost of making a complete inventory of his stock in hand, decides to estimate his stock from a sample group of items selected from the whole. Imagine about $200,000 worth of stock, made up of about 4,000 different types of items, the price of which varies from $\frac{1}{2}$ cent to over $100 each. Because of the large variation in the price of the different items, a random sample would certainly have to be very large if it was going to be truly representative of the parent stock. Considering the time it would take to select such a sample properly, it would be just as easy to do a full count in the old-fashioned way.

But suppose that the stock is classified into 4 groups –

(1) items under $1.00;
(2) items ranging from $1.00 to $4.99;
(3) items ranging from $5.00 to $19.99;
(4) items $20.00 or more (a small group).

Then an ordinary random sample could be taken from each of the first 3 groups, the fourth group being small enough and important enough to warrant being counted in full. In this way, only 700 of the 4,000 lines would have to be tallied, and yet the risk of under-representing or over-representing any particular items in the under $20 groups is kept to a minimum.

This example is based on a true case described by Raymond Obrock in the *Journal of Accountancy*, March 1958, pp. 53–9. The estimated inventory proved perfectly satisfactory, and the saving in cost was considerable.

This method of achieving randomness by taking a random sample from different levels (strata) of the parent group is called *stratified random sampling*. For maximum efficiency, each stratum should be as homogeneous as possible (in respect of the feature being studied); a preliminary pilot survey may be necessary to decide on the strata. Because of its in-built safety mechanism, it provides an accurate way of dealing with any parent group which possesses considerable internal variability, and is economical because for this type of case the sample size needed to give a specified degree of precision is very much smaller than would be the case with ordinary random sampling.

Stratified random sampling is widely used in market research and opinion polls, for it is fairly easy to classify people into occupational, economic, social, religious, and other strata.

Other Sampling Methods

There are yet other situations which require other sampling techniques. Ordinary random sampling needs a complete list of individuals in the group to be sampled. Stratified random samp-ling needs a knowledge of the relative numbers of individuals in each stratum. These requirements simply cannot be met in some circumstances.

Consider the case of trying to select samples for inspection from a continuously moving production line in a factory. The parent group from which the sample group will be extracted is literally on the move. This situation is handled by what is called *systematic sampling*. This entails selecting individual items at regular specified intervals along the line. Suppose that the degree of precision required has dictated that 10% of the parent group should be inspected, this would be done by examining every 10th individual along the line. A Table of Random Numbers is used to choose the 1st item out of the initial 10 in the line; suppose this turns out to be 3, you would then sample the 3rd, 13th, 23rd, 33rd, and so on, item along the line. If for any reason it is sus-pected that the parent group may vary in some regular cyclical way, it would be necessary to select individuals at irregular intervals, for example at random intervals out of each group of 10 (or whatever proportion is being sampled).

A common kind of sampling in medical research is the accumu-lation of what may be termed a *presenting sample*. This consists merely of a consecutive series of patients who present themselves for treatment of a certain complaint. For instance, a doctor may report on the last 50 cases of appendicitis that he has seen. Such a sample can often be accepted as being equivalent to a truly random sample, such as might be drawn from a parent group consisting of all possible cases of the disease in question. But this is not always the case. For one thing, the patients must be genuinely consecutive; the exclusion of any cases from the series would certainly destroy the randomness, and it would obviously be unwise to draw any general conclusions from a hand-picked sample. Another consideration is that there may be a strong element of selection at work, unwittingly, as in the case of specialist and hospital practice. The clientele in such cases is often far from random. For instance, skin specialists report that adolescent acne (pimples) is twice as common in females as in males, but this may reflect nothing more than a greater concern about their appearance on the part of young ladies, so that they report for treatment in greater numbers, rather than indicate the true relative incidence in the population as a whole. Or again, a hospital report on the incidence of measles complications may only be applicable to those cases of measles which are severe enough to need admission to hospital; it is quite likely that the hospital never gets to see any of the milder cases of measles which are ordinarily treated at home.

Bearing these reservations in mind, the fact that a presenting sample is often equivalent to a random sample is no excuse for using the word 'random' in a loose and incorrect manner, as in the following excerpt from a well-known American medical journal in 1960 –

The trial with this new drug was carried out on a random sample of patients suffering from respiratory illnesses. No attempt whatever was made to select patients.

By definition, a random sample requires a very careful selection of individuals, according to a definite plan and using a Table of Random Numbers. Presumably the above trial was carried out on a presenting sample of patients; it therefore may or may not have been equivalent to a random sample.

To select a random sample from a truckload of coal or from a sack of flour, it is best to take a number of portions from various parts of the parent group, pool these portions and mix well, divide into two parts, take either half and again mix well (discard the other half), and keep on halving and mixing in this way until the sample has been reduced to the size required. Such a sample is random because every part of the parent material has an equal chance of getting into the final sample.

There are still other statistical sampling methods, with special applications, but we have covered the main ones and have emphasized the need for correct sampling. Remember, a badly chosen sample is worse than useless – *it is positively misleading*!

CHAPTER 4

AVERAGES AND SCATTER

'MANY AUSTRALIANS LOVE swimming.' This is true, but vague. It might equally well mean that half a million Australians (5% of the population) love swimming, or even eight million (80% of the population). Similarly – 'A few men can run a mile in 4 minutes.' How many men, 10 or 10%? Words like 'many' 'a few', 'often', 'sometimes', and 'most', are useful in everyday language for conveying some idea of proportions, but actual figures are necessary for the precise specification of the frequency of occurrence of anything. Such figures are obtainable from a properly selected sample or from a total count of the parent group.

Now, when it comes to describing some measurable character-istic (weight, length, speed, etc.), we are faced with the fact that all things, even those of the same kind, tend to vary more or less from one individual to another. Think of houses, hats, horses, or husbands. In any group of such things, whether it be a parent group or a sample group, the measurements of the individuals will vary through a certain range. To deal with this situation, we need a single figure which will fairly represent the whole group. A figure which does this and which 'sums up' the group measure-ments is called an average, or mean.

Averages are very useful for descriptive purposes. If we read that the adults of a certain race of pygmies have an average height of 3 feet 6 inches, we can immediately form an accurate mental picture of their size.

Averages are also needed for making comparisons between different groups. One group of measurements might tell us that 'Prince Hairylegs' galloped a mile in 1 minute 40 seconds on Monday, 1 minute 53 seconds on Tuesday, and 1 minute 44 seconds on Wednesday; to compare such performances with those of another racehorse, we would need to work out the averages.

In fact, averages are so useful that they are used on practically everything that can be measured. In view of the great variety of applications, then, it has been found necessary to devise a number of different averages, each with its own particular use.

The Arithmetic Mean

The ordinary average which we learnt at school is properly termed the *arithmetic mean*. It is used so commonly that it is often simply called 'the mean' or 'the average', but it is generally best to give it its full title so as to avoid confusion with other averages.

The arithmetic mean of any set of numbers can be calculated by applying this formula –

$$Arithmetic\ mean = \frac{x_1 + x_2 + x_3, \text{etc.}}{n}$$

where x_1, x_2, x_3, etc., are the values of all the individual measurements forming the set, and $n =$ the total number of measurements in the set.

Thus the arithmetic mean of 20, 23, 25, and 26 is –

$$\frac{20 + 23 + 25 + 26}{4} = \frac{94}{4} = 23{\cdot}5$$

A shortcut (which is perfectly accurate) is to first subtract a fixed amount from each value, then find the arithmetic mean of the reduced values, and finally replace the fixed amount by adding it to the reduced mean. Thus in the above example we could subtract, say 20 from each value; the calculation would then be –

$$Arithmetic\ mean = 20 + \frac{0 + 3 + 5 + 6}{4} = 20 + 3{\cdot}5 = 23{\cdot}5$$

Whenever possible, we shall be using shortcuts like this, because they save time and also, by keeping the numbers small, they reduce our liability to arithmetical errors.

One point should be mentioned here. The arithmetic mean is only accurate as an indicator of the centre of data when that data is in units of an *equi-intervalled scale* (such as those of weight, length, area, temperature, or time). This is noteworthy because there are some scales in which the steps between successive units are not uniform. The units of such scales therefore behave like arithmetic numbers only in the respect that they both have a conventional order. Take the case of a measurement of human ability like an intelligence test. In such tests there can be no guarantee of a uniform grading of the difficulty of the questions posed, so that while it is valid to say that a person getting a score of 120 is more intelligent than one who has scored 100 at the same test, and that the first score is 20% higher than the other, it is not, however, to be claimed that the first person is 20% more intelligent than the other, for that would only be true if one knew that the units were equally spaced along a scale of increasing difficulty (and that the scale started from an absolute zero). It follows that the arithmetic mean of a group of such units will be erroneous to the extent that the intervals between units are unequal.

Measures of Dispersion

The arithmetic mean is a single number which 'sums up' a set of numbers by indicating their central tendency or location on a scale. However, it is incomplete as a descriptive measure, because it does not disclose anything about the scatter or dispersion of the values in the set of numbers from which it is derived. In some cases these values will be clustered closely about the arithmetic mean, while in other cases they will be widely scattered.

The importance of this can be seen from a simple example. Suppose we have tested a sample of 4 television tubes of Make A, and found that the tube life was (in turn) 20, 23, 25, and 26 months. (Tube life is ordinarily quoted as being so many hours; to keep the numbers small we shall stick to months, and assume that all tubes were used for 2 hours daily.) We saw above that the arithmetic mean of these numbers is 23·5. Now if we tested 4 tubes of Make B, and found that these tubes lasted 4, 10, 25, and 55 months, we may be somewhat surprised to discover that the arithmetic mean of these numbers is also 23·5. It is obvious that in such a case we need some way of specifying that the life of Make B tubes is much more variable than those of Make A.

We could simply state the *range*. This is the difference between the largest and smallest observations. The Make B tubes would then be described as having an arithmetic mean life of 23·5 months, with a range of $55 - 4 = 51$ months. This is very simple, but unfortunately it is a poor measure of the dispersion of observations around the arithmetic mean because, being determined solely by two extreme observations, it totally disregards the bulk of the observations. Furthermore, the range increases not only with greater dispersion in the data, but also with an increase in the number of observations; thus if you measure the height of 10 men, the range may be 6 inches, whereas if you measure 100 men from the same population, the range would probably be about 15 inches. So you can never be sure how much of the range is attributable to variability in the basic data, and how much is due to the size of the sample.

These disadvantages can be largely overcome by working out the *range between the lower and upper quartiles*. Given a set of measurements arranged in order of size, the lower quartile is that value which has a quarter of the observations below it and three-quarters of the observations above it. Likewise, the upper quartile has three-quarters of the observations below it and one-quarter above it. If the quartiles fall between observations, they are given a value which is the arithmetic mean of the numbers on either side of them. Here are the quartile positions in our first set of numbers –

$$20 / 23 \quad 25 / 26$$

The lower quartile is seen to fall between 20 and 23, so has a value of the mean of these numbers, viz. 21·5. The upper quartile falls between 25 and 26, so has a value of 25·5. The inter-quartile range is thus $25·5 - 21·5 = 4$.

Compare this with the second set –

$$4 / 10 \quad 25 / 55$$

in which the lower quartile has a value of 7, and the upper quartile a value of 40. The inter-quartile range of this set of numbers is therefore $40 - 7 = 33$. The greater dispersion of the data in this set as compared with the first set is shown by the larger inter-quartile range.

The inter-quartile range is less affected by extreme values than the range. For instance, if this second set of numbers had been 1, 10, 25, and 58, the arithmetic mean would be the same, but the total range has increased by 6, whereas the inter-quartile range has only increased by 3. Check my calculation for practice.

The inter-quartile range is easy to calculate, and provided that the data contains at least 50 or 100 observations, it is a reasonably accurate measure of the dispersion of that data. However, it is not used very much because it does not lend itself to further mathematical treatment in the way that the standard deviation does. In fact, the most useful thing about the inter-quartile range is that it actually provides an excellent shortcut for estimating the standard deviation in one type of case which we shall discuss later in this chapter; this shortcut depends on the fact that 1 standard deviation equals 0·741 of the inter-quartile range.

Without any doubt, the *standard deviation* has proved itself to

be the best measure of dispersion around the arithmetic mean.
It can be calculated in the following steps –

(1) all the deviations (differences) from the arithmetic mean
of the set of numbers are squared;
(2) the arithmetic mean of these squares is then calculated;
(3) the square root of this mean is the standard deviation.

In a phrase, it is the 'root mean square of the deviations'. Non-
mathematical people always find this a bit complicated at first,
but don't despair, there are shortcuts to take the sting out of it.
Meanwhile, refer back to Fig. 3 (p. 35) for a pictorial representa-
tion of the standard deviation – see how it is a measure of the
spread or dispersion of data around the mean.

Whenever calculating the standard deviation of a sample, a
better estimate of the standard deviation of the parent group
from which the sample has been drawn is obtained by a slight
modification of the second step, as follows. Instead of dividing
the sum of the squares by the number of measurements (n) in the
sample to get the arithmetic mean, the sum of the squares is
divided by ($n - 1$). Proof of this statement is beyond the scope
of this book.

Let us work out the standard deviation of the first set of numbers (20, 23, 25, 26). The arithmetic mean of these numbers was
found to be 23·5, so the deviations from the mean are 3·5, 0·5,
1·5, and 2·5 respectively. The squares of these deviations are
12·25, 0·25, 2·25, and 6·25. The sum of these squares is 21. This is
now divided by ($n - 1$), which is 3, to get 7. Finally, the square
root of 7 is 2·65; this, then, is the standard deviation. The
observations about the Make A television tubes can therefore be
described as having an arithmetic mean of 23·5 months, and a
standard deviation of 2·65 months.

In a like manner, the standard deviation of the data concerning the Make B tubes can be calculated, and turns out to be 22·8
months. You may care to check my arithmetic for practice at
doing this calculation. At the back of this book there are Tables
of Squares and Square Roots which will save you a lot of work.

The standard deviation is affected by extreme values to an even
smaller extent than is the inter-quartile range. We noted that
modifying the second set of numbers to make 1, 10, 25, and 58
had the effect of increasing the inter-quartile range by 3; the
standard deviation is only increased by 2·2. See if you agree
with my calculations. The net result of this is that the standard

deviation is less affected by chance fluctuations due to the sampling procedure than are other measures of dispersion.

Failure to appreciate the importance of dispersion is a fairly common fault. A typical example appeared in the *Medical Journal of Australia* in May 1964. A doctor had carried out a well-designed experiment to compare the duration of numbness produced by 2 local anaesthetic injections (one was old, the other new). The results were reported in this way –

Local anaesthetic	No. of patients	Arith. mean of duration of numbness
Lignocaine	110	90 minutes
Prilocaine	96	136 minutes

The average duration of numbness in these tests was 50% longer with Prilocaine than with Lignocaine, but the lack of any information about the degree of dispersion around these arithmetic means makes it impossible to assess the significance of the observed difference. Suppose, for instance, that the total range of variation with the Lignocaine had been from 85 to 95 minutes, and that the range with Prilocaine had been from 120 to 150 minutes. If this had been the case, it is obvious that there would be no question of the difference being due to chance. But what if the duration of numbness with the Lignocaine had ranged from 30 to 170 minutes, and that of the Prilocaine from 60 to 160 minutes? Your intuition will surely tell you that in this case the difference between the 2 averages might easily be due to nothing other than chance. So you see that we must be told something about the way the results are disposed around the means in order to make a meaningful comparison. This was therefore a good piece of research spoiled by lack of statistical knowledge.

Calculating an Arithmetic Mean and Standard Deviation

A simple shortcut for calculating the arithmetic mean was shown on page 52. There are many such shortcuts which apply to statistical calculations. We shall use them at every opportunity, for they simplify the arithmetic very considerably. In fact, shortcuts of various kinds have made it possible for anyone who can add, subtract, multiply, and divide to do absolutely every calculation in this book.

There will be only one rule to keep. *Copy the setting out of all calculations exactly as shown.* I learnt this rule the hard way, by making silly mistakes as a result of unsystematic layouts, so quite a bit of thought has gone into designing the layouts of all calculations, to ensure that they are as clear and as mistake-proof as possible.

In practice, the shortcut described previously for calculating the arithmetic mean can be carried a step farther, to bring it into line with a shortcut for calculating the standard deviation (which is often needed at the same time). There are 3 varieties of this combined calculation, each with its own particular use.

Method 1

This method applies to most cases in which there are less than about 25 observations.

Suppose we want to know the arithmetic mean and standard deviation of a set of 8 speed tests made by a racing car. The tests are carried out by timing the car over a measured distance, and converting the results into miles per hour.

Values of x (m.p.h.)	Differences $d = x - x_0$		Differences squared d^2
115·2	0·2		0·04
116·6	1·6		2·56
114·5		−0·5	0·25
115·8	0·8		0·64
115·1	0·1		0·01
113·3		−1·7	2·89
114·9		−0·1	0·01
113·0		−2·0	4·00
$x_0 = 115$	2·7	−4·3	10·40
$n = 8$	$\therefore A = -1·6$		$= B$

Calculation of the Arithmetic Mean

(1) Prepare a 3-column table as above, listing the observations in the first column as 'Values of x'. The variable, x, in the present example is the speed reached in each of the 8 tests.

(2) Look down the values of x, and choose a value which looks as though it will be about the group average. It will make no difference to the answer which value you choose, but the closer your guess is to the true arithmetic mean, the simpler will be the ensuing calculations.

The chosen value is called the *assumed mean*, and is represented by the symbol x_0. In the present example, I have chosen 115 as the assumed mean. It has been written at the bottom of the values of x.

(3) Subtract the assumed mean (x_0) from each value of x in turn. Some of these differences (d) will be plus values, and others will be minus; keep the plus values on the left of the second column and the minus values on the right, as shown.

(4) Add up the plus differences, and then the minus differences. Subtract the minus total from the plus total, and call this answer 'A'.

(5) Count the number of observations (n).

(6) The arithmetic mean can then be calculated from this formula –

$$Arithmetic\ mean = x_0 + \frac{A}{n}$$

Applying this formula to the present example we get –

$$Arithmetic\ mean = 115 + \frac{-1 \cdot 6}{8}$$

$$= 115 - \frac{1 \cdot 6}{8}$$

$$= 115 - 0 \cdot 2 = \underline{\underline{114 \cdot 8\ \text{m.p.h.}}}$$

Calculation of the Standard Deviation

(1) Square each difference (d) in turn, entering the answers in the last column (d^2). Use the Tables of Squares at the end of this book. Don't forget that squaring a minus value results in a plus answer.

(2) Add up all the values of d^2. Call this total 'B'.

(3) The standard deviation is then given by the formula –

$$Standard\ deviation = \sqrt{\frac{B - \dfrac{A^2}{n}}{n - 1}}$$

Do this calculation in 3 stages, thus –

(a) Calculate the value of $\dfrac{A^2}{n}$. First square A, using the Table of Squares, and then divide it by the number of observations (n). Don't calculate beyond 3 figures.

(b) Work out the fraction $\dfrac{B - \dfrac{A^2}{n}}{n - 1}$, to 3 figures.

(c) Look up the square root of this fraction in the Tables of Square Roots at the back of this book. This answer is the standard deviation.

Doing this with the racing car speed tests we get –

(a) $\dfrac{A^2}{n} = \dfrac{-1\cdot6^2}{8} = \dfrac{2\cdot56}{8} = 0\cdot32$

(b) $\dfrac{B - \dfrac{A^2}{n}}{n - 1} = \dfrac{10\cdot40 - 0\cdot32}{8 - 1} = \dfrac{10\cdot08}{7} = 1\cdot44$

(c) $\therefore$ Standard deviation $= \sqrt{1\cdot44} = \underline{\underline{1\cdot2 \text{ m.p.h.}}}$

This shortcut will probably strike you as being quite a bit different from the 'classical' method of calculating the standard deviation as described on page 55, but in fact it is algebraically identical with it. The result will therefore be the same with both calculations, but the shortcut method is simpler to do (except when there are only a few numbers to handle). Anyone interested in the derivation of this and other formulae given in this book should consult one of the textbooks of Mathematical Statistics, such as G. U. Yule and M. G. Kendall, *An Introduction to the Theory of Statistics* (Griffin), or M. G. Kendall and A. Stuart, *The Advanced Theory of Statistics* (Griffin).

Snedecor's Rough Check

A simple way of checking your calculation of a standard deviation is to estimate the approximate standard deviation from the following table –

If *n* is about	Standard deviation will be about
5	Range ÷ 2
10	Range ÷ 3
25	Range ÷ 4
50	Range ÷ 4·5
100	Range ÷ 5
200	Range ÷ 5·5
500	Range ÷ 6

(Adapted from G. W. Snedecor, *Statistical Methods*, Iowa State University Press, 1956, p. 44.)

In the example of the speed tests above, the slowest speed was 113·0 and the fastest was 116·6, so the range was 3·6. There were 8 observations, so we divide the range by 3 to get a rough estimate of the standard deviation.

$$\frac{3·6}{3} = 1·2$$

This confirms our formal calculation, although the correspondence is usually only approximate (this exact correspondence was just lucky).

Method 2

If there are more than about 25 observations, and if the values of the observations are whole numbers, it is usually possible to group the values, and this simplifies the calculations even further (without any loss of accuracy).

Think of a man selling washing machines. Some days he sells none, some days one, some days two, and so on. If a record is kept of his daily sales, his daily average (i.e. arithmetic mean) and standard deviation can be calculated as follows.

Suppose that on 42 successive workdays he sells the following number of machines –

2 5 3 1 0 2 2 2 1 5 0 0 1 2 2 3 2 4 6 4 5
2 3 1 3 3 2 2 1 2 3 4 1 4 1 3 3 4 1 3 3 4

Values of x (No. sold per day)	Tally	Frequency f	Differences $d = x - x_0$	Freq. × Diff. $f × d$	$d × (f × d)$
0	|||	3	−2	−6	12
1	++++ |||	8	−1	−8	8
$x_0 = 2$	++++ ++++ |	11	0	0	0
3	++++ ++++	10	1	10	10
4	++++ |	6	2	12	24
5	|||	3	3	9	27
6	|	1	4	4	16
		$n = 42$		35 −14 ∴ $A = 21$	97 $= B$

Calculation of the Arithmetic Mean

(1) Prepare a 6-column table as shown, and list the whole range of daily sales in the first column.

(2) Go through the observations and enter a tally mark for each one in the second column. On the first day he sold 2 machines, so put a stroke beside the value of $x = 2$. The next day he sold 5, so put a stroke beside $x = 5$, and so on. To simplify the addition of these tally marks, it is a good idea to group them in fives, drawing every fifth tally mark horizontally across a group of four vertical strikes.

(3) Add up each group of tally marks to get the frequency (f) with which each value of x occurred in the observations.

(4) Choose an assumed mean (x_0), and show it clearly. My guess is 2.

(5) Subtract the assumed mean (x_0) from each value of x to get the differences (d). Keep the plus differences on the left of the column, and the minus differences on the right.

(6) Multiply each frequency (f) by the value of its difference (d) from the assumed mean. Again keep the plus and the minus values separate.

(7) Add up the plus values in this $f \times d$ column, and then add up the minus values. Subtract the minus total from the plus total, and call this answer 'A'.

(8) Add up the total number of observations (n) in the frequency column.

(9) The arithmetic mean can then be calculated from the same formula as before –

$$\text{Arithmetic mean} = x_0 + \frac{A}{n}$$

The arithmetic mean of the number of washing machines sold each day is therefore –

$$2 + \frac{21}{42} = \underline{\underline{2 \cdot 5 \text{ machines.}}}$$

Calculation of the Standard Deviation

(1) Multiply each value of d (4th column) by its corresponding value of $f \times d$ (5th column), and enter the results in the final column. Remember that a minus number multiplied by another minus number gives a plus answer, so all the answers in this final column will be plus ones.

(2) Add up all the values in this final column, and call the answer '*B*'.

(3) The standard deviation can then be calculated from the same formula as before –

$$\text{Standard deviation} = \sqrt{\frac{B - \dfrac{A^2}{n}}{n - 1}}$$

In our example –

(a) $\dfrac{A^2}{n} = \dfrac{21^2}{42} = \dfrac{441}{42} = 10\cdot5$

(b) $\dfrac{B - \dfrac{A^2}{n}}{n - 1} = \dfrac{97 - 10\cdot5}{42 - 1} = \dfrac{86\cdot5}{41} = 2\cdot11$

(c) ∴ Standard deviation $= \sqrt{2\cdot11} = \underline{\underline{1\cdot45 \text{ machines}}}$

Snedecor's rough check: Our range was 6, and *n* about 50, so Snedecor's Table on page 59 indicates that our range should be divided by 4·5 to get an approximation of the standard deviation.

$$\frac{6}{4\cdot5} = 1\cdot3$$

Our calculated standard deviation is therefore of the right order of size.

Method 3

If there are more than about 25 observations, and the observations consist of measurements having a continuous spectrum of values, it is necessary to modify Method 2 by grouping the measurements into classes (such as 100–3, 104–7, 108–11, etc.). It is then assumed that the midpoint of each class will fairly represent all the observations in that class. This assumption involves a slight approximation, but the results will prove quite satisfactory for significance tests provided that the observations are divided into *at least 10 classes.*

Choice of the classes should be made with care. Divide the overall range by 10 (or more) to find an easy interval such as 1, 2, 5, or 10. The class boundaries must not overlap; intervals such as 70–80, 80–90, etc., would be no good, for where would you put an observation with a value of 80? We should therefore use 70–9, 80–9, etc. When the data shows a natural tendency to fall on certain values, it is best to make these values the midpoints of the classes, in view of the fact that we make the assumption that the class midpoints are the class averages. For example, annual salaries often have a definite tendency to settle around multiples of $500, so classes like $1,750–2,249, $2,250–2,749, etc., would be better than $1,500–1,999, $2,000–2,499, etc.

With the method described here, the class ranges must all be the same size.

Here is an example. A group of 40 students underwent a standard physical fitness test, at the end of which each one's pulse rate (heartbeats per minute) was noted. The results were –

136	110	121	108	⑦⑦	99	129	97	81	128
120	115	146	138	90	100	114	137	110	105
108	121	107	108	128	93	⑯⑧	144	125	111
137	100	92	101	108	125	112	116	129	115

Classes of x (Pulse rate)	Tally	Fre-quency f	Differences $d = x - x_0$	Freq. × Diff. $f \times d$	$d \times (f \times d)$
70– 79	I	1	−40	−40	1600
80– 89	I	1	−30	−30	900
90– 99	₩	5	−20	−100	2000
100–109	₩ IIII	9	−10	−90	900
$x_0 =$ 110–119	₩ III	8	0	0	0
120–129	₩ IIII	9	10	90	900
130–139	IIII	4	20	80	1600
140–149	II	2	30	60	1800
150–159		0	40	0	0
160–169	I	1	50	50	2500
		$n = 40$		280 −260 $\therefore A = 20$	12200 $= B$

Calculation of the Arithmetic Mean

(1) Prepare a 6-column table as shown. Select an appropriate class range, and list the classes in the first column.

In the present instance, the overall range is $168 - 77 = 91$. This lends itself nicely to 10 classes of 10 units each. There is no appreciable clustering of the observations around the tens (100, 110, 120, etc.), so it will be all right to start the classes at 70.

(2) Tally the observations into their appropriate classes.

(3) Add up each group of tally marks to get the frequency distribution of the various classes (f).

(4) Choose an assumed mean (x_0), and show it clearly. On the strength of the frequency distribution, I have chosen the 110–19 class, or rather the midpoint of this class. The class midpoint is found by adding the class range figures together, and dividing by 2. In our example, then –

$$x_0 = \frac{110 + 119}{2} = \frac{229}{2} = 114 \cdot 5$$

(5) Subtract the assumed mean from each class midpoint. Since the class ranges are of equal size, this simply entails listing the distances of each class from the class containing the assumed mean. As usual, keep the plus differences separate from the minus differences.

(6) Multiply each frequency value by each difference value, putting the answers in the next column ($f \times d$).

(7) Add up first the plus and then the minus values in the $f \times d$ column. Subtract the minus total from the plus total, and call this answer 'A'.

(8) The arithmetic mean can then be calculated from the same formula as before –

$$\text{Arithmetic mean} = x_0 + \frac{A}{n}$$

The arithmetic mean pulse rate after this fitness test is therefore –

$$114 \cdot 5 + \frac{20}{40} = 114 \cdot 5 + 0 \cdot 5 = \underline{\underline{115 \cdot 0}}$$

The exact arithmetic mean of this data (obtained by using the formula on page 52) is 115·225, so we see that the approximation is quite good.

Calculation of the Standard Deviation

(1) Complete the tabular part of the calculation by multiplying the values of d (4th column) by the corresponding values of $f \times d$ (5th column) to form a final column, $d \times (f \times d)$.

(2) Add up this final column, and call the answer 'B'.

(3) The standard deviation can then be calculated in 3 stages from the usual formula –

$$\text{Standard Deviation} = \sqrt{\dfrac{B - \dfrac{A^2}{n}}{n-1}}$$

In the case of the pulse rates –

(a) $\dfrac{A^2}{n} = \dfrac{20^2}{40} = \dfrac{400}{40} = 10$

(b) $\dfrac{B - \dfrac{A^2}{n}}{n-1} = \dfrac{12,200 - 10}{40 - 1} = \dfrac{12,190}{39} = 313$

(c) $\therefore$ Standard deviation $= \sqrt{313} = \underline{\underline{17{\cdot}7}}$ (approximately).

Snedecor's rough check: The observations ranged from 77 to 168, which is a range of 91; n was about 50, so the range divided by 4·5 gives us a rough idea of what the standard deviation ought to be (p. 59).

$$\frac{91}{4{\cdot}5} = 20$$

This does not agree with our result particularly well, so we should re-check our calculation (or better still, get someone else to calculate it independently). No fault being apparent, we accept the result of our formal calculation, and disregard the rough check on this occasion. At least it would have detected a gross mistake such as calling the square root of 313 a number such as 55·9.

Actually, the exact standard deviation of the data in this example, as calculated by Method 1, is 18·59. The shortcut of grouping the data into classes gave an answer (17·7) which was about 5% out. This error can be lessened by using smaller class intervals; with intervals of 5 (that is, 70–74, 75–79, etc.) the labour is increased, but the standard deviation then works out to be 18·3, a much better result. Even so, this method gives better approximations of arithmetic means than of standard deviations.

Before proceeding, you should now familiarize yourself with these shortcut methods by answering the following questions. Just follow the instructions step by step, and they won't take long to do. Answers are given on page 360. You will then feel confident to tackle any data that interests you.

Questions

Q14. A party of 10 young men went to see a strip-tease show. Afterwards they found that they had collected a fair number of fleas. In fact, their individual scores were –

$$1 \quad 3 \quad 0 \quad 4 \quad 4 \quad 5 \quad 1 \quad 2 \quad 4 \quad 3$$

Assuming this group of men to be a random sample of patrons, estimate the mean number of fleas per patron, and the standard deviation, using one of the shortcut methods. Apply Snedecor's check.

Q15. The number of pages in successive issues of a daily newspaper were as follows –

```
22  26  34  26  24  20  28  24  24  26  26  28
26  30  20  24  28  26  24  22  24  26  22  26
28  24  24  28  28  26  22  30  24
```

Estimate the arithmetic mean and the standard deviation of the number of pages in issues of the paper, using an appropriate shortcut method. Apply Snedecor's check.

Q16. Twenty-five schoolboys conducted a snail race. Each boy entered a snail, which had to race the length of a brick according to a complicated set of rules. The average speed of each snail (in millimetres per minute) was thus –

```
20  10    15    3·5  9·5  11  10  16  14  10   5    10  12
10  12·5  11    10·5  6   21  15  10  10  11  12·5  9
```

From this sample, estimate the mean speed of snails, and the standard deviation of this speed, using a shortcut method. A small approximation would be allowable here. Compare your result with Snedecor's rough check.

Combining Arithmetic Means

One of the virtues of the arithmetic mean is that it is mathematically very versatile. For instance, given a sample of 10 items selected at random from a certain parent group, and another sample of 50 items taken from the same parent group, the arithmetic means of each sample can be combined to give a single mean as follows.

Suppose the total value of the 10 items was 21, the arithmetic mean of this sample would be $\frac{21}{10} = 2\cdot1$. If the total value of the

50 items was 115, the arithmetic mean of this sample would be $\frac{115}{50} = 2\cdot3$. At first glance it might seem as if the combined mean would be $2\cdot2$, but this ignores the difference in sample sizes. Look at it this way: if we combine the 2 samples to make a combined sample of 60 items, the total of the values will be $21 + 115 = 136$, so the combined arithmetic mean will be $\frac{136}{60} = 2\cdot27$.

In other words, if one sample has n_1 items and a total value of X_1, and another sample has n_2 items and a total value of X_2, and so on, the arithmetic mean (M) of the combined samples will be given by the formula –

$$M = \frac{X_1 + X_2, \text{ etc.}}{n_1 + n_2, \text{ etc.}}$$

Another way of arriving at this answer is to 'weight' the means of the samples in proportion to the sample sizes. The smaller sample in the above example represents $\frac{10}{10 + 50} = \frac{1}{6}$ of the combined samples, while the larger one is $\frac{50}{10 + 50} = \frac{5}{6}$ of the pair. Then the combined arithmetic mean will be –

$$(\tfrac{1}{6} \times 2\cdot1) + (\tfrac{5}{6} \times 2\cdot3) = 0\cdot35 + 1\cdot92 = 2\cdot27$$

This technique is used for finding the arithmetic mean of a *stratified random sample*. Expressed in general terms, if the proportion of each stratum in the parent group is known $(p_1, p_2, \text{ etc.})$, and the arithmetic mean of each stratum is determined $(m_1, m_2, \text{ etc.})$, the combined arithmetic mean (M) of the whole sample will be –

$$M = (p_1 \times m_1) + (p_2 \times m_2), \text{ etc.}$$

For example, suppose the International Cat Society wants to know the average number of cats per house in a certain city. Suppose furthermore that poor people generally have more cats than rich ones. The city could then be investigated by means of a stratified random sample, with the houses divided into 3 strata – poor, medium, and rich. The relative proportions of these 3 categories might be known from some census, failing which it could be estimated by a pilot survey of a random sample from the city concerned. Suppose the proportions turned out to be 2:2:1 respectively. The results of our survey might then be as follows –

Strata	Strata proportions in population	Size of samples	No. of cats	Arithmetic mean of cats
Poor	$\frac{2}{5}$	100	12	$\frac{12}{100} = 0 \cdot 12$
Medium	$\frac{2}{5}$	100	6	$\frac{6}{100} = 0 \cdot 06$
Rich	$\frac{1}{5}$	80	4	$\frac{4}{80} = 0 \cdot 05$

To have kept their true relative proportions, the strata should have contained 100, 100, and 50 houses, but the latter may have been deemed an insufficient number to give the degree of accuracy required. Anyway, this does not affect the calculations.

We now have the following data –

$$p_1 = \tfrac{2}{5} \qquad m_1 = 0 \cdot 12$$
$$p_2 = \tfrac{2}{5} \qquad m_2 = 0 \cdot 06$$
$$p_3 = \tfrac{1}{5} \qquad m_3 = 0 \cdot 05$$

The arithmetic mean of the whole sample is therefore –

$$(\tfrac{2}{5} \times 0 \cdot 12) + (\tfrac{2}{5} \times 0 \cdot 06) + (\tfrac{1}{5} \times 0 \cdot 05)$$

$$= 0 \cdot 048 + 0 \cdot 024 + 0 \cdot 010 = \underline{\underline{0 \cdot 082}}$$

The arithmetic mean number of cats per house in this city is thus seen to be 0·082, or if you prefer, a bit over 8 cats per 100 houses.

Combining Standard Deviations

The standard deviation is a more complex measure than the arithmetic mean, and it would be wrong to combine the standard deviations of 2 or more random samples simply by weighting the individual standard deviations according to the sample sizes. This is because each standard deviation is calculated with reference to each sample mean, whereas the standard deviation of the combined samples must be referable to the arithmetic mean of the whole group. To combine standard deviations, therefore, it is necessary to re-calculate from all the original measurements.

If you are given the size, arithmetic mean, and standard deviation of 2 or more random samples, the standard deviation of the combined samples can be found by re-constructing the values of A and B for each sample, and adding these values to get a combined A and a combined B for substitution in the ordinary standard deviation formula (p. 58), for this latter formula works regardless of whether we use the shortcut of an assumed mean or not. Without that shortcut, A represents the total sum of the observed measurements, and knowing the mean (m) and size (n) of each sample, we can work back and derive each value of A because –

$$m = \frac{A}{n} \qquad \therefore A = n \times m$$

Let us call the value of A for the first sample A_1, that of the second sample A_2, and so on. The sum of the measurements of the combined samples (that is, the value of A for the combined samples) is then obtainable by simple addition, thus –

$$A = A_1 + A_2 + A_3, \text{ etc.}$$

Now, you will recall that the standard deviation of each sample (s_1, s_2, etc.) is calculated from –

$$s = \sqrt{\frac{B - \dfrac{A^2}{n}}{n - 1}}$$

and it is a simple bit of algebra to square both sides and then turn this equation around and show that –

$$\therefore B = [s^2 \times (n - 1)] + \frac{A^2}{n}$$

Knowing n, s, and A for each sample, we can therefore find the value of B for each sample (B_1, B_2, etc.). These values are the sums of the squares of the measurements in each sample, and these, too, are additive quantities, hence the value of B for the combined samples is –

$$B = B_1 + B_2 + B_3, \text{ etc.}$$

Finally, the size of the combined samples (N) is simply $n_1 + n_2 + n_3$, etc.

Having found the values of N, A, and B, we can now calculate the standard deviation of the combined samples (S) thus –

$$S = \sqrt{\frac{B - \dfrac{A^2}{N}}{N - 1}}$$

To show that this is not a difficult procedure, suppose that a random sample of 20 male white mice were found to have a mean length of life of 18 months, and the variation in the sample, the standard deviation, was 2 months. Another random sample, consisting of 30 female white mice, had a mean longevity of 16 months and a standard deviation of 3 months. By weighting the means, you can confirm that the mean length of life of the whole 50 white mice was 16·8 months. We now want to find the combined standard deviation of the 2 sample groups.

Start with the sample of male mice –

$$n_1 = 20 \qquad m_1 = 18 \qquad s_1 = 2$$
$$A_1 = n_1 \times m_1 = 20 \times 18 = 360$$

$$B_1 = [s_1{}^2 \times (n_1 - 1)] + \frac{A_1{}^2}{n_1}$$

$$= (2^2 \times 19) + \frac{360^2}{20} = (4 \times 19) + \frac{129{,}600}{20}$$

$$= 76 + 6{,}480 = 6{,}556$$

The data for the sample of female mice was –

$$n_2 = 30 \qquad m_2 = 16 \qquad s_2 = 3$$

To familiarize yourself with this procedure, confirm that $A_2 = 480$ and $B_2 = 7{,}941$.

Then $N = n_1 + n_2 = 20 + 30 = 50$
$$A = A_1 + A_2 = 360 + 480 = 840$$
$$B = B_1 + B_2 = 6{,}556 + 7{,}941 = 14{,}497$$

We now have the necessary data to calculate the standard deviation of the combined sample groups, thus –

$$B - \frac{A^2}{N} = 14{,}497 - \frac{840^2}{50} = 385$$

$$\therefore S = \sqrt{\frac{385}{49}} = \frac{\sqrt{385}}{7} = \frac{19\cdot6}{7} = \underline{\underline{2\cdot8 \text{ months}}}$$

Exactly the same method can be used to estimate the standard deviation of a parent group from 2 or more *stratified random samples*, but this is only valid if the sample sizes are truly in proportion to their relative numbers in the total parent group. In the case of the white mice, for instance, it is to be presumed that males and females occur equally often, so it would have been necessary for the 2 random samples to have been of equal size before one could say that the combined standard deviation referred to white mice in general. Our calculation above therefore referred only to the dispersion of the observations *in* the sample groups; to make the answer generally applicable would have required weighting the sample means and standard deviations in the proportions of the parent groups (say, by putting $n_1 = n_2 = 25$, instead of using the actual numbers in the samples).

The Harmonic Mean

Although the arithmetic mean is the type of average which is called for most commonly, it is certainly not the only one. The rest of this chapter will be concerned with some other important averages.

Consider this situation. It is 30 miles from town *A* to town *B* and uphill most of the way. If a man rides a bicycle from *A* to *B* at a speed of 10 miles per hour, and then rides back home to *A* at 30 miles per hour, what is his average speed for the whole trip?

The correct calculation here is as follows –

From *A* to *B*, 30 miles at 10 m.p.h. will take 3 hours.
From *B* to *A*, 30 miles at 30 m.p.h. will take 1 hour.
Total distance covered = 60 miles.
Total time taken = 4 hours.
∴ Average speed = 60 ÷ 4 = 15 m.p.h.

An average calculated in this way is called a harmonic mean.

If you imagined that the average was the arithmetic mean of 10 and 30 m.p.h., which is 20 m.p.h., the 60-mile trip would have taken 60 ÷ 20 = 3 hours, which was not the case.

The harmonic mean can be regarded as a specially calculated arithmetic mean, which is often appropriate when the data consists of ratios (miles per hour, calories per ounce, dollars per dozen, etc.).

Mode and Median

You may have noticed that in the cases presented so far, the observations have been disposed more or less symmetrically around the arithmetic mean. For example, if 5 observations had values of 20, 25, 30, 35, and 40, the arithmetic mean is 30, and the observations are disposed with perfect symmetry on either side of the mean.

With larger numbers of observations, such symmetry shows up in the frequency distribution. Refer back to the frequency column of the calculation table on page 60, and see how the

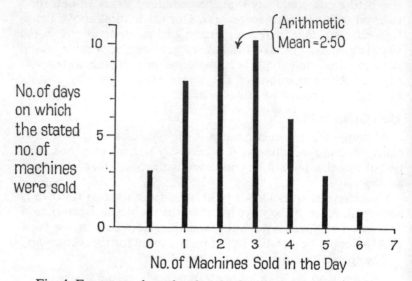

Fig. 4. Frequency chart showing the frequency of various daily sales of washing machines, from data on page 60.

frequencies form a pattern, with the daily sales of washing machines rising to the arithmetic mean (which was 2·5) and then tapering off again. This data is presented pictorially in Fig. 4. The tops of the bars form a curve which is approximately symmetrical, and bears a close resemblance to the shape of the Normal Probability Curve (p. 35).

Without any question, the arithmetic mean is the best representative of any data which is more or less symmetrically distributed.

Now there are certain types of data which are *not* distributed evenly around the arithmetic mean. Many types of biological measurements, such as measures of human achievements, come under this heading. Examples that readily come to mind are people's incomes, and the number of children in families. The asymmetry (= lack of symmetry) in such cases shows clearly when the data is presented as a frequency chart, of which Fig. 5 provides a good example. This chart was prepared from data obtained by asking a random sample of 550 married couples in a certain community how many children they had. All told, there were 1,660 children, which divided among the 550 families, gives an arithmetic average of 3·02 children per family.

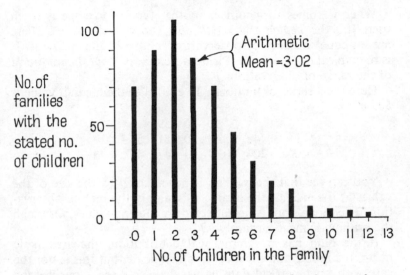

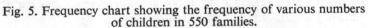

Fig. 5. Frequency chart showing the frequency of various numbers of children in 550 families.

Notice that, because the data is distributed asymmetrically, this arithmetic mean does not represent the data well. In fact, it is really misleading. If you started building houses for an 'average' family with 3 children, you would soon find that families with 1 and 2 children were commoner than those with 3.

In a case like this, the *most frequent value* (in the present case, 2 children) is generally the most representative single number. Being the most 'fashionable' value, it is called the *mode*.

A common use of the mode is when a doctor describes the 'usual' duration of a disease. Thus, most colds get better in a week (this is the usual or modal duration), although a few get better in less time, and some may last for 2 or 3 weeks.

An early example of the use of the mode was described by the historian of classical Greece, Thucydides (400 BC), in *The Peloponnesian*

War. He told of the Plataeans who, desirous of forcing their way over the enemy's walls, constructed ladders, the length of which was calculated by getting their soldiers to count, again and again, the number of layers of bricks forming the walls, which though they could not approach closely, they could yet see clearly. Accepting the number of layers which was counted most frequently, they multiplied this mode by the known height of the bricks to get an accurate estimate of the height of the enemy's walls.

When it comes to incomes, another type of average is often used. It is the *median*, the *middle observation* in the series. There are an equal number of observations above it and below it. It is the typical middle-man. (Please note that it is *not* the midpoint of the range of observations.)

Here are 3 series of numbers. In each case the median value is 50.

```
                    3   50   59
              48   49   50   51   51·1
      1,002  204  161   50   47   45   44
```

You can see that the median is not affected by the size of the values at the ends of the series; it is simply the middle observation. The mode, too, is not affected by the end values, although the arithmetic mean certainly is.

If the data has a symmetrical distribution, the arithmetic mean, the mode, and the median all coincide, but this is not the case with asymmetrical data as we shall now see. Consider the following salaries in an Australian factory –

1 general manager	@	$23,000 =	$23,000
2 sub-managers	@	$10,000 =	$20,000
3 salesmen	@	$4,000 =	$12,000
2 supervisors	@	$3,500 =	$7,000
4 clerks	@	$3,000 =	$12,000
10 craftsmen	@	$2,500 =	$25,000
1 apprentice	@	$1,000 =	$1,000

Total = 23 employees   Total = $100,000

The most frequently occurring salary, the mode, is $2,500. The middle salary, the 12th in the series, which is the median, is $3,000. The arithmetic mean is $100,000 divided by 23, which is $4,350. Each of these values can be called an average , so it is obviously necessary to specify which average is meant. Your choice will be governed by your purpose. If the employees want a raise, they will point out that the average salary is only $2,500, which will be rebuffed by the general manager – he will show them how to calculate the average salary 'properly' to get $4,350. The employees will then have fun pointing out that, if that is the case, nearly the whole staff are getting below-average salaries! You can see why the median is the appropriate average here, despite the fact that, as is often the case with grouped data, the median does not split the data into two equal parts (only 8 men earn more than the median salary whereas 11 men earn less than it).

The mode and the median are thus useful measures for describing the average location of asymmetrical data. They can also be used for data which is expressed in units of a scale which has unequal intervals between the units (p. 52); such units do no more than express an ordered scale of values, with 50 being larger than 49, which in turn is larger than 48, and so on. Thus we could satisfactorily compare two groups of children with respect to their IQ's if we found that the median of one group was 90, while that of the other was 105; this would enable us to state that a greater proportion of children in the second group were above a certain standard (say, IQ 90) than in the first group. (For reasons explained previously, the arithmetic means of such data can be misleading, and there is no way of telling just how much error, if any, is involved.)

However, modes and medians are not used much in statistical analysis, because unlike arithmetic means, they do not lend themselves readily to further mathematical treatment. For instance, you may find that the mode of one group is bigger than another, but there is no way of assessing whether this difference is due to chance or not. To do this, you need the geometric mean.

The Geometric Mean

The remainder of this chapter involves some simple work with logarithms. If your memory is a bit rusty on this subject, here is a brief refresher course.

$$
\begin{aligned}
\text{Log } 1 &= 0 \\
\text{Log } 10 &= 1 \\
\text{Log } 100 &= 2 \\
\text{Log } 0{\cdot}1 &= \bar{1} \ (\text{'bar one'}).
\end{aligned}
$$

Check these in your tables –

Log 2·011	= 0·3034	Antilog 0·0133 = 1·031
Log 20·11	= 1·3034	Antilog 0·1333 = 1·359
Log 201·1	= 2·3034	Antilog 1·3333 = 21·54
Log 0·2011	= $\bar{1}$·3034	Antilog $\bar{1}$·3333 = 0·2154

We can now proceed. A series of numbers, each of which is formed by multiplying the previous one by a constant number, is called a *geometric progression*. For example, starting from 1 and multiplying by 2 gives the geometric progression 1, 2, 4, 8, 16, 32, 64, etc. Or again, multiplying by 1·2 (that is, increasing at a rate of 20%) gives 1, 1·2, 1·44, 1·728, 2·0736, etc. Patterns like these occur with compound interest, growing populations, biological growth, and accelerations.

The same sort of thing happens 'in reverse' with depreciations and deteriorations, only here the process is one of progressive division instead of multiplication. Suppose a man buys 2 lb of chocolates on Monday. On Tuesday he eats half of them, and on each day thereafter eats half of what is left. His supply will diminish thus –

Monday	Tuesday	Wednesday	Thursday	Friday
32	16	8	4	2 oz

In a series like this, the most meaningful average is the amount of chocolate that he had in the midpoint of the period of observation, which is 8 oz. The arithmetic mean of these numbers is 12·4 oz, which as you see, tends to be unduly biased towards the largest numbers in the series.

Fig. 6 shows this data plotted on a graph; notice the bend in the curve. Now if this same data is plotted on a logarithmic scale, the curve becomes a straight line, as shown in Fig. 7. Because it is a straight line, the arithmetic mean of these logarithmic values will be a truly representative average for this data. Let us work this out.

$$\text{Log } 32 = 1·5051$$
$$\text{Log } 16 = 1.2041$$
$$\text{Log } 8 = 0·9031$$
$$\text{Log } 4 = 0·6021 \qquad \frac{4·5154}{5} = 0·9031$$
$$\text{Log } 2 = 0·3010$$

$$\text{Total} = 4·5154$$

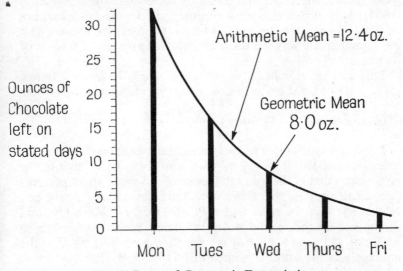

Arithmetic Mean = 12·4 oz.

Geometric Mean
8·0 oz.

Ounces of
Chocolate
left on
stated days

Fig. 6. Curve of Geometric Depreciation.

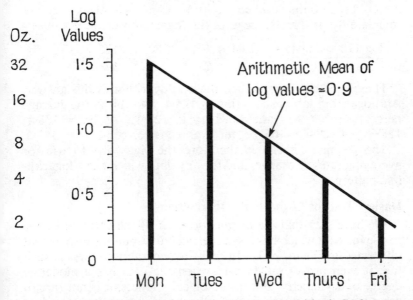

Arithmetic Mean of
log values = 0·9

Fig. 7. Same Data as Fig. 6, Plotted on a Logarithmic Scale.

As shown, the arithmetic mean of these logarithmic values is 0·9031. Now watch this for a conjuring trick: the antilogarithm of 0·9031 turns out to be 8·000, which is precisely the figure which we agreed above was the most meaningful average of this set of numbers.

This average is called the *geometric mean*. In general terms, given a set of n values, x_1, x_2, x_3, etc., it is calculated thus –

$$\textit{Geometric mean} = \text{Antilog}\left(\frac{\log x_1 + \log x_2 + \log x_3, \text{ etc.}}{n}\right)$$

For another example of the geometric mean, consider this. If sales increase to 110% in one year, and to 150% of that in the next year, what is the average increase? To make these percentages clear, suppose that there were 100 sales initially, at the end of the first year the sales had risen to 110% of 100 = 110, and at the end of the second year this figure would have risen to 150% of 110 = 165. Now if we were tempted to think that the average was the arithmetic mean of 110% and 150%, which is 130%, sales in the first year would theoretically have been 130% of 100, which is 130, and in the second year 130% of 130, which is 169, so this average is obviously wrong.

Now the geometric mean of 110% and 150% is what is required, because the numbers multiply upon themselves. First we calculate the arithmetic mean of the logarithms of these values –

$$\frac{\log 110 + \log 150}{2} = \frac{2·0414 + 2·1761}{2} = \frac{4·2175}{2} = 2·1088$$

The geometric mean is then the antilogarithm of this answer. Antilogarithm tables show this to 128·4, and this is the average percentage that we want. Let's try it. 128·4% of 100 = 128·4; 128·4% of 128·4 = 165, so our average is correct.

The geometric mean is therefore the right one to use for averaging rates of change, and for any data which has a logarithmic pattern.

Davies' Test for Logarithmic Distribution

We have seen that the arithmetic mean is the correct average to use for data which has a symmetrical distribution, such as that illustrated in Fig. 4, p. 72. In fact, its general usefulness is such that it even proves satisfactory when the data has a moderate degree of asymmetry. But in biological sciences, it is not rare to come across data which is markedly crowded over the low values of the variable (i.e. with a tall hump on the left of a graph, as in Fig. 5, p. 73). For these cases the geometric mean is a better average than the arithmetic mean.

This raises the question: just how asymmetrical must the distribution of the data be before we should use the geometric mean? At what point do we abandon the arithmetic mean?

Fortunately a simple test is available – it tells which mean to use. It was devised by Professor George Davies (*Journ. Amer. Statist. Assoc.*, 1929, pp. 349–66).

Davies' test entails calculating the following 'coefficient of skewness' –

$$\frac{(\log LQ + \log UQ) - (2 \times \log MQ)}{\log UQ - \log LQ}$$

where LQ = value of the lower quartile,

MQ = value of the middle quartile,

and UQ = value of the upper quartile.

The way to calculate the value of these quartiles will be shown in a moment.

If this coefficient works out to more than $+0·20$, the data is symmetrical enough to use the arithmetic mean. On the other hand, if the answer is $+0·20$ or less, the data is approximately logarithmic in distribution, so the geometric mean is appropriate.

Davies' test works well provided that –

(1) The data is definitely asymmetrical in its distribution (the results are misleading when the data is roughly symmetrical);

(2) the asymmetry is such that the high frequencies are over the low values (as in Fig. 5, p. 73); the answers are meaningless if the frequencies are heaped over the high values on the right of a graph; and

(3) the sample contains at least 50 observations (because quartile values are unreliable with small samples).

The Quartiles

We met the quartiles briefly on page 54. The time has come to have a closer look at them.

Given a series of measurements arranged in order of size, the median divides them into two equal halves. Thus the median of 2, 3, 4, 5, 6 = 4, there being two numbers on each side of it.

When the series consists of an even number of observations, the median will fall between the two centre numbers. Thus the median of 2, 3, 4, 5 = 3·5.

Likewise when a series of measurements is divided into four equal portions, the boundary values are called *quartiles*. Consider the following series –

$$1 \quad 2\,/\,3 \quad 4\,/\,5 \quad 6\,/\,7 \quad 8$$

The dividing strokes are at the quartile positions, for they divide the eight observations into four equal parts. The lower quartile is situated between 2 and 3; it has a quarter of the observations below it, and three-quarters of the observations above it; its value is 2·5. The middle quartile is situated between 4 and 5, with a half of the observations below it and a half of them above it; the middle quartile is thus synonymous with the median; its value in this instance is 4·5. The position of the upper quartile is between 6 and 7; it has three-quarters of the observations below it and one-quarter above it; its value here is 6·5.

To find the *position* of these quartiles in any set of measurements which have been arranged in order of size (from smallest to largest), use the following formulae in which n = the total number of measurements –

$$Lower\ quartile = \frac{n+2}{4}$$

$$Middle\ quartile = Median = \frac{n+1}{2}$$

$$Upper\ quartile = \frac{(3n)+2}{4}$$

To exemplify the use of these formulae, consider the above case where there are 8 measurements (that is, $n = 8$). The position of the upper quartile will therefore be –

$$uq = \frac{(3 \times 8)+2}{4} = \frac{26}{4} = 6\text{·}5$$

By 6·5 is meant a position midway between the 6th and 7th measurement in the set.

The Quartiles of Grouped Data

When dealing with large samples, it is usual to group the data into classes (as in the examples on p. 60 and p. 63). The quartiles will then fall somewhere *inside* one or other class.

Finding the value of the quartiles is easy when each class consists merely of a single value. For example, houses might be classified as having 4 rooms, 5 rooms, 6 rooms, etc. If each of these categories constituted a class, the value of any particular observation, say the 30th, in a set of say 500 observations of this kind is simply determined by finding the class in which this 30th observation occurs, for all the observations in any one particular class have the same value. If a quartile falls between two classes, split the difference and assign it an intermediate value.

But it often happens that the data is grouped into classes which do not possess single values but rather, have ranges. The estimation of the value of a quartile inside such a class is then made on the assumption that the observations are spread evenly through the class concerned, which allows an approximate value of the quartile to be estimated by simple proportions. The narrower the class range, and the more the observations, the more accurate will this estimate be.

The best way to understand this calculation is from an actual example.

A group of 221 patients with a certain skin disease was given a standard treatment, and the duration of the rash from onset to cure was noted in each case. As shown in the data table below, the results were grouped, partly for brevity and clarity, and partly because it was difficult to judge the duration more accurately than to the nearest week or so.

Duration (weeks)	No. of patients	Patient #
$\frac{1}{2}$– $2\frac{1}{2}$	20	1– 20
$2\frac{1}{2}$– $4\frac{1}{2}$	60	21– 80
$4\frac{1}{2}$– $6\frac{1}{2}$	58	81–138
$6\frac{1}{2}$– $8\frac{1}{2}$	31	139–169
$8\frac{1}{2}$–$10\frac{1}{2}$	26	170–195
$10\frac{1}{2}$–$12\frac{1}{2}$	9	196–204
$12\frac{1}{2}$–$14\frac{1}{2}$	5	205–209
$14\frac{1}{2}$–$16\frac{1}{2}$	4	210–213
$16\frac{1}{2}$–$18\frac{1}{2}$	2	214 and 215
$18\frac{1}{2}$–$20\frac{1}{2}$	1	216
$20\frac{1}{2}$–$22\frac{1}{2}$	2	217 and 218
$22\frac{1}{2}$–$24\frac{1}{2}$	1	219
$24\frac{1}{2}$–$26\frac{1}{2}$	2	220 and 221

(From Dr J. Crissey, *Journ. Invest. Dermat.*, August 1954.)

82   · AVERAGES AND SCATTER

To find the *values* of the quartiles, we must first determine the *position* of the quartiles, so we apply the positional formulae, as follows –

$$lq = \frac{n+2}{4} = \frac{221+2}{4} = \frac{223}{4} = 55 \cdot 75$$

$$mq = \frac{n+1}{2} = \frac{221+1}{2} = \frac{222}{2} = 111$$

$$uq = \frac{(3n)+2}{4} = \frac{(3 \times 221)+2}{4} = \frac{665}{4} = 166 \cdot 25$$

The lower quartile (*lq*) thus lies between the 55th and 56th patient in the series. We must now find the value (duration) corresponding to this level.

In this example, the observations are grouped into classes, each of which has a range. Thus the first class contains all those patients whose skin rash lasted anything between $\frac{1}{2}$ and $2\frac{1}{2}$ weeks. There were 20 such patients. To assign a duration value to any particular patient in this class, we have to assume that the 20 patients would have been spread more or less evenly through the range of this class, like this –

Duration $\frac{1}{2}$ week   1 week $1\frac{1}{2}$ weeks   2 weeks $2\frac{1}{2}$

Patient   1 2 3 4 5 6 7 8 9 10 11 12 13 14 15 16 17 18 19 20

This assumption results, of course, in an approximation, but it works quite well in practice so long as the class range is not ridiculously large.

We can now say that if the distribution is fairly even, 10 of the 20 patients could be expected to have a duration in the lower half of the class range, i.e. from $\frac{1}{2}$ to $1\frac{1}{2}$ weeks, while the other 10 ought to be in the upper half of the range, i.e. from $1\frac{1}{2}$ to $2\frac{1}{2}$ weeks. Becoming more specific, the 1st patient in this class can be assigned a duration of $\frac{1}{2}$ week (the lower boundary value of the class), and the 20th patient likewise a duration of $2\frac{1}{2}$ weeks (the upper boundary value of the class). Intermediate positions would then be assigned intermediate durations in proportion to their relative levels along the scale. Thus the 10th patient would be considered to have a duration midway between the class boundaries, i.e. $1\frac{1}{2}$ weeks. The 15th patient would be $\frac{15}{20}$ of the way between the lower and upper class boundaries. It is even possible to calculate fractional positions such as the value of the duration at a position midway between the 15th and 16th patient, which would be $\frac{15 \cdot 5}{20}$ of the range between the class boundaries.

Let us now apply these principles to finding the duration value of the lower quartile. We found its positional value to be $55 \cdot 75$, which is between the 55th and 56th patient. The data table on page 81 shows that there are 20 patients in the first class ($\frac{1}{2}$ to $2\frac{1}{2}$ weeks), so with the 21st patient we enter the second class ($2\frac{1}{2}$ to $4\frac{1}{2}$ weeks). This second class contains 60 patients. Allowing for the 20 patients in the first class, the second class therefore contains the 21st to 80th patient of the whole sample group. The lower quartile, being situated between the 55th and 56th patient, must therefore lie somewhere in this second class.

By subtracting the 20 patients in the first class, we see that the lower quartile (the '$55 \cdot 75$th' observation in the whole sample) will be the $55 \cdot 75 - 20 =$ the '$35 \cdot 75$th' observation in the second class. There are 60 patients in the second class, so this quartile position will be $\frac{35 \cdot 75}{60}$ of the way into this class. Just what duration value will such a position demand? Since the second class extends from $2\frac{1}{2}$ to $4\frac{1}{2}$ weeks, the class range is $4\frac{1}{2} - 2\frac{1}{2} = 2$ weeks, and $\frac{35 \cdot 75}{60}$ of 2 weeks is $1 \cdot 19$ weeks. The value of the lower boundary of the second class ($2\frac{1}{2}$ weeks) must be added to this to get the duration value of the '$55 \cdot 75$th' observation. Thus the value of the lower quartile, $LQ = 1 \cdot 19 + 2 \cdot 5 = 3 \cdot 69$ weeks.

This whole calculation can be condensed into a general formula which gives the quartile values of data which is grouped into classes which have a range –

$$Q = \frac{(q - p)r}{n_c} + b$$

where $Q = $ *value* of the lower, middle, or upper quartile corresponding to

$\quad q = $ *position* of the lower, middle, or upper quartile.

$\quad p = $ number of observations in all the classes *preceding* the class in which the quartile occurs.

$\quad r = $ range of the class in which the quartile occurs.

n_c = number of observations in the class containing the quartile.

b = lower boundary value of the class containing the quartile.

See how this formula is used to find the value of the middle quartile (the median) in the present example. You will recall that we found the position of the middle quartile (mq) to be 111.

$$MQ = \frac{(111 - 80)2}{58} + 4 \cdot 5 = \frac{31 \times 2}{58} + 4 \cdot 5 = 5 \cdot 57 \text{ weeks.}$$

Similarly –

$$UQ = \frac{(166 \cdot 25 - 138)2}{31} + 6 \cdot 5 = 8 \cdot 32 \text{ weeks.}$$

We have now got the necessary data to apply Davies' test to the distribution of the data in the table on page 81. The test is applicable because the distribution of patients is obviously asymmetrical, with crowding over the low values, and there are more than 50 observations.

$$LQ = 3 \cdot 69 \qquad \text{Log } LQ = 0 \cdot 5670$$
$$MQ = 5 \cdot 57 \qquad \text{Log } MQ = 0 \cdot 7459$$
$$UQ = 8 \cdot 32 \qquad \text{Log } UQ = 0 \cdot 9201$$

Davies' coefficient of skewness is therefore –

$$\frac{(0 \cdot 5670 + 0 \cdot 9201) - (2 \times 0 \cdot 7459)}{0 \cdot 9201 - 0 \cdot 5670} = \frac{1 \cdot 4871 - 1 \cdot 4918}{0 \cdot 3531}$$

$$= \frac{-0 \cdot 0047}{0 \cdot 3531} = \underline{\underline{-0 \cdot 013}}$$

This answer is less than $+0 \cdot 20$, so the geometric mean is the right one to use for this data.

The Logarithmic Mean

When Davies' test indicates that the geometric mean is the correct one to use, it is a sign that the data has a logarithmic distribution. Let us see what is meant by this.

Fig. 6 (p. 77) shows the pattern of a simple logarithmic distribution, and we saw in Fig. 7 (p. 77) how this curve became a straight line when the data was re-plotted on a logarithmic scale.

Now consider the distribution of the data we have been discussing above. This is illustrated in Fig. 8. It forms an asym-

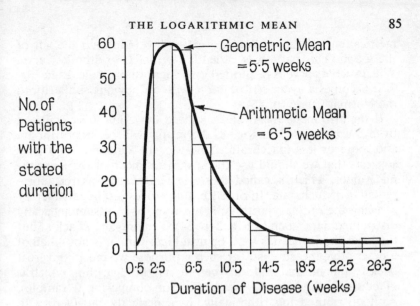

Fig. 8. Frequency chart showing the duration of 221 cases of pityriasis rosea. A frequency curve has been superimposed; it is a Normal Logarithmic Curve.

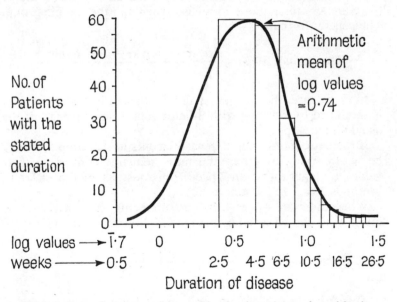

Fig. 9. Same frequency data as Fig. 8, plotted on a logarithmic scale. Doing this makes the frequency curve symmetrical.

metrical curve with its hump over the low values on the left of the graph. This pattern is called a Normal Logarithmic Curve. When such a curve is re-plotted on a logarithmic scale, as in Fig. 9, it becomes a symmetrical curve with an obvious similarity to the Normal Curve (p. 35).

If the arithmetic mean and standard deviation are calculated from the units used in Fig. 8, the results will be inappropriate and more or less inaccurate. However, the symmetry of Fig. 9 suggests that we should use the arithmetic mean of the logarithmic values, which is called the *logarithmic mean*, as the representative of such data. In practice, this proves just as satisfactory as using the ordinary arithmetic mean in cases of symmetrically distributed data (such as Fig. 2, p. 34). The only drawback about the logarithmic mean is that, being a logarithm, it is not much of a descriptive measure, so for such a purpose we must convert it back to our normal scales by getting its antilogarithm, which is of course the geometric mean. But for comparing 2 samples, both distributed logarithmically, it is perfectly satisfactory to compare the logarithmic means as such.

The way to calculate the logarithmic mean has already been demonstrated on page 76. This method should be used whenever there are less than about 25 measurements in the sample group. The general formula is –

$$Logarithmic\ mean = \frac{\log x_1 + \log x_2 + \log x_3,\ \text{etc.}}{n}$$

In many cases the accuracy will be sufficient if this calculation is simplified by working with 3-figure logarithms (instead of the usual 4-figure ones).

If there are more than 25 measurements in the sample, it may be worth trying to group the measurements, converting their values to logarithms, and proceeding just as in the shortcut Methods 2 or 3 (p. 60 and p. 62).

When there are at least 50 measurements to be averaged, a lot of work can be saved by using the following approximation –

$$Logarithmic\ mean = \frac{\log LQ + \log UQ + (1{\cdot}255 \times \log MQ)}{3{\cdot}255}$$

This formula is especially convenient because the logarithms of the quartiles have already been determined for Davies' test.

To show the application of this formula, consider the data we have been discussing. We saw that the data table on page 81 yielded the following logarithms of the quartile values –

$$\text{Log lower quartile } = 0.5670$$

$$\text{Log middle quartile } = 0.7459$$

$$\text{Log upper quartile } = 0.9201$$

The logarithmic mean of this data is therefore –

$$\frac{0.5670 + 0.9201 + (1.255 \times 0.7459)}{3.255} = \frac{1.4871 + 0.9361}{3.255}$$

$$= \frac{2.4232}{3.255} = \underline{\underline{0.7445}}$$

Anyone interested in the descriptive value of this mean can take its antilogarithm and the answer will be the geometric mean (antilog $0.7445 = 5.55$ weeks).

Incidentally, note that the logarithmic mean is the arithmetic mean of the logarithmic values; it is *not* the logarithm of the arithmetic mean.

The Logarithmic Standard Deviation

The standard deviation of logarithmically distributed measurements is needed for certain significance tests. Whenever the logarithmic mean is used, the standard deviation must be similarly logarithmic.

In these cases the distribution of the measurements is symmetrical when plotted on a logarithmic scale (Fig. 9, p. 85), so the calculation of the logarithmic standard deviation is exactly the same as for an ordinary standard deviation except that all measurements must first be converted into their logarithms.

When there are 50 or more measurements, there is another simple formula which gives a satisfactory approximation of the logarithmic standard deviation. It depends on the fact that, as was noted on page 54, 1 standard deviation equals 0.741 of the inter-quartile range. Therefore –

Logarithmic standard deviation $= 0.741 (\log UQ - \log LQ)$

In the example under discussion, the logarithmic standard deviation is –

$$0.741(0.9201 - 0.5670) = 0.741 \times 0.3531 = \underline{\underline{0.2616}}$$

For descriptive purposes, the antilogarithm of this is 1·83 weeks, which can then be called the geometric standard deviation.

This shortcut, and the one for the logarithmic mean, are mentioned in Professor Davies' article (cited on page 79); their derivation is discussed in his *Methods of Statistical Analysis* (Wiley & Sons, 1933), pp. 303–4.

Incidentally, the geometric standard deviation is quite different from the standard deviation calculated without logarithmic transformation; the ordinary (arithmetic) standard deviation of this data is 4·28 weeks, a figure which obviously has been drawn out by the long right-hand tail of this distribution.

Questions

Q17. In the following set of numbers, find (*a*) the arithmetic mean, and (*b*) the median.

$$1 \quad 2 \quad 3 \quad 4 \quad 5 \quad 6 \quad 7$$

Q18. What is the median value of each of the following sets of numbers?

(*a*) 8 3 1 2 5
(*b*) 14 12 9 10 13 11
(*c*) 38 42 60 70
(*d*) 10 17 14 18 13 11 14 13 18

Q19. In the following set of numbers, find (*a*) the mode, (*b*) the median, and (*c*) the lower and upper quartiles.

46 45 44 41 49 46 48 43 44 40 46 41

Q20. Determine the values of the lower and upper quartiles of the factory salaries in the example on page 74. Would Davies' test for logarithmic distribution be applicable to this data, and why?

Q21. A set of 100 observations is found to have a lower quartile value of 2·37, a median of 2·87, and an upper quartile of 3·99. Apply Davies' test and say which mean would be best for this data.

Q22. Suppose that a random sample of 80 people who owned a television set were asked to keep a record of the time they spent watching television for a period of 1 week, with the results tabulated below. It is desired to compare these results with those obtained in another city, so determine the correct mean and standard deviation of this data.

Viewing time (hours)	No. of people
0– 3·99	5
4– 7·99	25
8–11·99	30
12–15·99	11
16–19·99	4
20–23·99	2
24 or more	3

Q23. The weekly pocket money of 80 schoolboys is tabulated below. Draw a frequency chart of this data (preferably on graph paper), and then say whether the arithmetic mean or the geometric mean would be the better in this instance. Would Davies' test be of assistance in making this decision? Do you think the class ranges used in this table are satisfactory, or not?

Pocket money	No. of boys
0–11$d.$	0
1$s.$–1$s.$ 11$d.$	5
2$s.$–2$s.$ 11$d.$	15
3$s.$–3$s.$ 11$d.$	25
4$s.$–4$s.$ 11$d.$	20
5$s.$–5$s.$ 11$d.$	10
6$s.$–6$s.$ 11$d.$	3
7$s.$–7$s.$ 11$d.$	1
8$s.$–8$s.$ 11$d.$	0
9$s.$–9$s.$ 11$d.$	1

Q24. Suppose that a survey of the teeth of 100 teenagers showed the following numbers of decayed teeth. Which average would best represent this data?

No. of decayed teeth	No. of teenagers
0	0
1	0
2	0
3	2
4	6
5	44
6	24
7	14
8	6
9	3
10	1

Q25. Calculate the arithmetic mean of the 50 measurements of the width of my desk, given on page 33, using one of the shortcut methods. How does this compare with the mode and with the median of these measurements? Why is this so?

Q26. A comedian made 3 films. The first was 30 minutes' duration, and caused the audience to laugh 24 times. The second lasted 90 minutes, and brought forth 53 laughs. The third caused 25 laughs in its 50 minutes' run. What is the average number of laughs per minute in these 3 films?

Q27. If you bought $1,000 worth of shares on the stock market at $10 each, and another $1,000 worth at $20 each, what is the average price of the shares you bought?

Q28. A tennis club wants to compare the average performance of its 3 competition teams with those of another club. In the past year, its *A* grade team played 45 games and won 73·3% of them; the *B* grade team played 40 games and won 75%; and the *C* grade team played 35 games and won 94·3% of them. What is the group average?

Q29. Thus spake Zarathustra: 'Gentlemen, we need more funds if we are to continue our research into breeding super-rabbits. Since we started 3 years ago, our rabbit stock has grown from the original 6 to precisely 6,020. In other words, we have looked after an overall average of 6 + 6,020 divided by 2, which is 3,013 rabbits.'

Then cried Quibbler: 'Your calculation is wrong, sire. The average number of your rabbits is only 190.'

Who is right?

Q30. Lindyville and Thornton both have an average September temperature of 63°F. What further information would you need about the temperature before deciding which of the 2 places would be better for a September camping holiday?

CHAPTER 5

DESIGN OF INVESTIGATIONS

OBSERVATIONS FORM THE basis of all factual knowledge. It is now time to look at the methods of organizing the search for facts along logical lines.

In the past, there have been many men of genius who have groped their way intuitively from the darkness of ignorance into the light of knowledge. In those days it was sufficient to observe with keen eyesight, to interpret the observations with keen judgement, and to prove the facts by means of simple direct experiments.

Consider the story of smallpox. Known since earliest times in Asia, it was observed to be contagious. It was furthermore noted that if a person recovered from smallpox, he never got it again. By AD 1000 we find that a Chinese physician, Yo-meishan, successfully inoculated the emperor's grandson with dried crusts from a mild case of the disease, with the idea of giving him a mild attack in order to render him immune to the disease in its more severe form. This practice of inoculation was introduced into England in 1721 by the wife of the British Ambassador at Constantinople, Lady Wortley Montague, who had her son inoculated in this way in Turkey. Despite opposition by the clergy ('it robs Providence of the power over life and death'), inoculation gradually gained acceptance throughout England and Europe. However, the severity of the induced attacks was rather unpredictable and, moreover, inoculated patients were contagious and so spread the disease more than ever. Smallpox became as common as measles is today, and it is estimated that 60,000,000 people died from it in eighteenth-century Europe.

Jenner's Experiments

This, then, was the state of affairs when along came an English country doctor called Edward Jenner. He himself had been through the very unpleasant business of smallpox inoculation as a child, and so was quite intrigued when he heard the local folk-lore that persons who had contracted a mild infection from cows, called cowpox, were subsequently immune to smallpox. He spent 20 years investigating the subject, and in 1798 published a booklet entitled, *An Inquiry into the Causes and Effects of the Variolae Vaccinae, or Cow Pox*. Here are some excerpts from this remarkable work –

Case I. Joseph Merret lived as a Servant with a Farmer near this place in the year 1770, and occasionally assisted in milking his master's cows. Several cows became affected with the Cow Pox, and soon after attending them Merret developed the Cow Pox on his hands.

In April, 1795, Merret was inoculated with the Small-pox along with his family. Though the Small Pox matter was repeatedly inserted into his arm, I found it impossible to infect him with it. . . . During the whole time that his family had the Small Pox, he remained in the house with them, but received no injury from exposure to the contagion.

Case IV. Mary Barge, of Woodford, had the Cow Pox when she lived in the service of a Farmer in this parish in 1760. She was inoculated with Small-pox matter in the year 1791. A palish red eruption appeared about the parts where the matter was inserted, but died away in a few days without producing any Small Pox symptoms. She has since been repeatedly employed as a nurse to Small-pox patients, without experiencing any ill consequences.

Case V. Mrs. H—, a respectable Gentlewoman of this town, had the Cow Pox when very young. . . . Soon after this event Mrs. H— was exposed to the contagion of the Small Pox, where it was scarcely possible for her to have escaped, had she been susceptible to it, as she regularly attended a relative who had the disease in so violent a degree that it proved fatal to him.

In the year 1778 the Small Pox prevailed very much at Berkeley, and Mrs. H— not feeling perfectly satisfied respecting her safety (no indisposition having followed her exposure to the Small Pox) I inoculated her with active Small-pox material. The same appearance followed as in the preceding cases – an eruption on the arm without any effect on the constitution.

Case XVII. Sarah Nelmes, a dairymaid at a Farmer's near this place, was infected with the Cow Pox from her master's cows in May, 1796. She received the infection on a part of the hand which had been previously in a slight degree injured by a scratch from a thorn. . . . Then accurately to observe the progress of the infection, I selected a healthy boy, James Phipps, about eight years old, for the purpose of inoculation for the Cow Pox. The matter was taken from a sore on the hand of Sarah Nelmes, and was inserted, on the 14th of May, 1796, into the arm of the boy by means of two superficial incisions, barely penetrating the skin, each about half an inch long. . . . On the ninth day he became a little chilly, lost his appetite, and had a slight head-ache . . . but on the day following he was perfectly well. . . .

In order to ascertain whether the boy, after feeling so slight an affection of the system from the Cow-pox virus, was secure from the contagion of the Small-pox, he was inoculated the 1st of July following with Small-pox matter. . . . The matter was carefully inserted, but no disease followed. . . . Several months afterwards, he was again inoculated with Small-pox matter, but no sensible effect was produced on the constitution. . . . To convince myself that the Small-pox matter made use of was in a perfect state, I at the same time inoculated a patient with some of it who had never gone through the Cow-pox, and it produced the Small-pox in the usual regular manner.

In all, Jenner described 20 patients who had been infected with cowpox, either naturally or artificially, and who remained unaffected by subsequent artificial inoculation with active smallpox virus (notice how he proved the potency of his material by using controls). From these observations and experiments he concluded that 'the Cow-pox protects the human constitution from the infection of the Small-pox.'

This publication brought forth a storm of protest. Not only the clergy complained this time, for in 1798 the idea of intentionally infecting human beings with an animal disease was utterly repugnant, even to the medical profession. As one doctor said, 'Can any person say what may be the consequence of introducing a bestial humour into the human frame, after a long lapse of years? Who knows but that the human character may undergo strange mutations from quadrupedal sympathy.'

Nevertheless, the logic of Jenner's experiments won the argument, and his vaccination procedure was gradually accepted by the civilized world. A hundred years later the results were reviewed by the President of the Royal College of Surgeons of England, Sir John Simon, in these words – 'Jenner's services to mankind in respect of the saving of life have been such that no other man in the history of the world has ever been within measurable distance of him.'

Nowadays smallpox only occurs in those parts of the world where vaccination is neglected.

*   *   *

This piece of medical research is a model of simplicity and clarity. The results speak for themselves. Jenner lived at a time when statistical procedures were in their infancy; they were certainly not being applied to medicine. So he makes no mention of the method by which he collected this sample of 20 patients. But if they represented a random sample of all those people who had had cowpox but not smallpox, we are now in a position to calculate the probability of the observed results coming about simply as a result of chance. We would not be far from the mark if we assumed that with people who had never had smallpox, inoculation with smallpox virus would produce the disease, on the average, 9 times out of 10; 1 in 10 times there would be no reaction because of inactive inoculation material or inadequate inoculation technique. The probability of a non-reaction is therefore 1 in 10, or $\frac{1}{10}$. It is as though each inoculated person took a 'lucky dip' into a large bag containing 90% black and 10% white marbles; he would have a reaction if he drew a black marble, but not if he drew a white one. Now Jenner had 20 people who had not had smallpox before, and not one of them developed smallpox from the inoculation. The likelihood of this happening by chance can be visualized by the 20 people picking a marble from that bag, and all of them drawing a white marble. By the Multiplication Law, this probability is $(\frac{1}{10})^{20}$, which is 1 chance in 100,000,000,000,000,000,000 times. Jenner was therefore quite safe in ignoring the possibility that his results could have been due to chance.

Most of the basic facts of science known today are derived from observations and experiments which are as clear as this. More often than not, however, men's experiments have not led to clear-cut results containing indisputable truths, and sorting out all the erroneous, conflicting, misleading, and doubtful results often takes many years. Even in the present century, for example, and on an issue as important as the testing of BCG vaccine against tuberculosis, a lack of proper controls delayed the emergence of the true answer for a period of 25 years.

For the art of finding the truth by means of experiments is by no means easy. It's not like a simple chemistry experiment at school. Take a farming example. Long ago men discovered the best time of the year in which to plant their crops, and somewhere along the line they noted that certain substances acted as growth stimulants (fertilizers) which increased the yield of their crops. Now suppose a man wanted to compare 2 fertilizers, to see if one was better than the other. He might try magic, or contemplation, or long discussions with his friends, but such methods only proved right 50% of the time. So he must put the fertilizers to a practical test. One way would be to use one of the fertilizers on his land for one year, and then use the other fertilizer on his land the next year, and compare the results. However, this would not be entirely reliable because different climatic conditions in each year could influence the results in a way, and to an extent, which would be impossible to assess accurately. Now he saw that this source of uncertainty could be eliminated by improving the design of the experiment; he would compare both fertilizers in the one year, treating half his land with one of them, and the other half of his land with the other. The only trouble now was that he could never be sure that regional differences in the soil were not interfering with the fairness of the comparison; some parts of his land might well be naturally more fertile than others, and it was difficult, if not impossible, to divide his land into 2 portions of exactly equal fertility. So again he improved the design, dividing his land into a number of small plots of equal size, like a chess-board, so that every second plot could be treated with one fertilizer, and the alternate plots with the other one. This would cancel out any regional differences over his land, but one more problem still remained – perhaps his results would not apply to other farms.

Some years ago the Irish Department of Agriculture performed a careful set of experiments which showed that one particular variety of barley, called Spratt-Archer, was superior to all others. Farmers all over Ireland were delighted with it, that is, except in one district where they claimed that it was not as good as their own native barley. Now the Department's experiments had demonstrated that this was not so, so to convince the farmers in the resisting district, the Department went out to that district and repeated the experiment right there. Imagine their surprise when they found that the farmers were quite right, because the weeds, which flourished in this district, overgrew and smothered the Spratt-Archer barley, whereas the native variety thrived by virtue of the fact that it sprouted more quickly. The original experiments had, of course, been carried out on ideal land, without the weeds. ('Student', *Nature*, March 1931, p. 405.)

And so, bit by bit, the methods of research improved. However, as men's knowledge of things increased more and more, it was natural that their questions and probings should become ever deeper, more pointed, more specific. Their experiments became more sensitive, their apparatus became more intricate, but somehow the truthfulness of their results still tended to be erratic. There seemed to be no way of ensuring the truth of their results. So that, in 1935, one man wrote, 'The waste of scientific resources in futile experimentation has, in the past, been immense.' (Fisher.)

Obviously the time had come when it was essential to take a long, deep look into the *methods* of research.

The man who began this inquiry was a genius, an English statistician named Professor Sir Ronald A. Fisher (1890–1962). He tackled the job at its very foundations, pointing out that the truth of any piece of research was necessarily and intimately bound up with Inductive Logic. It was always a question of, 'How far can the results of this particular experiment or observation be trusted as a generalization?' Now, whenever an argument proceeds from the particular to the general, there must always be an element of uncertainty, but the degree of this uncertainty is a thing which can be calculated quite precisely by the application of statistical procedures based on the Laws of Chance. This meant, of course, that the experiment would have to be designed in such a way that it would lend itself to statistical analysis, and this point had been overlooked through the years because of the historical accident whereby probability calculations had originally been worked out solely and specifically in relation to games of chance. The occasional attempts to apply such calculations to the more serious business of scientific research had failed as often as not, because the statistical procedure and the experimental design had not been put into partnership before the experiment was begun. Fisher then showed how such a fusion could be accomplished, and went on to devise an entirely new statistical procedure, called the *Analysis of Variance*, which was geared specially for use in scientific research. Since its introduction in 1924, this has proved easily the most useful mathematical technique in the whole statistical repertoire, with applicability in every imaginable field. (The variance of a set of results is the square of their standard deviation.)

In 1935, Fisher published his ideas in a book, *The Design of Experiments* (Oliver & Boyd), and from then on, scientific research was put on an entirely new footing. In every department, both in and outside the laboratory, research methods were overhauled, and many improvements were introduced. Any investigator today who fails to avail himself of this new knowledge is running a great risk of wasting time, money, and effort, unless his work is as clear-cut as that of Jenner.

Let us take a look at the important principles which have stemmed from Fisher's work.

All research passes through 3 stages of thought –

　(1) *Defining the problem* – what is being sought?

　(2) *Choosing the method* – how shall the answer be found?

　(3) *The interpretation* – what do the results really mean?

What Is Being Sought?

To define exactly what is being sought is sometimes easy, sometimes hard, and often overlooked. It calls for strict definitions of what is being sought, from which parent group, under what conditions, and by what criteria. All this may sound a bit pedantic, but experience has shown it to be extremely important.

The question of *what is being sought* needs to be clearly formulated right at the beginning. Any vagueness here will carry right through to the end. In particular, is the investigation aimed at some isolated piece of information (such as, 'What percentage of Frenchmen get chilblains?' or, 'Will this bomb explode?'), or is it aimed at comparing two or more things (such as, 'Are Frenchmen more prone to chilblains than Englishmen?' or, 'Is this bomb more powerful than last year's models?'). This simple, direct approach is nearly always better than asking compound questions such as, 'Are Frenchmen especially prone to chilblains, and if so, is it due to their clothing, their wine-drinking, or other factors?'

The *parent group* which provides the material for the investigation needs to be defined carefully. Exactly what sort of cases are being tested? The *Literary Digest* poll (p. 38) was a clear example of taking a sample from a parent group (the Digest and telephone subscribers) which was obviously only a portion of the entire parent group of voters. Whenever you come across the phrase, 'A random sample showed . . .', always ask, 'A random sample from which parent group?' From cases of measles in hospital, or at home (p. 50)?

An example of a good definition of a parent group was provided by the English Medical Research Council in its report on the effectiveness of streptomycin treatment for tuberculosis in patients with 'acute progressive bilateral pulmonary tuberculosis, of presumably recent origin, bacteriologically proved, unsuitable for collapse therapy, aged 15 to 30 years'. (*Brit. Med. Journ.*, October 1948.)

Imagine the chaos when such defining is neglected in a piece of research carried out by a *group* of investigators. Yet during 1957, of 24 such group investigations published in the *British Medical Journal*, 12 gave adequate diagnostic criteria, 7 gave vague criteria, and 5 gave none. (Dr K. Cox, *Med. Journ. Australia*, November 1963.)

Defining the *conditions* under which the investigation will be performed is especially important in all biological fields. It is not enough to say 'the ointment is to be applied twice a day'; each patient has to be instructed as to the exact method of application – is the treated area to be washed before each application, are scales or crusts to be removed, is the ointment to be smeared on gently and briefly or rubbed in for some minutes, is the ointment to be applied thinly or thickly, is a bandage or other dressing to be applied? And of course every patient in the trial must get the same instructions. In the case of tablets, is a fixed dose going to be used on all patients, regardless of their age and weight, or will the dose be varied to suit the individual, which introduces another variable in the form of the skill of the investigators?

The *criteria* by which success or failure of the investigation will be judged also needs consideration. It will often entail defining such things as the endpoint of a disease, which (unless it be death) can be difficult to specify except in terms such as 'duration of stay in bed' or 'until the patient's temperature has been normal for 3 days'. The same principles apply in non-medical fields too.

How Shall the Answer Be Found?

Having defined what is being sought, the next step is to decide how to go about getting the answer.

In broad terms, there are 2 kinds of scientific investigation –

(*a*) *Surveys*, in which observations are made on things as they are, without interference.

(*b*) *Experiments*, in which something is done to see what effect it has.

In most cases, the type of problem will itself determine the kind of investigation required. Thus if you want to know which flavour ice-cream will prove the most popular at your child's birthday party, a survey of those invited will reveal the answer. On the other hand, if you want to know whether an electric spark will ignite gunpowder, an experiment will decide the issue.

However, with the majority of questions being asked today, the difficulty in getting a true answer is very much greater than you may imagine. This applies not only to research into atomic structure, in which particles having a life of a millionth of a second are discovered, but even to such ordinary things as finding out which radio station is the most popular in your town. Here is an actual case –

The relative popularity of different radio stations has great importance in the advertizing trade, because the most popular stations can demand the highest advertizing fees. In one such survey, carried out by an established market research firm, a properly selected sample of people were asked to write down each program that they listened to for a period of a fortnight. Now it so happened that, quite independently, a university department was also conducting a survey on the relative popularity of radio stations in the same town and at the same time. The university department also chose a properly selected sample of people, but their method was to ask the selected individuals to list the radio stations in order of preference, according to their usual listening habits. The percentage points awarded by these 2 surveys were as follows –

Station	A	B	C	D	E	F
Research firm	23·6	23·4	17·3	15·8	15·4	4·5
University	20·1	19·9	13·3	16·5	23·4	6·8

Which are we to believe? According to one survey, Station E is the second bottom; according to the other it is the most popular. Both surveys selected their sample groups correctly, and both had samples of reasonable size. The difference in the results, then, must be due to the difference between the 2 ways of collecting the information. This is the kernel of the problem. You can see in this case how important it is to frame the question you want answered in very precise terms. Are you interested in the programs that people will listen to, and write down, during a specific fortnight, or are you interested in peoples' recollections of their favourite programs? How would you react if you were included in these surveys? Would you have any trouble in listing 6 radio stations in order, according to your listening habits? Would you tend to listen to your radio more, or differently, if you were under some kind of surveillance? Would you tend to listen to a certain station more than usual if that station happened to be giving substantial prizes to lucky listeners at frequent intervals during the survey period?

It should be clear, then, that to get the truth is not an easy matter, even in a case as simple as the one just described. In fact, it nearly always needs the combined knowledge of 2 experts – the expert in the field being investigated, and the expert in the design of investigations. The design expert is called a statistician. Many people are not aware that one of the basic aspects of the science of Statistics today is the study of how to design investigations in such a way that the conclusions will be correct and meaningful. Only too often the completed results of some investigation are presented to a statistician for his approval, and he finds that the design has been so inadequate that no amount of mathematical manoeuvring can make the results prove anything.

Our next example turned out to be a lucky exception to this rule. It comes from W. A. Wallis and H. V. Roberts, *Statistics – A New Approach* (Free Press, 1960).

A sociologist wished to study the effect of television upon the use of library facilities in a community. From the list of card holders at the public library, he selected a sample to whom he mailed a questionnaire. One question asked was whether the respondent owned a television set and, if so, when it was purchased. After the questionnaires had been returned, the sociologist started to organize his information. For all respondents who had bought television sets within the preceding two years, he compared the rate of borrowing from the public library before and after the purchase of television. He did this by averaging the rates before purchase, and comparing this with the average rate after purchase. He found a decline of about 10%.

Next, he computed the average rate of use for the same two-year period by nonpurchasers of television sets. The average for this group was about the same as for the other group before they purchased television. The sociologist tentatively concluded that his data showed that television had caused a decline in the rate of use of the library. He decided to discuss his results with a statistician before publishing them.

The statistician's opinion was that the averages presented were inconclusive, because (1) the average rate for the 'control group', those who had not purchased television, may also have declined during the two-year period, and (2) the method of calculating 'before' and 'after' rates for television owners left the possibility that a seasonal effect might distort the findings. To illustrate the latter possibility, suppose that people tend to use the library more in winter than at other seasons, and suppose a lot of people bought television sets two months after the beginning of the survey period (which began on January 1, 1950), then their 'before' rates would be based on two winter months, while their 'after' rates would include both winter and summer months. Thus, even if acquisition of television had no effect on library use, there would be an apparent decline in use after purchase, solely because of the seasonal pattern.

The statistician suggested the following procedure: For each television purchaser, compare the rate of library use during the twelve months prior to purchase with the rate during the twelve months after purchase. This could be done from the library's records. For each purchaser, choose a nonowner at random from those matching the purchaser in such characteristics as age, neighborhood, etc. For the nonowner, calculate separate rates of library use for exactly the same two time periods as for the purchaser with whom he was paired. Then compare the averages of owners and nonowners.

The sociologist naturally felt somewhat foolish, for the statistician's suggestions seemed to represent only common sense rather than technical knowledge; but they represent a kind of common sense that becomes highly developed in good statisticians through varied experience in analyzing data. In this particular study, the sociologist was unusually fortunate in being able to reorganize the data he had obtained, though he regretted the extra work and delay. Often, people are not so fortunate: no reorganization of data will bring out facts that were not collected in the first place.

The right time to consult a statistician, then, is at the very beginning, in the planning stage.

Here is an interesting contrast with the polio vaccine trial on page 46. The first properly-designed, large-scale medical experiment to be carried out in the USA was started in 1951, when it was desired to find out if gamma-globulin would prevent poliomyelitis. To help them plan the experiment, the doctors consulted a statistician, who estimated that the trial would need 50,000 children, half of whom would be given gamma-globulin while the other half would be given something which looked exactly the same but which in fact contained no gamma-globulin or other active substance. The effectiveness of the gamma-globulin could then be assessed by comparing the incidence of polio in the 2 groups. At first the number of children needed seemed to be excessive, but after the trial it was clear that anything less than this number would have led to inconclusive results. Bearing in mind the cost of such a large-scale experiment, one should at least be reasonably sure that it will lead to a definite answer, one way or the other. Admittedly no one can guarantee that the results of any piece of research will prove useful, but at least it should not be doomed from the start to prove nothing. (Dr W. Hammon *et al.*, *Journ. Amer. Med. Assoc.*, 1952, Vol. 150, p. 744.)

Let us therefore take a look at the ways in which statisticians can help in designing surveys and experiments.

Surveys

There are 3 types of survey –

    (*a*) Retrospective
    (*b*) Current
    (*c*) Prospective

Retrospective surveys look back over information recorded in the past. The usefulness of this type of investigation is often limited by records being incomplete in respect of the data being sought. It is hard to prevent this fault, for who knows what data may interest us in 10 years' time, and it is costly to keep records of things that will never be wanted. Nevertheless, good work has been done with this technique, as for instance –

During the Second World War it was necessary to keep planes in action as much as possible, so it was decided to see if the number of time-consuming engine overhauls could be reduced without risk. A retrospective survey was made of planes that were lost, and contrary to all expectations, it was found that the number of planes lost as a result of engine troubles was greatest right after overhaul, and actually decreased as the time since overhaul grew longer. This result led to a considerable increase in the intervals between overhauls, and needless to say, to important revisions in the manner of overhauling to make sure that all those nuts and bolts were really tightened up properly. (From W. A. Wallis and H. V. Roberts, *Statistics – A New Approach*, Free Press, 1960.)

A more extended example is discussed in Wallis and Roberts' book, #2.8.2. It concerns a retrospective survey of the incidence of insanity a century ago, to compare with the present rate, as part of a project to assess the influence of the pressures of modern living on mental health. A sound statistical approach has elevated this particular survey into a most important sociological document.

Current surveys are those which seek information about things as they are at the present time. The 2 surveys of radio audiences described on page 100 are typical examples.

Prospective surveys are a fairly new idea. They compare information obtained about some subject now, with a follow-up about the same subject obtained from the same individuals at a later date. The American Cancer Society's investigation described on page 37 is an example of a prospective survey.

Prospective and current surveys are beset with similar problems, so they will be discussed together. These problems concern the design of the questionnaire, the sampling, and the actual conduct of the survey.

Having decided on precisely what is being sought from the survey, the next thing is to prepare a question, or more usually a set of questions, designed to bring forth the desired information. The questions must be worded in simple, clear language. Standard answers should be provided whenever possible ('circle the answer which applies in your case'), otherwise you may find that in answer to a question such as, 'Where do you get chilblains?' one person may reply, 'Lower limbs', another may reply, 'On left big toe, $\frac{1}{4}$ inch from cuticle', while yet another may answer, 'Only at ski resorts'. After wording all the questions as carefully as possible, the next step is to try out the questionnaire on a number of people, for the express purpose of detecting unsuspected ambiguities or misunderstandings. Some people would be unsure about this question – 'What medicines, if any, do you take regularly?' They may be unsure as to whether 'medicine' refers only to a medicine prescribed by their doctor, or whether it also covers some mild patent remedy, such as throat lozenges or a laxative. Again, some questions to which a simple yes or no answer is expected may give trouble if the person finds that to be correct he should answer it 'sometimes yes, sometimes no'. An example of such a question would be, 'Do you usually take notes in lectures?' Perhaps the answer will depend on who the lecturer is, or on whether the subject is satisfactorily dealt with in textbooks, rather than on the habits of the person replying to the questionnaire. Maybe it would be better to change the question to read, 'Do you take lecture notes (a) always, (b) usually, (c) sometimes, (d) rarely, or (e) never?'

After this pre-testing, the questionnaire can be revised and, if necessary, re-tested to make sure it is faultless.

Even so, some of the answers will be wrong. Ladies will understate their age, gentlemen will overstate their occupation (the man who repairs electric toasters may elevate his status to 'electrical engineer'), and some adults might even be reluctant to admit that the main reason they buy a certain newspaper is because they like its comic strips.

Vance Packard, in *The Hidden Persuaders* (Penguin, 1962), quotes some nice cases. For example, a survey was carried out on behalf of a firm which sold kippered herrings, in order to discover why sales were lagging. It turned out that most of the people interviewed said they just didn't like the taste of kippers. However, under persistent and subtle probing, it was discovered that 40% of these people who said they didn't like kippers had never tasted kippers in their lives!

A manufacturer made 2 sizes of kitchen-range, one large and the other small. The large model proved much the better seller of the two, and a questionnaire revealed that housewives preferred it because it had much more working space on its top. Taking a tip from this, the firm then designed a medium-sized, lower priced stove with an exceptionally large working space. Imagine their disappointment when they found that this new stove just wouldn't sell. The manufacturer then consulted a market research firm to investigate the matter. The research firm's psychologists found that the housewives were not consciously telling fibs (as in the case of the kippered herrings, above), but the story about the large working area on the stove top was really only a rationalization of their unconscious desire for a big, expensive-looking stove.*

Let these examples be a lesson to anyone naïve enough to believe that people always tell the truth to opinion interviewers or questionnaires. In market research, these little 'fibs' can be very costly to the misled manufacturer, so psychologists are often brought into the affair. Their questions probe beneath the surface. Once they used to ask, 'We are thinking of marketing a new talc powder in a red tin, and would appreciate your opinion; do you think red would be a nice colour for the tin?' Nowadays a psychologist might ask exactly the same question in the following form: 'Do you associate the colour red with any particular part of the human body?'

To illustrate the potential value of depth probing, Vance Packard quotes the case of a psychologist who gave a Rorschach ink-blot test to 80 smokers who had a strong loyalty to one or other brand of cigarettes, and from this test was able to name their favourite brands in nearly every single case. The ink-blot test shows certain personality traits, which he was able to link with the various brand 'images' of the cigarettes.

However, most current and prospective surveys don't need any depth probes, just plain straight questions and answers.

Sampling is the next problem. As explained on page 41, the selecting of a proper sample, especially of people, is not easy. A stratified random sample is generally the best kind for surveys, particularly as the sample size required for a given degree of precision is quite a lot smaller than would be the case with an ordinary random sample (p. 48). But it's easier to talk about getting a good sample than it is to get one. It involves a continuous battle against factors which upset randomness. But the sample *must* be random.

Non-response is one of the main factors which will ruin the randomness of any sample, no matter how well-chosen the sample may be. All those people who are not at home, who would rather not answer that particular question, who are too busy to answer all those questions, who are too sick to answer any questions, or who simply don't know the answers to some of the questions, all these people had better be few and far between, or your sample could become seriously biased. Because the risk is always there that these non-respondents may differ significantly from those who do respond. Hence this fault cannot be rectified by simply questioning more people to allow for this 'shrinkage'.

The first Kinsey Report, a survey of the sexual habits of over 12,000 American men, published in 1948, came under considerable criticism for ignoring the problem of non-response. The authors of the Report say little about the method used to collect their sample, except to admit that randomness is hard to achieve. But we know that the mere size of their sample does not make it a random one (p. 38), so we are left to wonder whether those men who excluded themselves from the sample, i.e. the non-respondents, would have changed the sample averages or not.

The problem of non-response should never be ignored. It should be tackled by spending more time and money to reduce this pool of resistance to an absolute minimum; this may mean that the sample will be a smaller one, but the data obtained will be better.

The sample size needed depends on what is being sought and the degree of precision required. If the sample is random, its reliability increases with the square root of the number in the sample; if the sample is not random, its reliability remains a matter of luck, regardless of the sample size. We have discussed this matter on page 45. To save wasting time and money, a statistician should be consulted to make an estimate of the sample size; otherwise the survey may turn out to be too small for any reliance to be placed on its answers, or it may end up unnecessarily large.

Wallis and Roberts recall a statistical traffic jam which occurred in New York State in 1950. In order to find out the most suitable route for a new highway between New York City and Buffalo, a traffic commission decided to conduct a 24-hour survey of vehicles using the existing road. So, on a certain Sunday, every 4th vehicle was stopped, and its driver was asked 3 questions –

(1) Where did you come from?
(2) Where are you going?
(3) How often do you make this trip?

It seemed simple enough, but its object was largely defeated by acute oversampling. It caused a traffic jam 10 miles long, and many motorists, hearing of the jam, no doubt took alternative routes. The 25% sample size was much bigger than necessary. It would have been a better design to have taken a smaller survey, say every 100th vehicle, and spread the investigation over a whole week.

Despite a goodly number of formulae to help them estimate the sample size, statisticians will often ask for a small preliminary survey to be conducted; the results of this are then used to increase the accuracy of their estimate.

The actual conducting of the survey is generally done either by personal interviews or by mail. Mailing the questionnaires is cheaper, but the percentage of non-response may be high. It may take several letters, a telegram, or phone call, and finally sending an interviewer to get all the answers back. The *Literary Digest* didn't do these things (p. 38).

The efficiency of different interviewers varies quite a lot. Some have the knack of putting people at ease and getting the answers quickly and easily; others haven't. Sometimes the interviewer unwittingly affects the answers. For example –

Some years ago the National Opinion Research Centre in America found that when Negroes were asked, 'Do you think that Negroes are getting a fair deal in the Army?', 35% said no to Negro interviewers, while only 11% said no to white interviewers.

Of course, interviewers must be properly trained. For instance, they must never ask leading questions (i.e. questions worded in such a way that the answer is suggested) such as, 'After work, do you prefer fancy drinks like cocktails, or a man-size glass of beer?' It'd take a tough customer to admit to a cocktail under these conditions, wouldn't it? Or how about the following more subtle example –

Interviewer: What brand of toothpaste do you use?
Lady: Densol.
Interviewer: Why do you like Densol?
Lady: Ah ... um ... I've never really thought about it.
Interviewer: Well, try to give me an answer. (Pause). Is it because of its taste?
Lady: Yes, it's got a nice taste.

And the interviewer writes down, 'Uses Densol because she likes its taste.' But perhaps the lady wasn't prepared to admit the real reason, that it was the cheapest brand, or that she thought it counteracted her bad breath.

Any way you look at it, even a well-trained interviewer is still a human factor situated at a key point in all personal surveys.

These, then, are the kinds of things which must be dealt with in designing a survey.

Experiments

Experiments differ from surveys in that they are tests in which the effect of some deliberate act is observed. They may be carried out in a laboratory, or they may go out into the world and be carried out in the environment of everyday life (in the fields, in hospitals, on board ships, etc.). If you want to perform an experiment in a vacuum it would almost certainly need to be done in a laboratory; on the other hand, if you want to find out how a new road-surfacing compound stands up to heavy traffic, it would probably have to be done 'on location'.

Now regardless of where an experiment is performed, and regardless of its subject matter, all experiments have this in common: they are out to prove something. As mentioned earlier, the statement of what is to be proved must be framed in the clearest possible terms in the planning stage of any experiment, but in addition, we must now note a most important aspect of the logic of experimentation. It is that an experiment can only prove something which actually happens; no finite number of trials can ever prove that something won't happen, for there is always the possibility that it *will* happen on the very next trial. To permit a decisive conclusion, then, the question put to test by the experiment must therefore seek an affirmative (yes) answer. If a man wants to know whether gunpowder explodes when struck with a hammer, and asks the simple question, 'Does it?', he can nurse his wounds knowing that he has found the truth if the answer is 'Yes', but must forever remain in some doubt about whether he struck hard enough if the answer is 'No'.

In practice, the situation is handled as follows. Before the experiment is begun, it is tentatively assumed that the outcome will be negative ('It won't explode', or 'There is no difference between these 2 fertilizers'). This tentative negative assumption is technically called a *null hypothesis*. This assumption is then put to the test by the experiment. If it is proved wrong ('It did explode'), the result is clearly decisive; if it is not proved wrong

('There was no explosion'), the result is said to be 'not proven' under the conditions of the experiment. It is important to realize that a null hypothesis can never be proven or established by any experiment; it can only possibly be disproved.

All experiments are basically comparisons. This factor of comparison is quite obvious when 2 fertilizers are tested to see which is the better one, but it is also present (though hidden) in the above gunpowder experiment or any other similar one which is performed simply to see what happens. How? Because in such cases the outcome of the experiment is compared with the phenomenon of chance. Before any useful conclusion could be drawn from the experiment in which some gunpowder is struck by a hammer, one would have to know whether the gunpowder was likely to explode spontaneously (i.e. by chance) regardless of whether it was struck or not. If this has never been observed, the probability of it exploding by chance at the very moment when the hammer struck is so remote that we would be justified in concluding that the hammer blow rather than chance was the cause, if in fact it did explode at that moment. (Strictly speaking, chance is never a cause; it only refers to a happening which occurs in the absence of a cause.)

Consider another example, in which chance plays a larger part. Suppose that a lady claims that she can taste the difference between a cup of tea in which the milk has been added *before* the tea, and one in which the milk has been added *after* the tea. To test her claim, we could design an experiment in which 8 cups of tea, 4 made in one way and 4 made in the other way, were presented to her. She would be told the plan, and asked to taste them and divide the cups into the two categories. Now if the cups were presented to her in a random order, the probability of her doing this correctly by chance can be calculated, using the basic Laws of Chance. Assuming that chance alone determines the issue, the situation becomes exactly equivalent to blindly drawing 4 black marbles from a bag which contains 4 black mixed with 4 white marbles. For reasons described on page 26, the probability that the first marble drawn will be black is $\frac{4}{8}$, that the second one will be black is $\frac{3}{7}$, that the third will be black is $\frac{2}{6}$, and that the fourth one will be black is $\frac{1}{5}$. The probability of all these events happening, either one after the other or simultaneously, is governed by the Multiplication Law, so –

$$\frac{4}{8} \times \frac{3}{7} \times \frac{2}{6} \times \frac{1}{5} = \frac{1}{70}$$

From this we can see that if the lady had absolutely no ability to discriminate between the 2 kinds of tea, she could nevertheless divide them into 2 groups of 4, correctly, once in 70 trials (on the average) by pure guesswork.

This, then, is the chance factor against which we shall be testing the lady's performance. If she classifies each cup correctly, the chances are 69 out of 70 that she can really taste the difference, but there is 1 chance in 70 that she did it purely and simply by sheer good luck. If she can't really tell the difference, the odds are decidedly against her passing this test.

In a case like this, we start off with the tentative negative assumption, the null hypothesis, that she can't distinguish the two; if she fails in the test we haven't really proved anything except that she is not infallible, but if she succeeds we have an accurate yardstick by which to measure the degree of certainty that the result is truly as she claims.

This simple example, which incidentally comes from Fisher's *The Design of Experiments* (Oliver & Boyd, 1960), nicely demonstrates the need to incorporate statistical principles in the experimental design, which was one of Fisher's main points. For only by specifying that the cups of tea were to be presented in random order, as determined from a Table of Random Numbers, can the outcome of the experiment be properly assessed; the calculations above would all be quite meaningless in the absence of randomness.

Furthermore, it also serves to show how various designs affect the sensitivity of an experiment. Suppose it was desired to make the test tougher, to make it even less likely that the lady could pass the test by guessing. This could be done by repeating the experiment a number of times. We calculated that the probability of guessing the correct answer in one trial was 1 in 70. If the trial was carried out twice, independently, and the lady classified all the cups correctly, the probability of this happening by chance would be –

$$\frac{1}{70} \times \frac{1}{70} = \frac{1}{4,900}$$

which is a very remote chance indeed.

Alternatively, the test could be made more stringent by doubling its size; the lady would then have to pick two sets of 8 out of 16 cups. Using the same kind of calculation as before ($\frac{8}{16} \times \frac{7}{15} \times \frac{6}{14}$, etc.), we find the likelihood of this test being passed without a mistake, as a result of chance, becomes but once in 12,870 times. If the lady passed such a test, it ought surely convince the hardiest sceptic that she knew what she was doing.

Another way of altering the design would be to keep the size of the experiment unchanged (i.e. 8 cups), but to alter the internal structure. For example, we might change the predetermined proportions of the cups in each category from 4 and 4 to say 3 and 5. The lady would now be asked to distinguish the 3 cups in one category from the 5 in the other. The probability of doing this by chance is –

$$\frac{3}{8} \times \frac{2}{7} \times \frac{1}{6} = \frac{1}{56}$$

This is no good to us, for it actually renders the test less sensitive than it was before. But it does illustrate the rule that equal numbers in each category will ordinarily give any experiment its maximum sensitivity.

Another structural alteration would consist of not fixing in advance that 4 cups shall be of each kind, but rather allowing the treatment of each cup to be determined independently by chance, say by tossing a coin, so that not only would the cups be presented to her in random order, but also the number of each kind would be left to chance. As before, she would be told what the plan was. The design is now comparable with the drawing of 8 marbles from a large bag which contains 50% of black and 50% of white marbles. Correctly naming the 8 cups randomized in this way, without the help of true discrimination (i.e. purely by chance), would then have a probability of $(\frac{1}{2})^8 = 1$ in 256 chances. That this should be smaller than the 1 in 70 probability of the original version of the experiment is only to be expected when there is the additional uncertainty concerning the number of cups that will be in each category.

Now, when it comes to experiments in which 2 or more things are being compared, it is always essential to make sure that the test conditions are as alike as possible, or else the comparison may be quite unfair. If string A breaks at 65 lb, and string B breaks at 75 lb on the same testing machine under identical conditions, there is no question about the comparison; B is stronger than A. But in other cases, like that of the 2 fertilizers on page 96, it may take considerable ingenuity to ensure that the comparison is fair.

Suppose, for a further example, that you want to test some new nasal drops which you think may shorten the duration of common colds. This would be done by dividing a sample group of patients with colds into 2 halves, either using random numbers or by strict alternation of successive patients, and treating one group with the nasal drops and keeping the others untreated to serve as a *control group* for purposes of comparison. Any benefit from the nasal drops will then show up as a shorter arithmetic mean duration of the treated as compared with the untreated control group. Notice that the design of this experiment has ensured that the 2 groups will be genuinely comparable because, having been selected by a random process, the variations between the duration of individual colds can be expected to be the same in each group; there should be about the same number of 'long' and 'short' colds in both groups. The randomness is therefore a very important part of the design.

In the branch of medical research which seeks new treatments for human diseases, it is rarely possible to have a control group of *untreated* patients. Anyway, an untreated group would often be unsatisfactory as a standard for comparison, because it is known that the psychological effect of 'getting some treatment' can in itself influence the course of many diseases. This applies not only in nervous disorders and psychosomatic complaints, but even in some outright infections (e.g. warts). It is not uncommon to find reports in which 50% or more of patients given some bland or completely inert 'treatment' were quickly relieved of their symptoms.

It is thus unwise to draw conclusions from comparisons between treated and untreated groups of people, because any apparent benefit from the new treatment may be due to nothing but psychological factors. This could apply, for instance, to the case of the nasal drops discussed above. The same criticism also holds for research with new treatments for recurrent complaints like car sickness, sunburn, or herpes simplex ('cold sores'), in which the severity of the complaint under a new treatment is compared with the severity noted by the sample patients, on previous occasions, when they had no treatment (or at least none prescribed under the conditions of the research experiment).

The control group must therefore be given treatment of some kind or other. Well, then, what sort of treatment? When trying out a new remedy for the first time, it may occasionally be desirable to give the control group an inert 'treatment' (= *placebo*), but the shortcoming of this design is that it can only show if the new treatment is better than nothing. It is generally far better to try the new treatment on a small sample of patients (say, about 6), and if the results appear favourable, to proceed at once with an experiment to compare the new treatment with one or more standard treatments. This is done by randomly dividing a sample of patients into the required number of groups, and then the new treatment is given to one group, while standard treatments are given to the control groups. Not only is this design fairer to the patients, but it also provides a proper assessment of the status of the new remedy (i.e. better, same, or worse than standard). This approach is not used nearly as often as it should be. Large numbers of new treatments are constantly being devised, and unless such 'status' comparisons are made, it leads to a new kind of doctor's dilemma, in which he knows a lot of treatments for any particular complaint, but is not at all sure which one is best.

Because the investigator's enthusiasm can be 'infectious' to many patients, it is often desirable for the investigating doctor, as well as the patients, to be ignorant as to who are in the 'new treatment' group and who are in the control group. Only then can we be sure that all psychological factors are exactly the same for both 'new treatment' and control patients. As this design keeps both doctor and patients 'in the dark', it is called the *double blind* technique. Of course, proper records are kept by a clerk, but the results are not de-coded until the trial has been completed.

The only circumstance in which a control group may not be absolutely necessary is when the behaviour of the disease in question is thoroughly known from past experience. Thus when Jenner found 20 persons who had had cowpox and who were resistant to smallpox inoculations, he did not need a special control group of 20 persons who had never suffered from cowpox for making the comparison, for he was well aware that thousands of inoculations had amply demonstrated the rarity of such resistance in the general population. Or again, the data in the example on page 81 is perfectly adequate for comparison with any future treatment (there is no evidence that psychological factors can affect this particular disease).

An example which illustrates the application of a number of these principles was reported by some American doctors in the *Journal of Investigative Dermatology*, May 1950. A new antiseptic called hexachlorophene had been discovered. In laboratory experiments it had proved highly effective against a wide range of bacteria. It had also been shown to reduce the number of bacteria on normal skin to a very considerable extent. By rights, it ought to be an excellent cure for bacterial skin infections. But this had never been tried, so an experiment was designed to test it. 154 consecutively-encountered children suffering from certain specified skin infections made up the sample group. Each was given a jar of ointment, told how to apply it, and given strict instructions to wash the affected areas thoroughly with soap and water once a day. They were seen again a week later, at which stage 133 of the children were cured. The jars of ointment were then de-coded, and the following results emerged –

Ointment	No. of patients	No. cured	% cured
Base only	46	37	80%
Base + 2% hexa-chlorophene	108	96	89%

Notice that the doctors did not know which jars contained the ointment base alone, and which contained the antiseptic, so they could not influence one group any differently from the other. The difference in the cure rates between the actively treated and the control group was so small that it could easily be due to chance, so it was concluded that 2% hexachlorophene ointment was of no use for treating bacterial skin infections in children.

The above trial could have been improved in 2 ways. The first is that it would have been much more useful to use the best available remedy instead of the plain ointment base for the control group. The results would then have indicated which treatment should be used in future. The second point concerns the number of patients in each group. In testing a new product, it may seem logical to use it on as many patients as possible, and to restrict the control group to a minimum size ('after all, they're only controls'), but as was shown in the tea cup experiment (p. 111), it is more efficient to have an equal number of patients in each group. This is one occasion when our 'common sense logic' could mislead us. The mathematical proof in the present instance is somewhat harder than that of the tea cup experiment, but you will be able to prove it later with Yates' χ^2 Test.

The above remarks should suffice to emphasize the need for experiments to be designed in such a way that useful and proper comparisons are made. So far, we have only considered comparisons between independent groups of things. In some experiments it is possible to tighten up the design and make the comparison much more precise by what is called *paired comparison*. This entails dividing the sample group into 2 halves in such a way that every individual in one half has an exactly matching partner in the other half. The way in which this improves the sensitivity of the experiment will be seen most easily from an example.

Paired Comparisons

Suppose we have a certain kind of petrol (= gasoline), and want to find out if the addition of a new additive chemical affects the number of miles per gallon given by the petrol. Suppose, too, that it is decided to assess the matter on a variety of cars, in case the results from one type of car are not applicable to others. One way of doing this would be to make a list of all the types of car available, allot each type a number, and then select 2 groups, each containing 4 types of car, using a Table of Random Numbers in the manner described on page 42. Suppose that the first group turned out to be Volkswagen, Ford, Fiat, and Rolls-Royce, while the second group was MG, Austin, Chevrolet, and Morris. Now we would need to choose one of each of these cars; demonstration cars would suit our purpose if they were available for all 8 types, failing which individual cars could be selected by again using the Table of Random Numbers.

Thus we have got 8 cars, of 8 different makes, divided randomly into 2 groups of 4. We could now test (under standard conditions) the 4 cars in the first group with the plain petrol (P), and the 4 cars of the second group with the petrol containing the additive (P + A), and compare the averages of the 2 groups. Suppose the results were as follows –

P	m.p.g.	P + A	m.p.g.
Volkswagen	36	M.G.	31
Ford	20	Austin	26
Fiat	38	Chevrolet	19
Rolls-Royce	14	Morris	32
	Av. = 27		Av. = 27

The comparison reveals no difference between the averages of the 2 groups. Notice that the design of this experiment has ensured that this comparison is perfectly valid, for every make of car had an equal chance of being included in the experiment. At the same time, it must be admitted that the experiment as it stands is very insensitive. For it might easily have happened that the Rolls-Royce got into the second group, and the Morris into the first group. In this case the above results would have distributed thus –

P	m.p.g.	P + A	m.p.g.
Volkswagen	36	MG	31
Ford	20	Austin	26
Fiat	38	Chevrolet	19
Morris	32	Rolls-Royce	14
	Av. = 31·5		Av. = 22·5

The averages are now quite different; the additive appears to have reduced the average efficiency of the petrol by 9 miles per gallon. If you examine the figures in the above tables you will see what is wrong – the size of each group is much too small for accuracy, taking into account the amount of variation between individual cars (14 to 38 m.p.g.). We have previously noted that the sample size required depends in part on the amount of variation inherent in the parent group (p. 45), and this is a clear example of this. In the present instance it has resulted in any difference between the 2 groups *due to the additive* being swamped by the difference between the 2 groups *due to the degree of variability* inside each group.

It follows from this that it would have been better to have matched each car in the first group with another of the same make and model in the second group. This will make the comparison more precise, so that the results might be as follows –

P	m.p.g.	P + A	m.p.g.
Volkswagen #1	36	Volkswagen #2	33
Ford #1	20	Ford #2	22
Fiat #1	38	Fiat #2	42
Rolls-Royce #1	14	Rolls-Royce #2	15
	Av. = 27		Av. = 28

This is now an example of paired comparison using matched groups. Notice how, without increasing the size of the experiment, this design will show up much smaller differences between the 2 groups than was the case with independent groups which were not matched but merely selected at random. This increased sensitivity has been achieved by cancelling out the disturbing effect of the internal variations in each group, by making this factor the same in both groups.

Even so, the design is still imperfect, for the accuracy of the comparison will depend to some extent on the investigator's ability to match the pairs in respect of the characteristic being examined. Perhaps Volkswagen #2 was out of tune a bit. Can one be reasonably sure that the performances of all the #2 cars were identical with those of their #1 counterparts at the beginning of the experiment? There is a catch here, for there is no strict criterion as to just how identical the matching pairs must be. All that can be said is that they must be as identical as possible. If there is any doubt about this, as in the present case, it is necessary to handle the results (from the point of view of statistical analysis) as if they were independent, unmatched groups.

The design of the present experiment can, however, be changed into a perfect state of paired comparison by using the same 4 cars for testing both the plain petrol and the petrol plus additive. This will have the desired effect of completely eliminating any variation due to the cars, and any difference between the 2 groups will then be entirely due to the effect of the additive (within the limits of experimental error). The results might then look like this –

Car	With P (m.p.g.)	With P + A (m.p.g.)
Volkswagen	36	39
Ford	20	21
Fiat	38	41
Rolls-Royce	14	15
	Av. = 27	Av. = 29

By using exactly the same car for each pair of tests, the precision has been increased, although the experiment still consists of but 8 tests. In practice, of course, one would use a much larger sample to get even greater precision; the small groups were used here merely because they illustrate the principles more easily.

Paired comparisons are thus of 2 types –

(a) those using matched groups, and
(b) those using a single sample group.

As explained in the example above, the single sample type is the more accurate of the two. But it is not always attainable. For instance, it would not be applicable for comparing 2 operations for removing the appendix, because each patient has only got one appendix. We would therefore have to be content with comparing 2 independent randomly-divided groups of patients or, at best, 2 matched groups in which each patient had been carefully matched with a partner as regards age, sex, severity of illness, etc., with additional protection against bias by determining which partner shall have which treatment by the toss of a coin or other random process. It makes no difference whether the pairing is made after the whole sample group has been assembled, or whether the pairs are found first and then put together to make up the sample. However, you will realize that it is almost impossible to match patients accurately when they form a presenting sample, arriving at various intervals over a period of many months, so usually we are obliged to use randomly divided independent groups.

Paired comparisons can be used in quite a wide variety of ways, which can conveniently be classified as follows –

(a) *Paired comparison of matched groups*

(i) *Simple matching* of pairs according to the characteristic being investigated can be exemplified by an experiment to compare the effectiveness of 2 types of pills for treating obesity. Suppose that a small preliminary trial has made it seem that the efficacy of one or both pills may vary considerably depending on the initial weight of the patients; they might for example prove much more effective for slightly fat than for extremely fat people. If the whole sample group of patients was simply divided at random into 2 independent groups (as in *Q11*, p. 44), and one type of pill given to each group, the weight loss of different individuals inside each group might then vary quite a lot so that, exactly as in the first petrol-mileage experiment, any difference between the effectiveness of the 2 pills would tend to be hidden by the overall variation. This could be overcome by pairing off patients according to their weight at the beginning of the experiment, and then allocating one treatment to one of each pair and the other treatment to the other one, by a random process. Unless it is suspected that the pills could be affected by any variable other than the patient's initial weight, the pairs need not match as regards age, sex, nationality, or other respects, for the purpose of this design. However, error could arise here if this assessment was wrong, so in general it is best to choose pairs which are as identical as possible.

Identical twins are often ideal for this purpose. They are used fairly commonly in animal experiments, where litters are apt to be large, but with humans this technique is rather limited by the scarcity of such material. However, even with small numbers, this design can yield very convincing results. One such example will be found among the miscellaneous questions at the end of this book.

(ii) *Symmetrical matching* has many applications in biological work. It consists of comparing the effect of 2 different treatments applied to opposite sides of the body. This can be used, for example, to compare 2 ointments for treating skin rashes which are present on both arms or both legs; one ointment is applied to one arm, the other ointment to the other arm. The decision as to which arm gets which treatment is made, of course, at random so as to cancel out any tendency for the left arm (which is treated by the right hand) to get different applications than the right arm. Any difference between the ointments shows up much more markedly with this design than if 2 independent groups of patients are compared.

This idea has been extended to experiments on bilateral organs, such as kidneys.

(iii) *Split samples*, in which each individual making up the sample group is divided into 2 parts, provides another very good way of getting 2 matched groups. Inanimate things such as pieces of wood or paper, metal rods, and batches of chemicals lend themselves readily to such division. Thus to compare a new process for tempering steel with an existing process, one could take a number of different kinds of steel, cut each piece in two, and submit one half to the new process and the other half to the old.

(b) *Paired comparison of single sample groups*

All cases coming under this heading are basically the same, in that all the individuals of the sample group are examined on 2 or more occasions with reference to the same characteristic (e.g. miles per gallon). The circumstances may be –

(i) *Different treatments*, of which the petrol-mileage experiment would be a typical example. The 2 kinds of petrol represented different treatments, all other variables having been kept constant throughout the experiment. This is an excellent design, but care must be taken (particularly in biological work) to ensure that the first treatment does nothing which could affect the response of the individuals to the second treatment; it would hardly be suitable for comparing the efficacy of 2 rat poisons!

(ii) *Different methods of testing* for the characteristic under examination. This would apply to an experiment designed to compare 2 methods of examining students, say written versus oral exams.

(iii) *Different observers*, for instance comparing the results obtained by 2 examiners who have independently judged the same group of students in the same subject (Latin, Art, or whatever it is).

(iv) *Different occasions*, before and after some event which might affect members of the sample. Suppose you wanted to find out if the body weight of soldiers was affected by their first 6 months in the Army. One way would be to determine the average weight of a random sample of newly recruited men, and compare this with the average weight of another random sample taken from the ranks of those who had been in the Army for 6 months. This would be a case of comparing 2 independent samples, comparable with the first petrol-mileage experiment, and like it, rather insensitive. A much better way would be to compare the average weight of a random sample of new recruits with the average weight of exactly the same sample of men taken 6 months later, which brings the design into line with the last petrol-mileage experiment.

Factorial Designs

So far, we have been dealing with experiments which compare 1 factor with the effect of chance, or of 2 factors with each other, and have seen how proper design of experiments improves the reliability of the results and keeps wastage to a minimum.

Now it was first shown in experimental agriculture, and since confirmed and applied in many other fields, that these principles of efficient design could be extended, with profit, to more complex experiments. We saw, on page 96, how men learnt to compare the effects of 2 fertilizers, balancing the design so that any factors which might interfere with the comparison were cancelled out. The secret of success there was to alternate the treatments over multiple plots of ground.

There is, of course, absolutely no reason why this principle cannot be used to compare more than 2 fertilizers at the one time. Suppose there were 4 fertilizers, *A*, *B*, *C*, and *D*. Each one might be a single substance, or a mixture of substances. A number of $\frac{1}{4}$-acre blocks of ground, preferably spread over several districts, could each be divided into 4 plots, either strips or squares, to which the fertilizers would be allocated by strict random process. Several of the blocks could thus be represented like this –

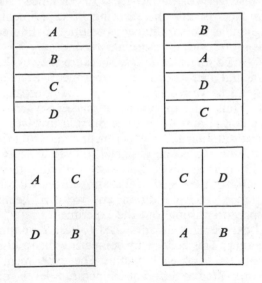

The same design could be used to compare the effect of 4 different strengths of the one fertilizer, or to compare 4 different varieties of wheat under uniform agricultural treatment. By extension of the use of the words 'block' and 'plot' to other things, this method could equally well be used to compare 4 ways of sterilizing hospital blankets, or 4 ways of washing motor cars, or 4 ways of learning to read faster, or 4 ways of fattening pigs. But remember that the data can only be analysed meaningfully if the treatments have been allotted at random.

However, as Professor Fisher pointed out in 1935, excessive stress has often been laid on the idea of varying only one experimental condition at a time. The very nature of research is such that the investigator is really in no position to know which particular factor will turn out to be the most important one in any given experiment. Exploring the unknown differs greatly in this regard from the demonstration of known physical and chemical laws in experiments at school. Furthermore, it can happen that the rigid control of all variables except the one being investigated can result in a highly artificial state of affairs, the conclusions from which may not then be applicable in the outside world. And, to top the argument, there is the practical consideration that different factors may interact, and either boost or inhibit each other's effects.

Fisher's discovery of the Analysis of Variance technique (p. 97) did, in fact, permit the devising of experimental designs which can handle the simultaneous testing of different factors, and what is more, with extraordinary economy.

They are called *factorial designs*, because they evaluate different experimental factors. In any experiment, each component part which can be varied is a *factor* in the construction of the experiment. Thus an experiment to make oxygen by heating potassium chlorate and manganese dioxide involves 3 factors – the 2 chemicals and heat. An experiment comparing the strength of 2 weight-lifters would be designed to evaluate just 1 factor – the weight lifters themselves. Likewise, a running race evaluates the single factor of the speed of the competitors, all other factors such as distance, track condition, and weather, being constant for all competitors throughout the experiment.

Factorial experiments are designed to test 2 or more factors at the same time. This is done by comparing the results obtained with different *levels* of each factor. The experiment may, for example, be repeated at different temperature levels, or at different voltages, or with different doses of a drug, or for different durations. Each of these factors (temperature, voltage, medication, duration, etc.) is a thing which can be varied, and the word 'level' is used for the various strengths, quantities, or concentrations of each factor. Thus if a factor like temperature is tested at 100° and 150°, it is tested at 2 levels. If the covering power of 1 coat, 2 coats, and 3 coats of a paint are compared, the thickness of paint is tested at 3 levels. If the taste of steak cooked with and without salt is compared, the salt factor is being tested at 2 levels (present and absent). It is important to understand these terms.

Let us now examine the principles of factorial design, starting with an experiment involving 2 factors (i.e. 2 factors will be varied while all others are kept constant). Each factor will be tested at 2 levels, present and absent. As this subject began in agriculture, it will be appropriate if we continue with reference to fertilizer experiments. Suppose, then, that we have 2 fertilizers, A and B (one nitrogenous, and the other phosphatic), and we want to evaluate their separate and combined effects on the growth of potatoes. One might imagine that this could be done by dividing a number of $\frac{1}{4}$-acre blocks of land into 3 plots each, so that each block would accommodate (by random selection) the 3 treatments – A, B, and a combined treatment with A and B (which we shall call AB), thus –

A	B	AB
300	400	500

The numbers quoted in the plots represent the total pounds of potatoes grown in the various blocks. We see that B gives a greater yield than A, and AB gives the best of all. But with this design there is no way of telling whether the effect of AB is one of simple summation, or whether there is *interaction* between them. Now interaction is an important phenomenon in many experiments. Two or more factors are said to interact if their combined effect is different from the simple addition of their separate effects. If a glass of water weighs 6 ounces, dissolving a $\frac{1}{2}$ ounce of sugar in it will increase the total weight to exactly $6\frac{1}{2}$ ounces; each weight factor remains independent of the other in their combined state, so there is no mathematical interaction in this case. On the other hand there are experiments in which any 2 out of 3 factors will be quite impotent, but the combination of all 3 will be effective. Making oxygen by heating a mixture of potassium chlorate and manganese dioxide to 240°C is an example of this; no 2 of these 3 factors will liberate oxygen; it takes the combination of all 3. These 3 factors are therefore said to interact in a positive way. Interaction may also be negative, when the combination of different factors results in more or less inhibition of their separate effects by some process of inter-ference or antagonism. Thus penicillin kills bacteria while they are in the act of dividing; tetracycline inhibits bacterial division; if penicillin and tetracycline are given at the same time, therefore, their combined effect will be less than a simple addition of their separate effects, so there is said to be a negative interaction be-tween them.

Returning to our fertilizer experiment, we noted that the 3-plot design did not give enough information to say whether fertilizer A had interacted with fertilizer B, positively or negatively, or not at all. Suppose, for instance, that A had no effect, that the total pounds of potatoes grown on the A plots would have been 300 lb even without A; if this was the case, the yield of the AB plots shows that A has stimulated B (or vice versa) to produce an additional 100 lb of potatoes – a positive interaction. On the other hand, suppose the effect of A was to increase the produc-tivity of the land from a basic 100 lb to the observed 300 lb, which is a stimulation of 200 lb. In this case one would have expected the effect of AB to be 200 lb greater than that of B on its own, i.e. a total yield of 600 lb. This was not achieved, so one would conclude that there was a partial negative interaction between A and B, if these were the circumstances.

This 3-plot design is therefore not very satisfactory. You can see that the thing that is missing is a knowledge of the potato yield of the untreated land. An untreated control plot (which we shall call O) should therefore be included in each block, to provide this information. Note carefully that O refers to 'no treatment', not to 'no yield'.

See what happens when each block of land is divided into 4 plots, 1 for each treatment (A, B, and AB) and 1 for the control (O). Sorting the results from the randomized plots, the experi-ment can be summarized in the following diagram –

A 300	B 400
AB 500	O 200

The difference in the yields of the plots treated with A com-pared with the untreated control plots, that is $A - O$, tells just what effect A has had. In the present case A has stimulated the yield of potatoes by $300 - 200 = 100$ lb. By the same reasoning, B has increased the crop by $400 - 200 = 200$ lb. Therefore if there was no interaction between A and B, the result of AB should just be a plain summation effect of the untreated soil (200 lb), plus the effect of A (100 lb) plus the effect of B (200 lb), which gives a total of 500 lb. This is, in fact, just what occurred, so we conclude that there was no interaction between A and B.

The above calculation can be expressed in the form of an equation; if there is no interaction between the 2 fertilizers –

$$O + (A - O) + (B - O) = AB$$
$$200 + \quad 100 \quad + \quad 200 \quad = 500$$

If this equation does not balance, interaction is present between A and B.

Another way of looking at the question of interaction is as follows. Interaction is a special effect obtained only when the interacting factors are present together. It can therefore be detected by comparing the effect of A on its own, with the effect of A when it is mixed with B. If there is no interaction, the effect of A will be the same in both cases. If there is interaction between A and B, the effect of A in the presence of B will differ from its effect in the absence of B. Now, our 4-plot factorial design provides this information, because –

$$AB - B = \text{effect of } A \text{ in the } \textit{presence} \text{ of } B, \text{ and}$$

$$A - O = \text{effect of } A \text{ in the } \textit{absence} \text{ of } B$$

By subtracting the one from the other, then, we can find out whether the presence of B has any effect on the yield of A. And this is precisely the interaction effect that we are after. Putting this into proper shape, we get a formula for determining the interaction effect of A with B, thus –

$$I_{AB} = (AB - B) - (A - O)$$

where the symbol $I_{AB} = $ interaction between A and B.

If this formula gives a zero answer, it means there is no interaction between A and B. If the answer has a positive ($+$) value, there is a positive interaction to the extent indicated. If the answer has a negative value, negative interaction is present.

Let us apply this formula to the data of our experiment.

$$I_{AB} = (500 - 400) - (300 - 200) = 0$$

That this answer should be the same as before is not surprising, for if you juggle the first formula around a little, you will soon find that it is algebraically identical with the second one.

Note well that our calculation of the interaction effect between A and B makes use of information from all the plots. It is as though the experiment was designed purely for determining this particular piece of information, and nothing else. But, now, come to think of it, 2 out of every 4 plots were treated with A (namely, A and AB), while the other 2 plots (B and O) were not treated with A. Perhaps ... yes, perhaps we could also calculate the effect of A utilizing information provided by the whole experiment. If this could be done, it would clearly be an advantage over merely using $A - O$, which makes use of only half of the data provided by the experiment; it would be tantamount to doubling the sample size, and this would lead to a more accurate result.

You've no doubt guessed it. It can be done. By treating the matter as a simple piece of algebra, where AB is equivalent to treatment $(A + B)$, you can see that the effect of B cancels out –

$$(A - O) + (AB - B) = 2A$$

Substituting the observed yields for these symbols, we get –

$$2A = (300 - 200) + (500 - 400)$$
$$= 200$$
$$\therefore A = 100$$

Now, if you think about it for a moment, you will realize that this formula would need to be modified if interaction between A and B had occurred. If the interaction was positive, it would have to be subtracted from AB; if negative, it would have to be added to AB. This can conveniently be written into the above formula, thus –

$$2A = (A - O) + (AB - B) - I_{AB}$$

Similarly, the effect of B can be calculated thus –

$$2B = (B - O) + (AB - A) - I_{AB}$$

If the interaction is negative, remember that subtracting a negative value converts both signs to positive; thus $x - (-2) = x + 2$.

Here, then, is the remarkable ingenuity of this factorial design: the effect of each single factor can be calculated from the total information provided by the experiment, and at the same time the effect of their interaction is also calculated from the total information. Each of these effects is thus determined with the degree of precision that would obtain if the entire experiment had been devoted to its exclusive elucidation!

Now this experiment has concerned 2 factors, A and B, each at 2 levels, present and absent. It is easy to see that this idea could be extended to 2 positive levels (quantities) of each fertilizer. Thus fertilizer A could have been tested in 2 different concentrations, a heavy application which we shall call A, and a light application (the smaller test level) which we shall call a. The other fertilizer could likewise have been tested at 2 levels, B and b. The 4 plots would then be treated as follows –

	A	a
B	AB	aB
b	Ab	ab

The structure of the design is just the same as before; the inbuilt symmetry is still there, conferring the same benefits. The marginal code-letters help to explain the design, but in practice the allocation of treatments in each block must be decided by a random process, so that each plot has an equal chance of getting any 1 of these 4 treatments.

The calculations follow the same lines as before. Firstly, note that –

$$AB - aB = A - a, \text{ in the presence of } B, \text{ and}$$
$$Ab - ab = A - a, \text{ in the presence of } b$$

Now, in the absence of interaction between the 2 fertilizers, the difference between the effect of the first fertlizer in its 2 concentrations (that is, $A - a$) will be the same whether it is based on observations in the presence of B, or in the presence of b. If there is a difference between these 2 situations, it would signify interaction between the fertilizers. Therefore –

$$I_{AB} = (AB - aB) - (Ab - ab)$$

Knowing this, we can calculate the difference in effectiveness, if any, between the 2 levels of A (that is, $A - a$), again using all the data provided by the experiment, as follows –

$$2(A - a) = (AB - aB) + (Ab - ab) - I_{AB}$$

Similarly, the difference between the 2 levels of fertilizer B is given by –

$$2(B - b) = (AB - Ab) + (aB - ab) - I_{AB}$$

These formulae remain applicable if either of the lower test levels, a and b, are actually zero (meaning the absence of the test factor). If both a and b are zero, these formulae simply reduce to the form in which we first met them (using the symbols A, B, AB, and O).

It does not need much imagination to see how this kind of design can be applied to things other than fertilizers. A and a could, for instance, be 2 varieties of wheat, while B and b remain as 2 concentrations of a fertilizer. Alternatively, A could be mumps with complications, a could be mumps without complications, B could be an experimental treatment, and b could simply be bed rest. A could be older children, a younger children, B one method of teaching, b another method of teaching. And so on.

The real beauty of factorial designs hardly begins to show until we reach designs for 3 and 4 simultaneous factors. Consider, for example, an experiment for evaluating 3 factors (say, fertilizers again, for simplicity), each at 2 levels. We shall call the fertilizers A, B, and C, and shall represent their presence and their effects at each level by the symbols A, a, B, b, C, and c. It is immaterial whether some or all of the lower levels represent 'absent' or merely lower levels than the levels symbolized by capital letters. As shown in the following diagram, the blocks of land would have to be subdivided into 8 plots each, and each plot would be allotted a particular treatment at random. The basic design is on the following pattern –

		A	a
B	C	ABC	aBC
	c	ABc	aBc
b	C	AbC	abC
	c	Abc	abc

You can see at a glance that half of the plots have received A, and the other half have received a, and since all the other factors are the same on both sides, we have again got the kind of design in which we shall be able to use every bit of the contained information for determining each of the effects in turn.

We must begin with investigating the triple interaction between A, B, and C. The question is, when these 3 fertilizers are used together, in the concentrations of their upper test levels, is the yield of the crop greater or smaller than one would expect from the simple addition of their separate effects? Does the combination result in some special boosting, or inhibiting, effect? To arrive at the formula which will give us the answer, notice in the first place that –

$ABC - aBC$ = effect of $A - a$ in the presence of B and C

$Abc - abc$ = effect of $A - a$ in the absence of B and C

$ABc - aBc$ = effect of $A - a$ in the presence of B

$AbC - abC$ = effect of $A - a$ in the presence of C

Now if there is interaction between A, B, and C, specifically requiring the presence of all 3 factors for its development, it will be demonstrable by comparing the effect of $A - a$ in the presence of B and C, together with the plain effect of $A - a$ (i.e. in the absence of B and C) on the one hand, as against the effect of $A - a$ in the presence of B alone and of C alone, thus –

$$I_{ABC} = [(ABC - aBC) + (Abc - abc)] \\ - [(ABc - aBc) + (AbC - abC)]$$

As before, if this works out to be zero, there is no triple interaction. If it gives a positive or negative value, there is positive or negative interaction respectively.

There are 3 possible double interactions to be studied: A with B, A with C, and B with C. The interaction between A and B is to be sought by comparing the effect of $A - a$ in the presence of B, with the effect of $A - a$ in the absence of B. This is given by the following formula –

$$I_{AB} = \underbrace{[(ABC - aBC) + (ABc - aBc)]}_{A - a, \text{ with } B} \\ \underbrace{- [(AbC - abC) + (Abc - abc)]}_{A - a, \text{ without } B} - I_{ABC}$$

Notice that it is necessary to subtract any triple interaction effect to compensate for the first item in the formula.

The interaction between A and C can be worked out along the same lines; likewise the interaction between B and C is –

$$I_{BC} = [(ABC - AbC) + (aBC - abC)]$$
$$B - b, \text{ with } C$$
$$- [(ABc - Abc) + (aBc - abc)] - I_{ABC}$$
$$B - b, \text{ without } C$$

After these have been worked out, one can proceed to calculate the separate effects of A, B, and C. Actually what we calculate is the difference between the effect of each factor at its 2 test levels.

For example, the effect of $A - a$ is calculated thus –

$$4(A - a) = (ABC - aBC) + (ABc - aBc) + (AbC - abC)$$
$$+ (Abc - abc) - (I_{ABC} + I_{AB} + I_{AC})$$

The last part of the formula is, of course, the necessary subtraction of any interaction effects. I_{BC} is omitted because it cancels out in the first step of the formula.

The effects of $B - b$, and of $C - c$, are calculated in a similar way. A numerical example is provided later ($Q39$, p. 135).

So once again this factorial design permits the calculation of the effect of each factor singly and in all combinations, and every individual result is based on the total data of the experiment. Amazing, isn't it?

Exactly the same principles apply to 4-factor, 2-level designs. The 4 single factors, A, B, C, and D, the 6 double and 4 triple interactions, and even the quadruple interaction (A, B, C, and D all together), can all be calculated with the degree of precision and reliability that would apply if the entire facilities of the experiment were devoted to each particular issue on its own. The general plan is as follows (though don't forget that in practice each block has its treatments allotted to its plots by random process) –

		A		a	
		D	d	D	d
B	C	ABCD	ABCd	aBCD	aBCd
	c	ABcD	ABcd	aBcD	aBcd
b	C	AbCD	AbCd	abCD	abCd
	c	AbcD	Abcd	abcD	abcd

Many more designs have been devised along these lines. Back in 1935, Professor Fisher summed up the situation in these words –

The efficiency with which limited resources can be applied is capable of relatively enormous increases by careful planning of the experimental program.

One more small point is worthy of notice. It sometimes happens that, in a factorial experiment, part of a growing crop is injured, or a test individual dies, or some technical error occurs, with consequent loss of portion of the total data. This need not render the whole experiment useless, as one might expect, for calculations have been devised which permit an estimate of the missing value to be made. The repaired experiment will not, of course, be quite as good as it would otherwise have been, but useful results may still be obtainable from it.

In discussing factorial designs, we have concentrated on the principles involved. We have made no mention of the fact that experimental error must always be taken into account; observed differences (say, between A and a) may represent nothing other than the vagaries of chance. Every such difference must be tested statistically to decide whether or not it is likely to be due to chance. And this is exactly what Fisher's Analysis of Variance was designed to do. We shall not in this book be dealing with this procedure, for it is essentially a tool for statisticians (we will, however, describe some simpler alternative methods). But those of you who are mathematically minded may be interested to see the full calculations of a 4-factor, 3-level factorial experiment concerning the manufacture of electrical condensers in M. J. Moroney, *Facts From Figures* (Penguin, 1962), pp. 394–418.

Sequential Design

Before concluding this section, one other important advance in the design of experiments must be mentioned. The story is an interesting one. In 1938, a 36-year-old mathematician called Abraham Wald came from Austria as a refugee and settled in the USA. There he was given a professorial post at Columbia University. Up to this point he had no particular knowledge or interest in Statistics. Then war broke out, and Professor Wald was put in charge of a Statistical Research Group at Columbia

University. In this capacity he turned his attention to problems of Statistics, which he approached with a fresh vision and a brilliant mind. The outcome was that by 1943 he had devised a novel idea which he called *Sequential Design*. Its immense value was recognized at once by the authorities, and it was classified as 'Restricted' within the meaning of the Espionage Act. It was

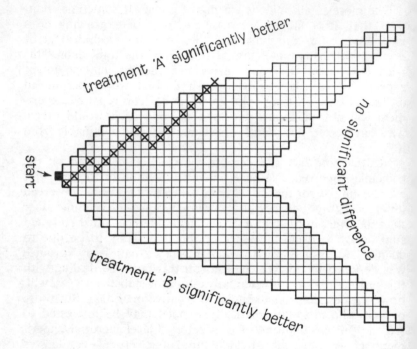

Fig. 10. Sequential Chart (in the style of Armitage). A cross is entered upwards and to the right for each case preferring Treatment '*A*', and downwards and to the right for each case preferring Treatment '*B*'. No entry is made for cases showing no preference. The trial is stopped as soon as one of the boundaries is crossed.

used extensively, both for wartime research and for checking the quality of goods made by over 6,000 wartime factories. In 1945, a few months before the war ended, it was released to the public, and has since revolutionized many aspects of statistical practice. Professor Wald was killed in an aeroplane crash in India in 1950.

Wald's basic idea was simple. He conceived the notion of continuously reviewing results as they came to hand, so that an experiment could be stopped just as soon as the results showed a predetermined degree of significance or non-significance. Statistically reliable conclusions could thus be wrung from the barest minimum of data. The sample size is thus not determined in advance, but the experiments are continued until the moment when the results are conclusive. Apart from keeping costs down, this approach has obvious advantages in medical research, where patients can benefit from the results at the earliest possible moment.

In practice, a sequential chart is prepared by a statistician to suit the circumstances of the experiment, and all that the research worker has to do is to plot the results as they occur. Fig. 10 shows the type of chart used for comparing 2 treatments. There are 3 boundary lines, and as soon as the plotted results pass any one of these lines, the experiment is ended. Depending on which boundary is reached, the result may indicate that treatment A is better than B, or that B is better than A, or that there is no significant difference between the two. Instructions for drawing a sequential chart suitable for medical research are given by Dr Peter Armitage, statistician to the Medical Research Council of England, in *Biometrika Journal*, 1957, pp. 9–26, and in his book, *Sequential Medical Trials* (Blackwell, 1960). The practical application of this technique is illustrated by Drs E. Snell and P. Armitage in *The Lancet*, 1957, i, p. 860.

Do you recall H. G. Wells' remark about statistical thinking (p. 20)?

Questions

Q31. 'Here are the figures you asked me to get, Mr Smith. I questioned over a thousand people, and found that 81·678% of people who own one or more cars have paid for them in full. The rest are still paying them off. Believe me, this is an accurate figure. I questioned people at random all over the place – at my golf club, in hotels, on street corners, at petrol stations, in a supermarket, and even out at the airport.'

Comment on –
    (a) the figure 81·678%,
    (b) the sampling technique,
    (c) the possibility of some answers being untrue.

Q32. A presenting sample of 200 patients complaining of various kinds of headaches was used to compare a new remedy with aspirin. The treatments were strictly alternated, so that the 1st, 3rd, 5th, and all the odd-numbered patients were given the new remedy, while the 2nd, 4th, 6th, and all even-numbered patients (in order of arrival) were given aspirin. The trial extended over a period of 3 months, at the end of which time the results were analysed, and it was found that satisfactory relief was produced in 90% of the aspirin-treated group, compared with 20% of those taking the new remedy.

(*a*) What do you think of the idea of alternating patients on arrival as a means of dividing the presenting sample into 2 comparable groups?

(*b*) What would be wrong with the systematic alternation of treatments for successive patients if it has been decided to keep the doctors ignorant of which treatment each patient was getting, by having a nurse give out the tablets according to the predetermined plan?

(*c*) Can you pick another shortcoming of the design of this clinical experiment?

Q33. A radio station advertisement in a trade journal stated, 'A survey has been conducted to find out exactly what kind of music listeners want to hear. The survey was made on a substantial cross-section of people aged 12 years and upwards. The result proved that 50% more people want to hear classical music than hit tunes.'

What further information, if any, would be needed to convince you that these findings were correct?

Q34. 'In tonight's lecture, I am going to present you with clear-cut evidence that horror films do not promote crime in children or adolescents. As you know, I have had considerable experience in this field, having been official psychologist at the Children's Court for the past 16 years. For the purpose of proving my point, I have chosen from my records the case histories of 40 young people who have habitually watched horror films, and will proceed at once to describe the pertinent points of each case.'

Assuming that his evidence is convincing, is it reasonable to draw general conclusions as in his opening sentence, from a highly selected sample such as this?

Q35. I have just determined, with the help of an ordinary ruler, that the pages of my Bible are 0·00114 inch thick. How did I do it?

Q36. Horse trainer: 'I have improved the performance of all 12 racehorses in my stables by giving them injections of an iron compound to improve the quality of their blood.' Comment on the validity of this experiment.

Q37. 'We are proud to announce that our JJJ car tyres have proved to be almost puncture-proof. Below are listed the names and addresses of 20 people who have tested our tyres for 2 years. Out of the entire group, Mr James was the only one to get a puncture. Contrast this with the national average, which would have amounted to a total of 14·2 punctures for the group over the same test period.'

Assuming that the 20 people named were bona fide, can you think of a way that such figures could be produced if the tyres were really no better than any others?

Q38. 'The dilapidated old farm-house pictured above is one of the homes selected by the XYZ Organization for their surveys which determine our TV program ratings. The occupants of this farm represent the television tastes of 52,000 homes. No wonder the accuracy of these program ratings has been doubted, when one household is expected to speak for so many others.'

Is this argument valid, and why?

Q39. You have no doubt heard that hypnosis can be used as an anaesthetic for operations. Now suppose that 2 doctors called Dr Ay and Dr a'Beckett (whom we shall simply call A and a) decided to see if they could speed up the induction of hypnotic anaesthesia by administering a sedative to allay anxiety, and/or by playing soothing music during the induction. As a sedative they chose Butobarbitone (B), and to serve as a control, other patients were to be given some ordinary bicarbonate of soda (b). For the soothing music they chose some Classical music (C), the effect of which would be compared with other patients given no musical background whatever (c). There were thus 3 factors, each at 2 levels, A a, B b, and C c, so naturally the investigation was carried out on a factorial design. It was decided to use 80 consecutive patients, divided evenly between the doctors (A and a), between the sedative and control (B and b), and between the presence or absence of music (C and c). A random list of 80 numbers was drawn up, so that successive patients were allotted, at random, to one or other of the treatment combinations. The average induction time for the 10 patients in each category is given (in minutes) in the following table.

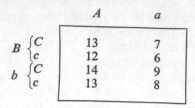

		A	a
B	C	13	7
	c	12	6
b	C	14	9
	c	13	8

Determine –

(i) Is there any interaction between the combined effect of the Butobarbitone and the Classical music and either of the doctors (i.e. between *A*, *B*, and *C*)?

(ii) Is there any interaction between either doctor and the Butobarbitone (i.e. between *A* and *B*)?

(iii) Is there any interaction between either doctor and the Classical music? Perhaps it disturbed or helped one doctor more than the other.

(iv) Is there any interaction between the sedative and the music (i.e. between *B* and *C*)? Perhaps they form a specially good, or a specially bad, combination.

(v) Allowing for any possible interactions, how does Dr Ay's speed of inducing hypnotic anaesthesia compare with Dr a'Beckett's?

(vi) How does the effect of Butobarbitone compare with bicarbonate?

(vii) How does the effect of Classical music compare with no music at all?

Q40. A farmer has 3 cows; we shall simply call them 1, 2, and 3. He wants to test 2 different food supplements, vitamin A and broken biscuits, each in 2 different quantities, to see if they increase the yield of his cows' milk. See if you can devise a suitable factorial experiment which will permit comparison of each cow's yield over the duration of the experiment, and at the same time will evaluate the effect of each level of the 2 food supplements. Disregard the possibility of interactions. Call the food supplement levels *A*, *a*, α, and *B*, *b*, β, the Greek letters representing the absence of the supplement.

CHAPTER 6

SIGNIFICANCE TESTS

What Do The Results Really Mean?

THIS, THE THIRD and final question posed on page 98, shall be the subject of the rest of this book. All that has been learnt about Statistics so far will now converge to provide the answer to this question.

We said (p. 109) that all experiments were comparisons. Consider the case of a man who wants to compare the accuracy of 2 rifles, *A* and *B*. He shoots 10 times at a small target with each rifle. If he hits the target 5 times with each rifle, one would be fairly safe in concluding that there was no appreciable difference between the rifles. But what if he got 5 hits with *A* and 6 hits with *B*? Is he then entitled to conclude that *B* is better than *A*? Our experience in such matters tells us that there is always an element of luck (i.e. chance) in tests of this kind, so perhaps the apparent superiority of rifle *B* was just a matter of luck on this occasion. Admitted. So he repeats the experiment, again shooting 10 times at the target with each rifle. Would you be surprised if the results were reversed this time, with a score of 6 hits with *A*, and 5 hits with *B*? Certainly not. Such variation is well within the scope of chance. Well, what if the trial had resulted in 5 hits with *A*, and 7 hits with *B*? Or 5 hits with *A*, and 8 hits with *B*? Just when would he be justified in concluding that *B* was really the better rifle?

There are 3 possible ways of answering this question –

(*a*) by guessing, or

(*b*) by drawing on past experience; an expert rifle tester could be expected to have some intuitive idea of the answer, or

(*c*) by statistical methods.

It should hardly need to be said that only this last way can give a precise and reliable answer. Sometimes its answer will, and sometimes it won't tally with what your judgement might lead you to expect, but at least it is completely objective and capable of exact mathematical proof. Here, then, is a way of arriving at conclusions which is quite unaffected by human prejudice, or by the memory of outstanding or unusual cases which may distort our judgement.

We shall later encounter a number of instances in which the statistical conclusion turns out to be very different from our 'guesstimate' of the situation. Here is a typical example. There are 3 shifts at a factory; each shift is of 8 hours, each has the same number of workers, and each involves the same work and same machines. In one particular month there are 7 accidents among the workers in the first shift, and 7 accidents among those in the second shift, but only 1 accident in the third shift.

Do these figures indicate that the workers in the third shift are more careful than those of the others? It certainly sounds like it, but later on (p. 277) you will see that this amount of variation could easily be explained by chance. There *may* be a genuine difference between the 3 shifts, but these figures simply do not prove it. (Example from K. A. Brownlee, in M. J. Moroney, *Facts From Figures*, Penguin, 1962.)

In short, the effects of chance must be taken into account in every piece of research.

The way that chance works is like this. Consider a large bag of marbles. If all the marbles are black, and you draw batches of say 10 marbles at a time, blindly from the bag, each and every sample group will be the same – all black. But if there is 1 white marble among all the black ones in the bag, every now and again a sample of 10 will contain that white marble. This means that the situation is now such that not all samples will be the same. Again, suppose there are 3 white marbles among all the black, samples of 10 could be drawn having 4 different compositions – all black, 1 white and 9 black, 2 white and 8 black, and 3 white and 7 black. Of course, these different compositions would not be encountered with the same frequency; all black would be the commonest, and 3 white with 7 black would be the rarest. But the main point to notice is that, with the exception of the situation in which the parent group is completely homogeneous (as when all the marbles in the bag were black), samples taken from the same parent group will ordinarily vary from one another as a result of chance.

We are now in a position to compare the trial of the 2 rifles, with which this chapter began, with the drawing of black and white marbles from a bag. The scoring of 5 hits out of 10 with rifle *A* is equivalent to drawing 5 white marbles (hits) and 5 black marbles (misses) in a random sample of 10 from bag *A*. Notice that we do not know the proportion of blacks and whites in the parent group; all that can be said is that our sample suggests that the proportion ought to be about 50:50. Now if rifle *B* scores 7 hits out of 10, this is equivalent to dipping into another bag of marbles, also of unknown composition, and drawing out 7 white and 3 black marbles. By the same reasoning as before, this suggests that bag *B* contains 70% white and 30% black marbles. But we know that, because samples drawn from the same parent group are liable to vary from one to another, there is a possibility that we could draw a sample consisting of 7 whites and 3 blacks from a parent bag containing 50% whites and 50% blacks. This would not occur as often as would be the case if the sample proportions matched the parent group proportions, but the fact that it can happen at all deserves consideration. We are therefore entitled to ask, '*What is the probability that both of the observed results could arise by chance from the same parent source*?' As we shall see, this probability can be calculated by significance tests based on the Laws of Chance, provided that the design of the experiment has ensured that our sample results are random (and not merely hand-picked ones).

If the answer turns out to be that the observed results could arise by chance from the same parent source once in 1,000 times, we are justified in concluding that, as this is very unlikely, the 2 parent sources are almost certainly different. Thus we judge the parent groups on the basis of our samples, and indicate the degree of reliability that our results can be applied as generalizations. We are not, after all, concerned only with what happened in the 10 shots with each rifle, but in what will happen in the long run, if we keep on shooting.

If, on the other hand, our calculations indicate that the observed results could be expected to arise from the same parent group once in every 5 times, we could easily have encountered one of those occasions, so our conclusion would be that a significant difference between the 2 rifles was not proven, in spite of the observed difference between the sample results. Further testing might reveal a genuine difference, so it would be wrong to say that out test *proved* that a significant difference did not exist; rather, we say that a significant difference was not demonstrated (on the strength of the evidence presented), which of course leaves the door open to reconsider the situation if further evidence comes to hand at a later date.

Now turn back to page 108 and refresh your memory about that tentative negative assumption (the null hypothesis). You can now see that the italicized question above contains this assumption. It is as though we had said, 'Let's assume for the moment that the observed results both derive from the same parent group; the results admittedly are not the same, but perhaps they are nothing more than chance variations such as can occur whenever 2 sample groups are taken from the same source, so let us find out just what the likelihood is of this being the case.' The negative assumption refers to 'let's assume that there is no difference in the parent source of these results' – it is always a case of *assuming no difference*. For only·in this way is it possible to develop a logical structure to our argument, regardless of how back-to-front it may seem at first. It just cannot be worked from the other direction. Thus if we had said, 'These results are decidedly different, so let us presume that they derive from different parent groups, and find the probability that this is not so', we would then be faced with the impossible task of trying to prove that 2 samples might have come from the same parent group, solely on the evidence of certain features or dimensions. It is obviously asking too much to try to prove identity of origin with incomplete data, and samples are, by their very nature, necessarily incomplete portions of their parent source. Samples can therefore only prove the likelihood of differences in their origins, but never the unity of their origins. It's a kind of one-way traffic.

There is no other way of getting the truth out of numerical data.

Probability Levels

If our calculations indicate that the observed results with rifle *A* and rifle *B* are very unlikely to have originated from a common parentage as a result of chance, the difference between the 2 rifles is said to be *statistically significant*.

Now, who is going to draw a line as to what is likely, what is unlikely, and what is very unlikely? Over many years, the following probability levels have been used satisfactorily by statisticians.

If the observed difference is likely to occur by chance with a frequency of more than once in 20 times (e.g. once in 5 times), it is not accepted as being a significant difference. Any probability larger than 5% is considered insufficient to deny our tentative assumption that the results have stemmed from the same source. In some quarters it is customary to call this answer 'not significant', but this implies a finality which is never justifiable. It is far better to describe this situation as *significant difference not proven*. Remember, it is always possible that a larger experiment may reverse this verdict.

If the observed difference could be expected to occur by chance only once in 20 times, which is a probability of 5%, this is considered to be unlikely to be due to chance. After all, the odds are 19 to 1 in favour of the difference being a real difference (in the sense that the samples came from different parent groups). Differences showing this probability level are generally said to be *probably significant*.

If the observed difference is only to be expected as a result of chance once in 100 times, which is a probability of 1%, it must be admitted that the likelihood of our no-difference assumption being correct is, in fact, very unlikely. We will then be right 99 times out of 100 if we reject our original assumption, and accept instead that the observed difference is due to the sample groups originating from different parent groups. The verdict here is that the difference is *significant*.

The 5% and 1% probability levels are thus standard critical levels. They are, of course, quite arbitrary, and it is up to the person interpreting the results to decide in advance of his calculations just what level of probability he would be prepared to accept as signifying a difference worth knowing about, taking into account the cost of the experiment and the seriousness of the consequences if the conclusion later proves to be wrong. One could naturally expect a different standard to apply to deviations in an astronaut's orbit, as against deviations in the automatic delivery of 1 lb packs of butter from a machine. But do realize that the solution is not necessarily that of simply insisting on small probability levels (1% or less) in all cases, for if the differences are small, such levels may only be attained with very large and costly samples. In general, however, it is better to be conservative and conclude that 'no difference has been proven by the experiment' than to accept that a difference exists when the evidence is not really adequate, and this even at the risk of sometimes concluding 'no difference' when a real difference would be shown by further experimentation.

We are, of course, dealing with probable rather than absolute truth (p. 19), so our conclusions must always be looked upon as provisional, in the nature of progress reports. As further evidence accumulates, our probable truths may be confirmed and thus become more certain, but there is always the possibility that a provisional conclusion may be contradicted in the light of additional observations. We should never lose sight of the strength of our evidence, for even if we found a result which could only occur by chance once in a million times, that chance will undoubtedly occur, with no less and no more than its due frequency, no matter how surprised we may be that it should have happened to *us*.

Furthermore, these interpretations apply only to data which is examined in accordance with the aim of the experiment or investigation. It has sometimes happened in the past that an experiment has failed to show what was being sought, but an examination of the results has revealed some unexpected finding. The application of statistical tests may show this finding to have a very remote probability of being due to chance, but it would be quite wrong to attach any special importance to such a result. For any group of numbers is likely to show some oddity if examined from enough points of view. So when you notice something 'worth testing' in a set of results, your choice is but one of many that might equally well have been chosen, and the calculated probability will therefore be false. The only thing to do in such a case is to carry out an entirely new investigation for the specific purpose of testing the significance of this unexpected finding; this may entail repeating the same experiment, or reviewing the results of previous experiments of the same kind, or perhaps performing an improved version of the experiment. Anyway, beware of conclusions based on 'post-mortem' examination of results, unless they have been tested properly in their own right.

Summarizing, it is valid to compare 2 or more sets of observations by determining the probability that the observed differences are only chance variations, arising naturally in the course of taking different sample groups from the one parent group, always provided that the sample groups have been selected in a random manner, and that it was decided to make the comparison before the results were known. The answer to this calculation is interpreted as follows –

Significance of Critical Probability Levels	
Probability (P) of differences being due to chance	Interpretation
P = more than 5%	Significant difference is not proven.
P = 5% or less	Difference is probably significant.
P = 1% or less	Difference is almost certainly significant.

In the significance tests which follow, additional guidance will often be given in the form of a level greater than 5% and another less than 1%. Statisticians do not ordinarily attribute any special significance to these outer levels, but they have been put in whenever possible in order to give perspective to answers which are greater than 5% or less than 1%.

Previously, for example on page 95, we have calculated the exact probability of a certain result being due to chance. Henceforth, with few exceptions, we shall be dealing with critical probability levels, for it is the significance of the result which is our real concern, rather than the knowledge that such-and-such could occur by chance once in 26·93 times. The latter probability will simply be expressed as being less than 5%.

One- and Two-sided Probabilities

It is to be noted that all probability levels for significance tests are the probability of the observed result itself plus any alternative results of equal or smaller probability. The need for this approach is most clearly seen in a simple example such as throwing 2 dice: if, as in all tests of significance, it is assumed tentatively that there is no real difference between 2 dice, the probability of getting a score of 3 and 1 must be balanced by adding the probability of getting a 1 and 3 (as on page 24), for the nature of the trial is such that both results are to be considered equally likely.

It is important to realize that in significance tests we do not ask, 'What is the probability of finding a man 6 or more inches *taller* than average?' (which may be a probability of 2%), but rather, 'What is the probability of finding a man 6 or more inches *different* from average?' (which would be a probability of 4%, half of which is contributed by those taller than average, and half by those shorter than average). The latter is a 'two-sided' probability, and is the correct one to use for such comparisons. After all, we begin by assuming that the man in question did derive from the parent population whose average is known, so it is equally likely that he will be either taller or shorter than average by this amount.

Now, there is a lot of muddled thinking about this particular aspect of statistical inference, even in high places, so some further discussion is warranted.

The argument to use a two-sided probability level applies even when we are only *interested* in deviations in one direction.

For instance, if the water supply of a town is known to have an average of 60 bacteria per millilitre, we would not be interested in any samples which had less bacteria than average, but would only want to test the significance of any sample which had more bacteria than average. In such a case it might seem reasonable to ask, 'If the true average of the water supply is unchanged, what is the probability of finding a sample with a bacterial count as large as this?' (as distinct from 'as different from average as this'). The answer to such a question, like the probability of finding a man so many inches taller than average, is a one-sided probability, and you will notice that for any given deviation from average, the one-sided probability is smaller than the two-sided level (e.g. 2% as against 4% in the example in the paragraph above). This creates two sets of probability levels, which is apt to be confusing, but the critical thing is that this duality of levels is brought about in these cases by the fact that the statistical test is only applied when the deviation is in a certain direction. In other words, the decision as to whether to apply a test or not depends on a 'post-mortem' examination of the sample results. Now surely it is ridiculous that such a decision should be able to affect the answer of the significance test. Yet it does. For unless we stick consistently to two-sided probabilities, we will find that a probability of 5% from one man's calculations may be equivalent to someone else's probability of 10% (the first being one-sided, the second being two-sided). The need for sorting out this problem is emphasized by the fact that published results usually fail to mention whether a given probability level is one-sided or two-sided. So let it be agreed that we shall use two-sided probabilities, and then we can carry out the appropriate significance tests whenever we like, adjusting our decision as to the probability level we shall accept as being significant according to our purpose. Naturally we will not change to using a new drug or other product which is significantly less effective than the one we have been accustomed to use, but the decision to change to the new product (if it proves more effective) can safely be based on the two-sided probability level of the comparative trial; the probability level should not be influenced by the direction of the outcome of the trial.

The problem arises again when there is some reason to believe that if a comparison between 2 things is going to show a significant difference between them, that difference will almost certainly be *in a predictable direction*. For instance, a small pilot study may show high hopes for a new drug, so that we may be tempted to believe that in a full scale experiment comparing this new remedy with the best established one, if any difference between the 2 drugs is demonstrated, it will presumably be the new one which will prove superior. Such a presumption may be nothing more than an impression or a hunch. But in an experiment on learning, if a group of people were given the same test twice, many people would probably feel that if a difference between the two tests was going to show up, it would be a fair thing to expect that the subjects would do better (not worse) at the second test. Or again, suppose you are going to repeat an experiment which proved statistically significant on the first occasion, for the purpose of confirming the original result. The presumption that a difference will show in a specified direction has now advanced from the 'hunch' stage to one of measurable 'certainty'. Some statisticians would use one-sided probabilities in all of these cases. But again this argument is unsound, for apart from perpetrating the confusing duality of probability levels mentioned above, it introduces an entirely subjective factor into the statistical analysis. The degree of certainty as to the direction of the outcome of any piece of research (which, after all, is only going to be carried out when the outcome is not known for sure) is obviously liable to vary from one person to another; if, then, a one-sided probability is allowable when a person *feels* or *believes* that the direction of the outcome can be predicted with some assurance, the one experiment may lead Tom to conclude $P = 4\%$, while Dick, who was not so convinced about the outcome before he saw the results, will conclude $P = 8\%$. Tom will then say the difference in the predicted direction is probably significant, while Dick will say that the observed difference is not proven to be significant.

When it all boils down, the idea of using a one-sided probability in these cases is essentially one of betting on the outcome. And what happens if you bet wrongly, and the outcome turns out to be in the opposite direction to that which you predicted? You are then faced with the unanswerable question, 'What is the probability if A is different from B, that it is better than B by the observed amount?', when the observed result in fact turns out to be that A is worse than B. A one-sided test is therefore meaningless in these circumstances, so you must either throw your results in the rubbish basket (and don't tell a soul!), or else slyly pretend that you really had in mind to ask, 'What is the probability of finding the observed *difference*?', in other words, that you had intended all the while to use a two-sided probability. The whole aim of statistical tests is to eliminate guesswork and to put inductive logic on a mathematical footing, and this means that these tests must remain completely objective. This impartiality, this freedom from human foible, is only possible if we stick to two-sided probabilities.

As a neat example to sum up these principles, imagine you have invented a safety device which makes cars automatically dip their headlamps at the approach of other cars, and you are going to compare the accident rate of a group of 2,000 cars fitted with this device with a control group of another 2,000 cars which do not have it. Of course, you will only be interested if the safety device lessens the accident rate, and your general knowledge suggests that if it is going to affect the accident rate at all, it will almost certainly be a reduction which will ensue. Here we have a combination of interest in one direction of outcome, and prediction of that direction. As always, we start by assuming that there is no significant difference between the accident rates of the two groups. If the observed results contradict this hypothesis at an acceptable two-sided probability level (say, $P = 5\%$), we interpret the results as meaning that the safety device either lessens or increases the risk of accidents, depending on the direction of the results. It's as simple as that. However, the use of a one-sided probability level involves acceptance of a different probability percentage (on the two-sided probability scale), which is unnecessarily confusing, and furthermore will invalidate the whole inquiry if the automatic device perchance increases the number of accidents (which is contrary to what was expected, but might still lead to some useful discovery, and therefore turn out to be of interest after all).

Now what do you think about the following statement, written by a statistician recently? 'The question of when to use two-sided and when to use one-sided probabilities for testing significance is apt to be a difficult one. . . . If there is the slightest doubt, it is far safer in the interests of truth to use a two-sided probability as a criterion. This is a matter of statistical ethics.' My answer is that in all research there must be at least some doubt about the direction of the outcome, so the problem is no longer 'difficult'. It is solved simply by using two-sided probabilities as a routine for all significance tests. This is not a question of good manners, but of logic.

Please understand that one-sided probabilities are not wrong in any arithmetical sense. The fault is that, for significance tests, they are answering a biased question. It is the structure of the question which is illegitimate. The error is fairly obvious in a case like this: 'In this experiment, I reckon A will prove better than B. However, I will start with the usual tentative negative assumption that there is no real difference between them, and will then calculate the probability of A being better than B by the amount that will be observed, as a consequence of chance alone.' As it stands, this calls for a one-sided probability answer; rather, the experimenter should have asked the probability of A being different from B (i.e. in either direction), which is a two-sided probability. But in experiments where a thing is tested 'against chance', the problem is more subtle. Suppose you want to find out if a person can tell the difference between the flavour of his favourite brand of cigarettes and that of other brands. You might then design a simple experiment in which the subject is blindfolded and is presented with a series of pairs of cigarettes, one of each pair being his favourite brand. The subject is asked to identify his favourite out of each pair. If he can't tell the difference, he is due to give the right answer in half of the tests, purely by chance. And if he can tell the difference, he ought to get more than half of them right. At this point the devotees of one-sided probabilities would put their foot in it by saying: 'We needn't allow for the possibility that he might get more than half of them wrong, for he can't be worse than chance.' But this is not true. He could be systematically wrong in every single test. And so long as this is physically possible, we must refrain from betting on the direction of the outcome. For we begin by assuming tentatively that he can't tell the difference, and if so, it is equally likely that he will guess either more, or less, than half of them correctly. We are therefore only entitled to ask the probability of getting the observed *difference* from that which could be expected by chance alone (50%), regardless of the direction of that difference, and this is a two-sided probability. In other words, instead of asking, 'Can he pick his own brand?' we must phrase it, 'Can he distinguish between his own and other brands?' If we get a satisfactory answer to this second question, and if the direction is right, we can finally say with measurable confidence that he can pick his own brand.

Bearing in mind that one-sided probabilities are smaller and hence more highly significant than their two-sided counterparts, let us reflect as follows: O ye little gods of Statistics, lead us not into temptation, lest we use one-sided probabilities in order to make our results appear as impressive as possible!

There are, however, some occasions when a probability is absolutely and unavoidably one-sided. This can happen in certain asymmetrical situations. Consider the case of a parent group which is distributed asymmetrically, as in the example given on page 81. The data Table there shows the probability of that skin disease lasting 1 to 2 weeks is, by the Proportionate Law, $\frac{20}{221}$, which is 9%. However, adding up the other end of the scale we find that 17 patients had a duration of 13 weeks or more; the probability of such a duration is thus $\frac{17}{221}$, which is 8%. The probability of having the disease for as little as 1 or 2 weeks, or for as long as 13 weeks or more is thus 9% + 8% = 17% (by the Addition Law). If a patient with that disease was treated with some new treatment, and the duration turned out to be 1 week, the result would not be proven to be significantly different from average, because a deviation from the average as great as this could be expected purely as a result of chance 17% of the time. On the other hand if the new treatment had resulted in a duration of 19 weeks, the deviation is probably significant, for the probability of getting such a result by chance is only $\frac{6}{221}$, which is about 3%. The data Table shows us that in this case there is no alternative result of equal or lesser probability on the other side of the average, so the probability level is unavoidably one-sided.

All tables of probability levels in this book give the two-sided probabilities, except where a one-sided probability is unavoidable because of asymmetry.

Statistical versus Practical Significance

In normal usage, the word 'significant' means 'important', as in the statement, 'He got a significant rise in salary.'

In Statistics, 'significant' means 'beyond the likelihood of chance', as in the sentence, 'The difference in performance of the 2 rifles was statistically significant.'

Whenever a difference between sets of observations has a practical significance, one expects to be able to demonstrate the difference to be statistically significant. The reverse, however, does not necessarily apply. A difference may be statistically significant, and yet have no practical significance or importance. Thus occasionally one will meet instances like this –

Take the average pulse rate in health to be 70, with a standard deviation of 5. Now suppose a group of 40 patients with 'Acute Terribilitis' have an average pulse rate of 73. We will show later (p. 155) that this difference is highly significant, being liable to occur as a chance variation from normal less than once in 500 times. Yet it would have no diagnostic value whatever, because a pulse rate of 73 is quite common in perfectly healthy people (as evidenced by the standard deviation of 5).

In other words, whenever you come across a result which is statistically significant, you must still assess whether the difference is of any practical consequence or not. There are times when a tiny, but real, difference can save millions of dollars, and other times when such a difference is simply not worth knowing about.

Introducing the Significance Tests

There are a very large number of different statistical tests. Each test is designed to deal with a certain type of situation. Thus a test which will compare the performance of 2 rifles will not necessarily be suitable for comparing the performance of 3 rifles; a test which will compare accident rates in different factory shifts cannot be used for comparing the yields of a crop under the influence of different fertilizers. Tests vary, too, in the broadness of their applicability, in their accuracy, in their efficiency, and in the amount of arithmetic which they entail.

The tests included in this book have been selected very carefully. They are more or less standard tests which cover a comprehensive range of situations; they are reasonably easy to do, and are sufficiently accurate and sensitive for all practical purposes. They are tests which, because of their simplicity, are ideal for checking results of investigations, particularly those which are published without statistical analysis. Practising statisticians do not, of course, limit themselves to any such basic group of tests, but beginners have to start somewhere, and the group presented herein can confidently be recommended to anyone whose work or interest brings them into contact with the results of research in any field.

As this book is designed primarily for non-mathematical people, the theory behind these tests will be kept to but brief explanations of the principles, sufficient to permit them to be used intelligently, in the manner of working tools. If any excuse is needed for this approach, let it be that many people will forever be denied access to these working tools if they are forced in the first place to digest the complicated mathematics under-lying these tests. It is possible to drive a car correctly without knowing all about the car's internal workings; it is even possible to use logarithms usefully and accurately without knowing how they work. So it is with these tests. Of course, there is nothing to prevent anyone from delving into the subject further, either using textbooks such as those recommended on page 59, or by going to the original sources which are quoted in the case of all the more modern tests.

A word of warning, though. In the absence of a complete mathematical understanding of these tests, you will have to be very careful indeed to use them strictly in accordance with the instructions provided, both as regards the type of situation to which each test may legitimately be applied, and also to the exact setting out of the calculations. Failure to do either of these things will often lead you silently and unknowingly into error. This advice, to stick to the instructions to the letter, is especially necessary because a great deal of effort has been spent in making the procedures as simple as possible, and you may therefore be misled into imagining that the tests are simpler than they really are. Take no liberties with them, and they will serve you well.

As in previous chapters, many of the examples used to illus-trate the tests are fictitious, in order to keep the calculations simple and clear. In this way the structure of each test (i.e. what is actually being done) becomes easy to grasp. Then you should have no difficulty in doing the set questions yourself, and then you will be able to apply these tests to your own particular field of interest. Do not neglect to answer all the questions as you proceed, for no one can say that they know how to do these tests until they have actually done them. They are not designed merely for schoolwork, but for familiarization.

Statisticians use calculating machines; copy them by using every mathematical aid that you can. This decreases both your labour and the risk of computational errors. Tables of squares and square roots are provided at the back of this book; they will give a sufficient number of digits for the degree of accuracy re-quired for significance tests. A small adding machine is handy. For multiplications and divisions, use logarithms or a slide rule. With the latter you can multiply or divide numbers in a matter of seconds; a 10-inch student's model is simple to use, sufficiently accurate, and inexpensive.

However, far and away the greatest labour-saver of all is the provision of special tables for interpreting the test results. These tables are provided for every test. They obviate the need to carry your calculations right back to the basic Laws of Chance by meeting you more than halfway along the road. The tables themselves have been calculated from these Laws, often with the help of electronic computers. They are wonderful time-savers. To ensure that they are easy to use, all these tables have been modified from the originals in order to produce a format and style which is as uniform as possible throughout. In some cases the transformation has resulted in tables which will scarcely be recognizable beside their parent source; this applies especially to those which have been converted from one-sided to two-sided probabilities.

Finally, after all the tests have been described and, I trust, mastered, there remains the problem of knowing which test to use for any set of data. It is one thing to know *how* to do these tests, and another thing to know *when* to use them. The only way to learn this is by practice. Accordingly, a fairly extensive set of miscellaneous questions is provided, specifically to enable you to become proficient with handling the sort of situations you will meet in real life. The guide to these miscellaneous questions is given in the form of a Table of Significance Tests at the very back of this book, where it serves as a kind of index. But you will need practice at using this Table, and the miscellaneous questions are there to provide just that, and at the same time will raise a number of additional points of interest. As always, answers and comments are provided.

The time has now come to let the statistical cat out of the bag.

zM TEST (1733)

Purpose

The *z* Test for Measurements (hence '*zM*') compares a random sample of 1 or more measurements with a large parent group whose mean and standard deviation is known.

Principle

All things that can be measured have an unlimited number of possible measurements, which together form a potential parent group of measurements. The 50 measurements of my desk (p. 33) were a sample from such a parent group. The frequency distribution of the measurements making up a parent group usually follows the pattern of the Normal Curve (p. 35). The most important dimensions of this pattern are the mean (the measure of central tendency) and the standard deviation (the measure of dispersion around the mean).

The difference between the mean of a parent group and the mean of a random sample is given by simple subtraction. Dividing this difference by the standard deviation of the parent group converts the units of measurement (inches, volts, etc.) into standard deviation units. This 'standardized difference' between the parent group mean and the sample mean is the value called *z*. The further the sample mean is from the parent mean, the larger will be the value of *z*.

In the case of parent groups whose measurements have a logarithmic distribution (p. 78), the above calculation can be carried out using logarithmic means and the logarithmic standard deviation (pp. 84 and 87). This does not distort the value of *z*.

The probability of encountering various values of *z* by chance has been calculated using the Normal Probability Formula (p. 34). This enables a simple Table to be drawn up showing the values of *z* which correspond to important probability levels.

To determine whether a certain sample differs significantly or not from a specified parent group whose characteristics are known, we begin by tentatively assuming that the difference between the measurements of the sample and those of the parent group are simply the result of chance. In other words, we assume that the sample has really come from the parent group concerned, and the probability of this being so is then determined by calculating the value of *z*.

If this indicates that the sample could arise by chance from the parent group with a probability of greater than 5% (i.e. more than once in 20 times), we cannot deny that the sample may well belong to the large group concerned. On the other hand, if the probability turns out to be 5% or less (i.e. once in 20 times or less often), we reject our tentative assumption of no real difference, and accept instead that the sample probably (or almost certainly) has come from a different parent group.

Data Required

n = number of measurements in the sample.

m = mean of the sample measurements, determined as on page 52 (or p. 86 for logarithmic means). Note that if the sample consists of a *single measurement*, the value of that measurement is the sample mean.

M = mean of the large parent group.

S = standard deviation of the large parent group.

Procedure

Calculate z from the following formula –

$$z = \frac{\sqrt{n} \cdot |M - m|}{S}$$

The straight brackets enclosing $M - m$ are a useful mathematical symbol meaning 'subtract the smaller from the larger value'. This ensures that the answer will be a positive value. E.g. $|20 - 30| = +10$. Remember the meaning of these straight brackets, for we shall be using them frequently.

Don't forget to use the Tables of Square Roots at the back of this book for finding the value of $\sqrt{n}$.

For data which has a logarithmic distribution, use the logarithmic means (M_{log} and m_{log}) and logarithmic standard deviation (S_{log}), as described on pages 86 and 87. The z formula then becomes –

$$z = \frac{\sqrt{n} \cdot |M_{log} - m_{log}|}{S_{log}}$$

Interpretation

The larger the value of z, the less the likelihood of our no-difference assumption being correct.

The exact values of z corresponding to important probability levels (as calculated from the Normal Probability Formula) are as follows –

z TABLE

Probability of no significant difference between M and m			
P = 10%	P = 5%	P = 1%	P = 0·2%
$z = 1·64$	1·96	2·58	3·09

Adapted from E. S. Pearson and H. O. Hartley, *Biometrika Tables for Statisticians*, Vol. 1, Table 4 (CUP, 1966).

If z is less than 1·96, the probability (P) of encountering a sample mean of m by chance from a parent group whose mean is M and whose standard deviation is S, is a thing to be expected on more than 5% of occasions. In this case, a significant difference is *not proven.*

If z is 1·96 or more, P = 5% or less, which is to say that the observed difference between the sample and the large group could be expected as a result of chance only once in 20 times, or less often. This possibility is reasonably remote, so in such a case we accept that the difference is *probably significant.*

If z is 2·58 or more, P = 1% or less. Such a probability is so slight that we reject our tentative no-difference assumption, and accept that the difference between the means is *statistically significant.*

The outer columns of the z Table (P = 10%, and P = 0·2%) are given to add perspective to the issue.

Examples

Ex. 1. The melting point of gold is known to be 1,060° C. This is, of course, an average figure, for unavoidable 'experimental error' causes more or less variation from this figure whenever the test is actually performed. The best measure of these variations is the standard deviation (S). Suppose this has been calculated from a large series of tests, according to Method 3 (p. 62), and found to be 3° C.

Now imagine that you are analysing an unknown metal, and

a test shows its melting point to be $1,072°C$. Is it likely that this unknown metal is gold? In other words, what is the probability that a sample of gold would show a melting point as different from its average as $1,072°C$?

Data: $n = 1$; $m = 1,072$; $M = 1,060$; $S = 3$.

Calculation:
$$z = \frac{\sqrt{n} \cdot |M - m|}{S}$$
$$= \frac{\sqrt{1} \cdot |1,060 - 1,072|}{3}$$
$$= \frac{1 \times 12}{3}$$
$$= \underline{\underline{4·0}}$$

Referring this answer to the z Table opposite shows that the probability of the sample measurement being merely a chance variation from the parent group of gold is less than $0·2\%$, which is a very remote chance indeed. The difference in melting points is statistically highly significant, so we reject the possibility that the tested metal is likely to be gold.

Ex. 2. Here is the calculation of the probability that the pulse rate in 'Acute Terribilitis' is within normal limits, as discussed on page 149.

Data: $n = 40$ patients with this hypothetical disease

   $m = 73$, the average of their pulse rates

   $M = 70$, the average pulse rate in healthy people (the parent group)

   $S = 5$, the standard deviation of the pulse rate in health

Calculation: $z = \dfrac{\sqrt{n} \cdot |M - m|}{S} = \dfrac{\sqrt{40} \cdot |70 - 73|}{5}$
$$= \frac{6·32 \times 3}{5}$$
$$= \underline{\underline{3·79}}$$

Reference to the z Table shows the probability of the observed difference (between the average pulse rate in health, and in these patients) being due to chance is considerably less than 1 in 500. This indicates that, although the difference is small, it is statistically highly significant. It is therefore almost certain that the observed difference is a genuine difference, and not just a variation due to chance.

Ex. 3. An example of a logarithmic distribution was given on page 81. It concerned the duration of a certain skin disease, under standard treatment, based on the observation of 221 patients. The data was presented in a table on page 81, and in diagrammatic form on page 85. Davies' test indicated that the distribution was logarithmic (p. 84). The logarithmic mean duration of the disease was found to be 0·7445 (p. 87), and the logarithmic standard deviation was found to be 0·2616 (p. 88).

These dimensions, being derived from a fairly large group of patients, can be looked on as the dimensions of the parent group of this disease (under the influence of standard treatment). It would have been more accurate to use a group of 2,000 patients, but it might take 20 years to collect such a group, so we accept the slight loss of accuracy in the interest of arriving at a provisional conclusion. So, here we have a parent group of known dimensions, against which sample groups given other treatments can be compared.

Such a comparison was, in fact, carried out by the doctor who reported this particular piece of research. He treated 14 other patients suffering from this disease with a different treatment, namely antihistamine tablets. The duration of the disease in this small group varied from 2 to 10 weeks. To compare this result with the parent group we shall compare the mean duration of each group. To do this, we must first find the logarithmic mean of the sample group, for we cannot compare a logarithmic mean with an arithmetic mean, and the examination of the parent group has told us that, in this instance, the logarithmic mean is the more accurate one to use.

We could find the logarithmic mean of the 14 durations of the sample group by adding the logarithms of each duration, and dividing by 14, as described on page 86. However, the values fell into natural groups (2 patients had a duration of 2 weeks, 2 had 3 weeks, 4 had 4 weeks, etc.), so we shall find this mean by using

a perfectly accurate, slight modification of Method 2 (p. 60) as
follows –

Values of x (duration in weeks)	Frequency (number of patients with the stated duration)	Log x	Freq. × Log x
2	2	0·301	0·602
3	2	0·477	0·954
4	4	0·602	2·408
6	4	0·778	3·112
7	1	0·845	0·845
10	1	1·000	1·000
$n = 14$			Total = 8·921

The logarithmic mean duration is the arithmetic mean of the
logarithms of the durations, so –

$$m_{\log} = \frac{8 \cdot 921}{14} = \underline{\underline{0 \cdot 637}}$$

The mean duration of the sample is thus less than that of the
large group. But are we entitled to infer that antihistamine
tablets shorten the duration of this disease? No, not until we
have found out the likelihood of the observed difference being
due to chance. To do this, we start by making the assumption
that there is no significant difference between the duration with
the standard treatment and with the antihistamine treatment.
Then we ask: What is the probability that a parent group whose
logarithmic mean is 0·744 and whose logarithmic standard deviation is 0·262 could yield, by chance, a sample group of 14
patients with a logarithmic mean as different from the parent
mean as 0·637?

The answer is found by calculating z. Before proceeding, however, let us make it a firm rule to always write down the symbols
and data values first, so that there will be no errors when it
comes to substituting the data values for the symbols in each
formula.

Data: $n = 14$ $M_{\log} = 0 \cdot 744$

   $m_{\log} = 0 \cdot 637$ $S_{\log} = 0 \cdot 262$

Calculation: $z = \dfrac{\sqrt{n} \cdot \mid M_{\log} - m_{\log} \mid}{S_{\log}}$

$$= \dfrac{\sqrt{14} \cdot \mid 0 \cdot 744 - 0 \cdot 637 \mid}{0 \cdot 262}$$

$$= \dfrac{3 \cdot 74 \times 0 \cdot 107}{0 \cdot 262}$$

$$= \underline{\underline{1 \cdot 53}}$$

The z Table (p. 154) shows that this value of z does not even reach the 10% probability level; it has a probability of more than 10%. This means that random samples of 14 patients taken from the large parent group could be expected to show this much difference from the parent mean more than once in every 10 such trials. A significant difference between the 2 means is thus not proven by this experiment. It is not proven that the antihistamine tablets reduce the duration of the disease, in comparison with the standard treatment.

If we had ignored the asymmetry of the parent distribution in this example, and had simply calculated the means and standard deviation without logarithmic transformation, we would find $M = 6 \cdot 53$, $S = 4 \cdot 28$, and $m = 4 \cdot 79$. Then

$$z = \dfrac{\sqrt{14} \cdot \mid 6 \cdot 53 - 4 \cdot 79 \mid}{4 \cdot 28} = 1 \cdot 52$$

The closeness of this to the logarithmic answer $(1 \cdot 53)$ is unusual; ordinarily with logarithmic distributions the difference will be greater than this, and the logarithmic answer will be the more trustworthy one. But this does illustrate how the zM Test tends to be relatively robust with respect to handling asymmetrical distributions.

Questions

Q41. In Example 1 (p. 154) it was stated that an unknown metal was found to have a melting point of 1,072°C. Is the unknown metal likely to be copper, the average melting point of which is 1,080°C, with a standard deviation of 5°C? (First write down the symbols and data values, then the formula to be used, and then proceed with the calculation.)

Q42. Continuing *Q41*, suppose the melting point of this unknown metal is determined 3 more times, with the following results –

    1,072° (original result),   1,071°,   1,072°,   1,073°.

Calculate the arithmetic mean of these 4 results (using a short-cut), and then find out whether this additional evidence increases or decreases the likelihood of the unknown metal being copper.

Q43. Suppose that, as part of an investigation into ladies' hair styles, the maximum length of hair of over 1,000 London women was measured, and the measurements were found to have a logarithmic distribution, with a logarithmic mean of 0·954 inches and a logarithmic standard deviation of 0·600 inches.

Now, suppose that the hairdresser who made this investigation went to Paris for a holiday, and while there carried out a similar investigation on a random sample of 40 Parisian women, and found their logarithmic mean to be 0·764 inches.

Obviously the Parisian sample average is quite a bit shorter than the London average, but would it be justifiable to claim that, on the evidence presented, Parisian women *in general* had shorter hair styles than those of London? In other words, is there a significant difference between these means, or could the observed difference be reasonably attributed to the kind of chance variation that always tends to be present when random samples are drawn from a parent group?

'STUDENT'S' t TEST (1908)

Purpose

To compare a random sample consisting of 3 or more measurements with a large parent group whose mean is known, but whose standard deviation is not known.

Principle

This test is a modification of the zM Test in which the standard deviation of the sample measurements is used instead of the standard deviation of the large parent group.

In making this substitution we are really using the standard deviation of the sample measurements as an *estimate* of the standard deviation of the parent group. Accordingly, it is necessary to make allowance for the fact that, in the same way as the means or proportions of different samples drawn from the same parent group show variation from one to another as a result of chance, so the standard deviation of different sample groups taken from the same source will also vary from one sample to another. This allowance for chance variations is conveniently made by modifying the z Table to take the sample size into account, for the smaller the sample, the larger the eccentricity is likely to be.

The way of working out this allowance was discovered in 1908 by William Gosset, a 32-year-old research chemist employed by Guinness, the famous Irish brewery. He reported it in *Biometrika Journal*, 1908, Vol. 6, pp. 1–25, and the story behind this report is very unusual. You see, it was a strict rule at Guinness' that employees were never allowed to publish their discoveries. However, because of the exceptional nature and importance of this statistical computation, Gosset was granted permission to publish it, but only on condition that he remained anonymous and used a pen name. He chose 'Student'. His real name was released in later years, but it is his pen name which endures as one of the most renowned names in Statistics. Gosset died in 1937.

As usual, the tentative negative assumption is made that there is no significant difference between the mean of the sample group and the mean of the large parent group, and the probability of this being the case is determined by calculating the value of 'Student's' t, and referring this answer to the Table provided.

Data Required

n = number of measurements in the sample group. This must be at least 3.

m = mean of the sample measurements.

s = standard deviation of the sample measurements, calculated at the same time as the mean, see pp. 57–65.

M = mean of the large parent group.

Note: (1) The sample must be an ordinary random one; the test as described here is *not* suitable for use with a stratified random sample.

(2) This test applies to groups of things whose measurements are distributed in the pattern of the Normal Curve (p. 35), and this, of course, covers the vast majority of situations. If, from the nature of the investigation, it was suspected that the measurements were distributed in some other pattern (e.g. logarithmically), it would be necessary to examine the parent group by Davies' test (p. 78), and you would then have sufficient information to proceed with the ordinary zM Test.

Procedure

Calculate 'Student's' t from the following formula –

$$t = \frac{\sqrt{n} \cdot |M - m|}{s}$$

The only difference between this and the z Formula (p. 153) is that this one uses the standard deviation of the sample (s), whereas the z Formula uses the standard deviation of the parent group (S).

Interpretation

The interpretation of the value of 'Student's' t is a similar process to that of interpreting z. The larger t is, the less the probability of our no-significant-difference assumption being correct.

However, with the t Test, the critical values of t vary with the number of measurements (n) in the sample, so you must use the row of t-values corresponding to the sample size; see the t Table on page 162.

The significance of these probability levels was discussed on page 140.

SIGNIFICANCE TESTS

'STUDENT'S' t TABLE

No. in sample n	Probability of no significant difference between M and m			
	P = 10%	P = 5%	P = 1%	P = 0·2%
3	$t = 2·92$	4·30	9·92	22·33
4	2·35	3·18	5·84	10·21
5	2·13	2·78	4·60	7·17
6	2·02	2·57	4·03	5·89
7	1·94	2·45	3·71	5·21
8	1·89	2·36	3·50	4·79
9	1·86	2·31	3·36	4·50
10	1·83	2·26	3·25	4·30
11	1·81	2·23	3·17	4·14
12	1·80	2·20	3·11	4·02
13	1·78	2·18	3·05	3·93
14	1·77	2·16	3·01	3·85
15	1·76	2·14	2·98	3·79
16	1·75	2·13	2·95	3·73
17	1·75	2·12	2·92	3·69
18	1·74	2·11	2·90	3·65
19	1·73	2·10	2·88	3·61
20	1·73	2·09	2·86	3·58
21	1·72	2·09	2·85	3·55
22	1·72	2·08	2·83	3·53
23	1·72	2·07	2·82	3·50
24	1·71	2·07	2·81	3·48
25	1·71	2·06	2·80	3·47
26	1·71	2·06	2·79	3·45
27	1·71	2·06	2·78	3·44
28	1·70	2·05	2·77	3·42
29	1·70	2·05	2·76	3·41
30	1·70	2·05	2·76	3·40
40	1·68	2·02	2·70	3·31
60	1·67	2·00	2·66	3·23
120	1·66	1·98	2·62	3·16
∞	1·64	1·96	2·58	3·09

Adapted from E. S. Pearson and H. O. Hartley, *Biometrika Tables for Statisticians*, Vol. 1, Table 12 (CUP, 1966).

Example

Ex. 4. A certain printing press is known to turn out an average of 45 copies a minute. In an attempt to increase its output, an alteration is made to the machine, and then in 3 short test runs it turns out 46, 47, and 48 copies a minute. Is this increase statistically significant, or is it likely to be simply the result of chance variation?

This question raises a couple of important preliminary points. First, notice that we are dealing with *measurements*. Any quantity or quality which is capable of being measured on some continuous scale is to be considered a measurement. Measurements usually have some units, such as inches, ounces, kilowatts, miles per hour, and so on. But even though the scale of units may be continuous (in the sense of being uninterrupted), the coarseness of our measuring devices often has the effect of breaking the continuous scale into a number of tiny discrete jumps, as in the jerky movement of the second hand of a watch, or in the ability of a person to count the number of copies coming out of a printing press only to the nearest whole number. Do not be misled, then, into thinking that the output of this press is not a measurement, simply because there are no technical units (such as horsepower) or because the output is quoted in whole numbers. Well, what about a count of the number of flowers on a bush, or the number of dogs roaming in a certain street, are these to be considered measurements too? Yes, they are, in spite of the fact that such measurements can never be anything but whole numbers (the number of dogs may be 3, 16, or 102, but could never be 6·15). This note has been called for because it is very important not to apply tests for measurements to things which, by their nature, are not measurements. Tests for proportions and for things divisible into mutually exclusive classes, and tests for things which arise as isolated occurrences, will be dealt with further on in this book.

The other point of interest raised by this example is that you might at first imagine that after the machine had been altered, the 3 test runs should be considered as 3 samples. Well, they are in a sense, but, if so, each sample consists of but 1 measurement. It is therefore usual in Statistics to group such sample measurements together, and to look upon them as 1 sample group containing so many measurements. Failure to appreciate this point would leave you wondering what the value of n (the number of measurements in the sample) would be.

We can now symbolize and summarize the data of this example, and proceed.

$$Data: \quad n = 3$$

$$m = \frac{46 + 47 + 48}{3} = 47$$

$$s = 1 \text{ (see below)}$$

$$M = 45$$

The standard deviation (s) of the sample measurements is calculated according to the method described on page 57, as follows.

x	d	d^2
46	-1	1
47	0	0
48	1	1
$x_0 = 47$	$1-\ \ 1$	2
$n = 3$	$\therefore A = 0$	$= B$

$$s = \sqrt{\frac{B - \dfrac{A^2}{n}}{n - 1}} = \sqrt{\frac{2 - \dfrac{0}{3}}{3 - 1}} = \sqrt{\frac{2}{2}} = 1$$

$$Calculation: \quad t = \frac{\sqrt{n} \cdot |M - m|}{s}$$

$$= \frac{\sqrt{3} \cdot |45 - 47|}{1}$$

$$= 1 \cdot 73 \times 2$$

$$= \underline{\underline{3 \cdot 46}}$$

Referring this value of t to the t Table on page 162 shows that for $n = 3$ (the top line), the probability of there being no significant difference between the mean of the sample group and the mean of the parent group is greater than 5% (actually it is between 5% and 10%). Such a difference as was observed would therefore be likely to occur by chance more than once in 20 times, so we conclude that a significant difference is not proven.

Questions

Q44. Continuing the above example, 2 more trials are made with the press, and each one turns out 47 copies a minute. Does this additional evidence alter the significance of the results? Would this have any practical significance?

Q45. Over the period 1945 to 1962, my telephone bill averaged £48 per half-year. Then I got a new secretary, and my telephone bills became, successively, £56, £51, £63, and £60. Should I sack my secretary for over-using the 'phone?

Q46. The average annual income of adult males in North Utopia is known to be $20,560. It is reported that a random sample of 49 adult males in South Utopia had an annual income which averaged $18,505. Is this difference statistically significant?

WILCOXON'S SUM OF RANKS TEST (1945)

Purpose

To compare 2 unmatched random samples of measurements, as for instance 2 samples taken from different sources.

Principle

The usual tentative assumption is made that there is no significant difference between the 2 sets of measurements – we imagine them to be independent samples drawn from the same parent group (or more specifically, from parent groups whose arithmetic means are the same), and attribute any difference between the samples to the whims of chance. The probability of this being the case is then determined along the following lines.

We test our tentative assumption by first pooling the measurements of both samples. Then we arrange the combined set of measurements in order of size. This is called *ranking* the pooled measurements. If each measurement is now awarded a rank value according to merit (from the largest to the smallest measurement, or vice versa), it should be clear that if the 2 samples possess the same number of measurements and are really derived from the same parent group, each sample group ought to gain about the same total of rank values.

The probability of getting unequal rank totals as a consequence of chance variation is determined by the present test, which was first described, for samples of equal size, in 1945 by an American industrial statistician, Dr Frank Wilcoxon, in *Biometrics Bulletin*, 1945, pp. 80–2. It was later extended to samples of unequal size by H. Mann and D. Whitney (*Ann. Math. Statist.*, 1947, pp. 50–60).

This test really amounts to an indirect way of comparing the arithmetic means of the 2 sample groups. It is insensitive to the dispersion of measurements in the samples, so that if Sample *A* consists of three measurements, 9, 10, and 11 units, while Sample *B* consists of 7, 10, and 13 units, this Wilcoxon's Test will show no difference between them, because they both have a mean of 10 units. However, Sample *B* has a greater dispersion or spread of measurements, and there may be occasions when you want to know the probability of such samples being derived, by chance, from parent groups with the same standard deviation.

This would be the case if you were primarily interested in the *uniformity* of samples from different sources (e.g. 'Do Eskimo women marry at a more constant age than Mexican women?'). The way to handle this is by **Siegel and Tukey's modification** of Wilcoxon's Sum of Ranks Test (*Journ. Amer. Statist. Assoc.*, 1960, pp. 429–44). In this modification the ordered pooled measurements are given rank values as follows: assign rank 1 to the smallest measurement, rank 2 to the largest, rank 3 to the second largest, rank 4 to the second smallest, rank 5 to the third smallest, rank 6 to the third largest, and so on in 'paired alternation' (you leave the central observation unranked if there is an odd number of pooled measurements). In this way extreme observations get low rank values, whereas central observations get high values, so that if one sample receives a lower total of rank values than the other, it means that the samples differ in the dispersion of their measurements. The significance of any inequality of rank totals is given by exactly the same Tables and z Formula as used for the ordinary Wilcoxon's Sum of Ranks Test.

The idea of using rank values (in place of the measurements themselves) for the purpose of significance tests came from Professor Spearman in 1904 (p. 199). The use of ranks simplifies the arithmetic very considerably, and has the additional advantage of being uninfluenced by different patterns of distribution of measurements in the parent groups from which the samples are drawn. Such tests are called 'distribution-free', and are equally good for normal, logarithmic, or any other distributions. The price to be paid for such simplicity and wide applicability is generally some slight loss in power, so that they tend to give slightly higher probability levels than their much more complicated alternatives. For non-professional students of Statistics, however, their virtues far outweigh this shortcoming. If you would care to check this statement, refer to one of the textbooks of Statistics which describe the form of t Test used for testing 2 independent samples and compare it with this Wilcoxon's Test.

Data Required

n_A = number of measurements in Sample A.
n_B = number of measurements in Sample B.
Plus all the individual measurements in both samples.

Procedure

(1) Prepare a table for the calculation with the following headings –

Data values	Tally	Rank values	A ranks	B ranks

(2) Examine the measurements in both samples, and find the smallest and the largest measurement in the combined set.

(3) In the first column of the calculation table write down an ordered list of values in the observed range, starting with the smallest measurement and ending with the largest. We shall call these numbers 'data values'.

(4) In the second column enter the tally. Work systematically through Sample A, and for each of its measurements put an A beside the appropriate data value in the table. Then enter a B likewise for each of the measurements in Sample B. The net result of this tallying process is to arrange all the observations in order of size, that is, to rank the observations.

(5) Assign a rank value to each observation. If $n_A = n_B$, or if either is more than 20, you can rank in *either* direction (from smallest to largest will do). However, if $n_A \neq n_B$ and neither exceeds 20, the measurements must be ranked in *both* directions, from the smallest to largest and vice versa. (The sign '$\neq$' means 'is not equal to'.)

If 2 or more observed measurements are the same (i.e. if they have the same data value), each measurement has to be assigned an *average rank value*. Thus if 2 measurements share the 1st and 2nd place in the ranks, each is given a rank value of $\frac{1+2}{2} = 1\frac{1}{2}$. If 3 measurements happen to share the 9th, 10th, and 11th places, each sharer is given a rank value of $\frac{9+10+11}{3} = 10$.

Such tied ranks tend to weaken the power of this test a little, so it is better to go back to the original data and work out tied measurements to an additional one or two decimal places if possible in order to break such ties. (All remarks in this section apply also to Siegel and Tukey's modification.)

(6) Now transfer the rank values belonging to Sample A into the next column, and those belonging to Sample B into the final column.

(7) Add up whichever is the smallest of the 2 or 4 sets of rank values. Call this *smallest rank total 'R'*.

The fact that the sum of a series of consecutive numbers 1 to N equals $\frac{1}{2}N(N+1)$ provides a method of checking the calculation at this juncture, as follows: the sum of the 2 rank totals, $R_A + R_B = \frac{1}{2}(n_A + n_B)(n_A + n_B + 1)$. This applies regardless of the direction of ranking.

Interpretation

The smaller the value of R, the less the chance of our no-significant-difference assumption being correct.

(a) *With up to 20 measurements in each sample*, the critical values of the smallest rank total (R) are provided in the Wilcoxon's Sum of Ranks Tables on pages 170–1. These Tables show the probability of getting (by chance alone) a rank total in either sample as small as shown in the body of the Tables, if there is really no significant difference between the 2 samples.

The significance of these probability levels was discussed on page 140.

(b) *With more than 20 measurements in one or both samples*, the significance of the smaller rank total (R) is to be found by calculating z from this formula –

$$z = \frac{n_R(n_A + n_B + 1) - 2R}{\sqrt{\dfrac{n_A \cdot n_B(n_A + n_B + 1)}{3}}}$$

– where n_R = number of measurements in whichever sample possesses the smaller rank total; it thus equals n_A or n_B depending on the circumstances.

The significance of the value of z was given on page 154.

R TABLES FOR WILCOXON'S SUM OF RANKS TEST

n_A	n_B	P = 10%	P = 5%	P = 1%	P = 0.2%	n_A	n_B	P = 10%	P = 5%	P = 1%	P = 0.2%
2	8	4	3	—	—	5	6	20	18	16	—
2	9	4	3	—	—	5	7	21	20	16	—
2	10	4	3	—	—	5	8	23	21	17	15
2	11	4	3	—	—	5	9	24	22	18	16
2	12	5	4	—	—	5	10	26	23	19	16
2	13	5	4	—	—	5	11	27	24	20	17
2	14	6	4	—	—	5	12	28	26	21	17
2	15	6	4	—	—	5	13	30	27	22	18
2	16	6	4	—	—	5	14	31	28	22	18
2	17	6	5	—	—	5	15	33	29	23	19
2	18	7	5	—	—	5	16	34	30	24	20
2	19	7	5	3	—	5	17	35	32	25	20
2	20	7	5	3	—	5	18	37	33	26	21
3	5	7	6	—	—	5	19	38	34	27	22
3	6	8	7	—	—	5	20	40	35	28	22
3	7	8	7	—	—	6	6	28	26	23	—
3	8	9	8	—	—	6	7	29	27	24	21
3	9	10	8	6	—	6	8	31	29	25	22
3	10	10	9	6	—	6	9	33	31	26	23
3	11	11	9	6	—	6	10	35	32	27	24
3	12	11	10	7	—	6	11	37	34	28	25
3	13	12	10	7	—	6	12	38	35	30	25
3	14	13	11	7	—	6	13	40	37	31	26
3	15	13	11	8	—	6	14	42	38	32	27
3	16	14	12	8	—	6	15	44	40	33	28
3	17	15	12	8	6	6	16	46	42	34	29
3	18	15	13	8	6	6	17	47	43	36	30
3	19	16	13	9	6	6	18	49	45	37	31
3	20	17	14	9	6	6	19	51	46	38	32
4	4	11	10	—	—	6	20	53	48	39	33
4	5	12	11	—	—	7	7	39	36	32	29
4	6	13	12	10	—	7	8	41	38	34	30
4	7	14	13	10	—	7	9	43	40	35	31
4	8	15	14	11	—	7	10	45	42	37	33
4	9	16	14	11	—	7	11	47	44	38	34
4	10	17	15	12	10	7	12	49	46	40	35
4	11	18	16	12	10	7	13	52	48	41	36
4	12	19	17	13	10	7	14	54	50	43	37
4	13	20	18	13	11	7	15	56	52	44	38
4	14	21	19	14	11	7	16	58	54	46	39
4	15	22	20	15	11	7	17	61	56	47	41
4	16	24	21	15	12	7	18	63	58	49	42
4	17	25	21	16	12	7	19	65	60	50	43
4	18	26	22	16	13	7	20	67	62	52	44
4	19	27	23	17	13	8	8	51	49	43	40
4	20	28	24	18	13	8	9	54	51	45	41
5	5	19	17	15	—	8	10	56	53	47	42

R TABLES FOR WILCOXON'S SUM OF RANKS TEST

n_A	n_B	P = 10%	P = 5%	P = 1%	P = 0·2%	n_A	n_B	P = 10%	P = 5%	P = 1%	P = 0·2%
8	11	59	55	49	44	12	13	125	119	109	101
8	12	62	58	51	45	12	14	129	123	112	103
8	13	64	60	53	47	12	15	133	127	115	106
8	14	67	62	54	48	12	16	138	131	119	109
8	15	69	65	56	50	12	17	142	135	122	112
8	16	72	67	58	51	12	18	146	139	125	115
8	17	75	70	60	53	12	19	150	143	129	118
8	18	77	72	62	54	12	20	155	147	132	120
8	19	80	74	64	56	13	13	142	136	125	117
8	20	83	77	66	57	13	14	147	141	129	120
9	9	66	62	56	52	13	15	152	145	133	123
9	10	69	65	58	53	13	16	156	150	136	126
9	11	72	68	61	55	13	17	161	154	140	129
9	12	75	71	63	57	13	18	166	158	144	133
9	13	78	73	65	59	13	19	171	163	148	136
9	14	81	76	67	60	13	20	175	167	151	139
9	15	84	79	69	62	14	14	166	160	147	137
9	16	87	82	72	64	14	15	171	164	151	141
9	17	90	84	74	66	14	16	176	169	155	144
9	18	93	87	76	68	14	17	182	174	159	148
9	19	96	90	78	70	14	18	187	179	163	151
9	20	99	93	81	71	14	19	192	183	168	155
10	10	82	78	71	65	14	20	197	188	172	159
10	11	86	81	73	67	15	15	192	184	171	160
10	12	89	84	76	69	15	16	197	190	175	163
10	13	92	88	79	72	15	17	203	195	180	167
10	14	96	91	81	74	15	18	208	200	184	171
10	15	99	94	84	76	15	19	214	205	189	175
10	16	103	97	86	78	15	20	220	210	193	179
10	17	106	100	89	80	16	16	219	211	196	184
10	18	110	103	92	82	16	17	225	217	201	188
10	19	113	107	94	84	16	18	231	222	206	192
10	20	117	110	97	87	16	19	237	228	210	196
11	11	100	96	87	81	16	20	243	234	215	201
11	12	104	99	90	83	17	17	249	240	223	210
11	13	108	103	93	86	17	18	255	246	228	214
11	14	112	106	96	88	17	19	262	252	234	219
11	15	116	110	99	90	17	20	268	258	239	223
11	16	120	113	102	93	18	18	280	270	252	237
11	17	123	117	105	95	18	19	287	277	258	242
11	18	127	121	108	98	18	20	294	283	263	247
11	19	131	124	111	100	19	19	313	303	283	267
11	20	135	128	114	103	19	20	320	309	289	272
12	12	120	115	105	98	20	20	348	337	315	298

These Tables have been adapted from S. Siegel and J. Tukey, *Journ. Amer. Statist. Assoc.*, 1960, pp. 434–40, with corrections from D. B. Owen, *Handbook of Statistical Tables*, #11.5 (Addison–Wesley, 1962).

Examples

Ex. 5. Suppose we wish to compare the recovery times of patients after 2 different versions of some operation, say removing the gallbladder. Operation A is performed through a vertical incision; Operation B through an oblique incision. Each operation is performed alternately (A, B, A, B, etc.) on a consecutive series of patients suffering from gallbladder disease, and the recovery times (say, number of days in hospital after operation, including the day of operation and the day of discharge from hospital) are then collected as follows –

Patient #	Days to recover from operation A	Patient #	Days to recover from operation B
1	16	2	18
3	20	4	19
5	25	6	15
7	19	8	16
9	22	10	21
11	15	12	17
13	22	14	17
15	19	16	14

For this test it is not necessary to know the sample averages, but just out of interest the average recovery time of the 8 patients who had Operation A is $158/8 = 19.8$ days, while the 8 who had Operation B have an average of $137/8 = 17.1$ days.

Now the question is: Do these 2 sets of results differ significantly, or could the observed difference be reasonably attributed to chance?

Calculation: The smallest measurement in the combined results is 14, the largest is 25, so we shall list all numbers from 14 to 25 inclusive in the column of data values.

A number of measurements are tied (e.g. 2 patients recovered in 15 days), but it would be folly to try to make the recovery times more precise in order to break the ties (whether a patient leaves hospital in the morning or afternoon is a matter of convenience rather than of recovery time). We shall therefore award average rank values to such ties.

Now examine the following tabular calculation carefully, and you should be able to follow the various steps of the procedure as they take shape (listing of data values, tallying of the observations, etc.).

Data values	Tally		Ranks values	A ranks	B ranks
14		B	1		1
15	A	B	$2, 3 = 2\frac{1}{2}$	$2\frac{1}{2}$	$2\frac{1}{2}$
16	A	B	$4, 5 = 4\frac{1}{2}$	$4\frac{1}{2}$	$4\frac{1}{2}$
17		BB	$6, 7 = 6\frac{1}{2}$		13
18		B	8		8
19	AA	B	$9, 10, 11 = 10$	20	10
20	A		12	12	
21		B	13		13
22	AA		$14, 15 = 14\frac{1}{2}$	29	
23					
24					
25	A		16	16	
Smaller rank total =				(—)	52

The total of the *A* ranks is 84, but this is not shown in the calculation table because we are only concerned with the smaller rank total.

Thus we have $n_A = 8$, $n_B = 8$, and $R = 52$. We now refer to the Wilcoxon's Sum of Ranks Table on page 170 and find the row corresponding to these values of n_A and n_B; there we see that *R* would have to be as small as 49 to reach the 5% probability level. $R = 52$ has a probability of more than 10%, which means that if there was no difference between the 2 operations, a difference as large as that observed in our results could arise by chance more than once in 10 such trials. A significant difference between the 2 operations is therefore not proven by this experiment.

Ex. 6. A man who was sick in bed decided to count the number of advertisements delivered per day by 2 competing radio stations, WHO and WHY. Each morning he tossed a penny to determine which station he would listen to that day; WHO won the toss 5 times, and WHY won 3 times. Eight days of this made his illness fatal, but he left us the following results to analyse –

WHO (Sample A) = 341   326   360   305   326
WHY (Sample B) = 352   382   347

WHY's average is obviously higher than WHO's, but is the difference statistically significant?

Calculation: First notice that $n_A = 5$, and $n_B = 3$, and since these are unequal and less than 20 we must rank the observations in both directions (from smallest to largest, and vice versa). The smallest rank total will then be chosen from the 4 sets of ranks.

Data values	Tally		Rank values (i)	(ii)	A ranks (i)	(ii)	B ranks (i)	(ii)
305	A		1	8	1	8		
326	AA		2, 3	6, 7	5	13		
341	A		4	5	4	5		
347		B	5	4			5	4
352		B	6	3			6	3
360	A		7	2	7	2		
382		B	8	1			8	1
Smallest rank total =					(—)	(—)	(—)	8

We turn now to the Wilcoxon's Sum of Ranks Table on page 170 to determine the significance of getting a smallest rank total as low as 8 when one sample consists of 5 measurements and the other of 3. To save unnecessary repetition, the smaller sample is labelled A throughout these Tables, so we must look under $n_A = 3$, and $n_B = 5$ (this has no effect on the result). There we find that $R = 8$ could occur by chance in more than 10% of cases in which there is no significant difference between the samples. This is not sufficient to contradict our tentative negative assumption, so we conclude that a significant difference in the number of advertisements between the 2 stations is not proven.

Notice that it is only necessary to rank the observations in both directions when using the Wilcoxon's Tables to assess the significance of R; it is not necessary to do this when either n_A or n_B is more than 20, for the z Formula takes care of this and gives the same answer either way.

Even in a case as simple as this, the calculation of the t Test mentioned on page 167 takes a page of arithmetic; it yields a probability of almost exactly 10%, which is slightly stronger than the Wilcoxon answer (although it doesn't alter our conclusion in any way).

To illustrate **Siegel and Tukey's modification**, let us see if there is a statistically significant difference in the dispersion of the measurements from the 2 radio stations. Perhaps one station is appreciably more consistent than the other. Using the ranking technique described on page 167, the calculation proceeds as follows –

Data values	Tally		Rank values (i)	(ii)	A ranks (i)	(ii)	B ranks (i)	(ii)
305	A		1	2	1	2		
326	AA		4, 5	3, 6	9	9		
341	A		8	7	8	7		
347		B	7	8			7	8
352		B	6	5			6	5
360	A		3	4	3	4		
382		B	2	1			2	1
Smallest rank total =					(—)	(—)	(—)	14

Notice that tied measurements are given average rank values, so that each measurement of 326 is assigned a rank value of $\frac{1}{2}(3 + 6) = 4\frac{1}{2}$ in column (ii).

We see that the smallest rank total, $R = 14$. As mentioned earlier, this is to be interpreted exactly as if it was the result of a Wilcoxon's Sum of Ranks Test. So we turn to the Wilcoxon Table on page 170, and on the line for $n_A = 3$ and $n_B = 5$ we see that R would have to be as small as 7 to reach the $P = 10\%$ level. This tells us that the observed dispersions could readily occur by chance from parent groups having the same dispersion as each other. Our tentative negative assumption is not denied.

The efficiency of Siegel and Tukey's test of spread tends to be reduced if there is an appreciable difference between the arithmetic means of the 2 sample groups. However, this can be neutralized by first adding the difference between the sample means to each measurement in the sample possessing the smaller mean. This *equalizes the sample means*, but does not disturb the patterns of dispersion. Try this yourself with the above data. Use the shortcut described on page 52 to find the arithmetic mean of each sample, and confirm that the difference between these means is 28·7. Add this amount to each of the measurements in Sample A (because its mean is the smaller of the two), and repeat the above tabular calculation using these adjusted data values. The strengthening of the test by this refinement is shown by the reduction in the value of R from 14 to 12. However, this is still not small enough to indicate statistical significance in the present case.

Ex. 7. Random samples of 2 brands of ladies' stockings were examined by a quantitative test to see if one brand was superior to the other. Altogether 18 stockings of brand A and 22 stockings of brand B were tested, and the smaller rank total turned out to be brand A with a total of 274. Does this indicate a statistically significant difference between the 2 brands?

Data: $n_A = 18$; $n_B = 22$

$n_R = n_A = 18$; $R = 274$

Calculation: As one of the samples contains over 20 measurements, it is necessary to calculate z to assess the significance. This is not hard if the formula is calculated in 3 stages. Here again is the whole formula –

$$z = \frac{n_R(n_A + n_B + 1) - 2R}{\sqrt{\dfrac{n_A \cdot n_B(n_A + n_B + 1)}{3}}}$$

First, let us work out the top line of the formula –

$$n_R(n_A + n_B + 1) - 2R$$

$$= 18(18 + 22 + 1) - (2 \times 274)$$

$$= (18 \times 41) - 548$$

$$= 738 - 548$$

$$= 190$$

Secondly, let us work out the bottom part of the formula –

$$\sqrt{\frac{n_A \cdot n_B(n_A + n_B + 1)}{3}}$$

$$= \sqrt{\frac{18 \times 22(18 + 22 + 1)}{3}}$$

$$= \sqrt{\frac{18 \times 22 \times 41}{3}}$$

$$= \sqrt{\frac{16{,}236}{3}}$$

$$= \sqrt{5{,}412}$$

= 73·6 (from the Table of Square Roots, of course.)

Finally, we re-form the fraction to get the value of z –

$$z = \frac{190}{73 \cdot 6} = \underline{\underline{2 \cdot 58}}$$

Reference to the z Table on page 154 shows that such a value of z could occur by chance if there was no significant difference between the 2 brands of stockings on 1% of occasions. This is deemed very unlikely to have happened in our particular experiment, so we conclude that there is a significant difference between the 2 brands. Comparison of the original measurements will reveal which of the 2 is the better brand.

Questions

Q47. You remember Big George, the gangster? Well, his trigger man, Tommy Gunn, liked to dabble in Statistics to while away the hours between 'jobs'. For instance, when a new brand of cartridge came on the market a few months ago, he compared the weight of gunpowder in 6 cartridges of his favourite make with 6 of the new kind. Here are his results (in grams) –

Favourite brand	3·05	3·01	3·20	3·16	3·11	3·09
New brand	3·18	3·23	3·19	3·28	3·08	3·18

Is there a significant difference between the means of the 2 brands?

Q48. The number of passengers travelling each week on a certain air route was as follows –

Week No.	1	2	3	4	5	6	7	8
Passengers	3,204	2,967	3,053	3,267	3,370	3,492	3,105	3,330

At that point the airline concerned put a new type of aeroplane on the route with the same seating capacity, and in the following month the weekly figures were –

Week No.	9	10	11	12
Passengers	3,568	3,299	3,618	3,494

Start with the assumption that there is really no difference in popularity between the 2 types of aeroplane, and determine the probability of this being so. Is the difference significant?

Q49. The Loyal and Ancient Society of Cat Lovers of Greater France decided to conduct a survey of the stray cats of Paris to find out whether they were underfed or not. They selected a random sample of 30 domestic cats and weighed them; then they collected 15 stray cats from the streets (that's all they could catch in the time available) and weighed them too. The strays had the smaller rank total, which was 234. Is this sufficient to establish a real difference between the weights of the 2 groups?

WILCOXON'S SIGNED RANKS TEST (1945)

Purpose

To compare 2 random samples of measurements which are matched. The matching may be achieved by –

(*a*) pairing members of 2 sample groups, carefully, or symmetrically, or by splitting (see p. 118), or

(*b*) using 1 sample group for different treatments, different methods of testing, different observers, or on different occasions (see p. 120).

Principle

This important test was described by Dr Frank Wilcoxon in 1945, in the same 3-page article as his Sum of Ranks Test (p. 166). It is recommended here in preference to a form of *t* Test for the same reasons as before (p. 167).

The test depends on the fact that if there is really no significant difference between 2 sets of paired measurements, any chance differences which are present ought to consist of about equal numbers of plus and minus differences. But this test takes into account not only the *direction* of the differences, but also the *size* of the differences between matched pairs, and this feature increases the sensitivity of the test to a point where it compares very favourably indeed with the more complicated *t* Test. The way that both the direction and the size of the observed differences are taken into account is to assign rank values to the various sized differences, and then attach plus or minus signs to the rank values; then, if there is really no significant difference between the pairs, the total plus and the total minus rank values should be about equal. The significance of any inequality can be determined in the way described below.

Data Required

n = number of pairs of measurements which show a difference. This must be at least 6 before a 5% probability level can be reached.

Plus all the individual measurements in Sample or Set A and their corresponding partners in Sample or Set B.

Procedure

(1) Subtract each measurement in Sample or Set *B* from its partner in Sample or Set *A*. Mark any minus answers with a minus sign.

If any pairs of measurements are identical, and show no difference, they are excluded from the test on the ground that they contribute nothing to our search for a significant difference between the 2 sets or samples. This reduces the number of usable pairs of measurements from *N* (the total number of pairs) to *n* (the number of usable pairs). In turn, this reduces the power of the test somewhat, so whenever possible re-examine the original measurements with a view to working such measurements to an additional decimal place in order to break such ties.

(2) Now prepare a calculation table with the following headings –

Difference values	Tally	Rank values	+ Ranks	− Ranks

(3) In the first column write down an ordered list of all the observed differences, from smallest to largest, ignoring any minus signs for the moment.

(4) In the second column enter the tally. Work systematically through the differences calculated in (1) above, and for each observation in turn enter a plus or a minus sign (according to the sign of the observed difference) in the tally column beside the appropriate difference value. Thus if the first difference between the 2 sets is +3, put a plus sign in the tally column beside the difference value of 3; if the next difference is −5, put a minus sign beside the difference value 5, and so on.

(5) Assign rank values to each tally entry, still disregarding whether the signs are plus or minus. Start from the smallest difference and work down to the largest.

Tied ranks are given average rank values, as described on page 168. This reduces the efficiency of the test to a slight extent, so if possible break any such ties by taking the measurements concerned to an additional decimal place.

(6) Now transfer the rank values belonging to the plus signs into the next column (+ Ranks), and those belonging to the minus signs into the final column (− Ranks). You now have 2 sets of 'signed ranks'.

(7) Finally, add up the rank values of whichever set (+ or −) has the *smaller* rank total ($=R$). If you like, you can check the calculation at this point, for the sum of the smaller and the larger rank totals equals $\frac{1}{2}n(n + 1)$.

R TABLE FOR WILCOXON'S SIGNED RANKS TEST

No. of pairs n	$P = 10\%$	$P = 5\%$	$P = 1\%$	$P = 0.2\%$
6	2	$\frac{1}{2}$	—	—
7	$3\frac{1}{2}$	2	—	—
8	$5\frac{1}{2}$	$3\frac{1}{2}$	0	—
9	8	$5\frac{1}{2}$	$1\frac{1}{2}$	—
10	$10\frac{1}{2}$	8	3	0
11	$13\frac{1}{2}$	$10\frac{1}{2}$	5	$1\frac{1}{2}$
12	$17\frac{1}{2}$	$13\frac{1}{2}$	7	$2\frac{1}{2}$
13	$21\frac{1}{2}$	17	$9\frac{1}{2}$	$4\frac{1}{2}$
14	$25\frac{1}{2}$	21	$12\frac{1}{2}$	$6\frac{1}{2}$
15	$30\frac{1}{2}$	25	$15\frac{1}{2}$	$8\frac{1}{2}$
16	$35\frac{1}{2}$	$29\frac{1}{2}$	$19\frac{1}{2}$	$11\frac{1}{2}$
17	41	$34\frac{1}{2}$	$23\frac{1}{2}$	$14\frac{1}{2}$
18	47	40	$27\frac{1}{2}$	$18\frac{1}{2}$
19	$53\frac{1}{2}$	46	$32\frac{1}{2}$	$21\frac{1}{2}$
20	60	52	$37\frac{1}{2}$	$26\frac{1}{2}$

Adapted from D. B. Owen, *Handbook of Statistical Tables*, #11.1 (Addison–Wesley, 1962).

Interpretation

The smaller the value of R, the greater the difference between the 2 sets of measurements, so the less the likelihood of our tentative negative assumption ('there's no significant difference between them') being correct.

(a) *With 6 to 20 pairs of usable measurements*, the critical values of the smaller rank total (R) are given directly in the Wilcoxon's Signed Ranks Table above. This Table shows the probability of getting (by chance) a rank total of either plus or minus values as small as shown in the body of the Table if our tentative assumption of no significant difference is really true.

The significance of these probability levels was discussed on page 140.

(b) *With more than 20 pairs of usable measurements*, the significance of the smaller rank total (R) is found by calculating z from this formula –

$$z = \frac{\frac{1}{2}n(n+1) - 2R}{\sqrt{\dfrac{n(n+1)(2n+1)}{6}}}$$

The significance of z values was given on page 154.

Examples

Ex. 8. Suppose you are investigating a new sleeping pill called Nockout, and want to compare it with a standard sedative such as Phenobarbitone. Things like sedatives can vary quite a bit from one person to another, so it is best to try both drugs on every person taking part in the experiment. This kind of paired comparison will show if there is a significant difference between the 2 drugs in a much smaller trial than with independent tests in which one drug is given to one group of people and the other drug to another group. We saw this in the petrol–mileage experiment (p. 117).

So you collect, say, 10 people suffering from chronic insomnia, and on one night give half of them (selected at random, please!) Nockout pills, and the other half Phenobarbitone, and observe the number of hours each person sleeps. You won't be able to judge this to closer than the nearest $\frac{1}{4}$ hour (because of the difficulty in determining exactly when a person falls asleep), so any tied results will have to be accepted as such.

A few nights later, when you can be sure that the effect of the first sedative has worn off completely, each person is given his second pill, which is whichever pill he didn't get the first time, and the hours of sleep are again noted.

The results of such a trial, expressed in hours of sleep, can then be tabulated thus –

Patient	J.B.	R.A.	S.T.	S.L.	P.Q.	E.V.	J.T.	L.O.	E.M.	B.O.
With Phenobarb.	$7\frac{1}{2}$	7	7	$5\frac{3}{4}$	$4\frac{1}{4}$	$9\frac{1}{4}$	8	$7\frac{1}{4}$	$8\frac{1}{4}$	$7\frac{3}{4}$
With Nockout	8	6	$6\frac{3}{4}$	5	$4\frac{1}{2}$	8	$7\frac{1}{2}$	$6\frac{1}{4}$	8	$7\frac{3}{4}$
Difference	$-\frac{1}{2}$	1	$\frac{1}{4}$	$\frac{3}{4}$	$-\frac{1}{4}$	$1\frac{1}{4}$	$\frac{1}{2}$	1	$\frac{1}{4}$	0

Do these results indicate a significant difference between the 2 drugs, or could they have arisen by chance with fair probability if there was absolutely no real difference between them?

Calculation: We are obliged to eliminate the last patient (B.O.) from the test because he showed no difference between the 2 drugs; this reduces the number of usable pairs of measurements (*n*) to 9, but it can't be helped in this instance.

Difference values	Tally	Rank values	+ Ranks	− Ranks
$\frac{1}{4}$	+ −	$1, 2 = 1\frac{1}{2}$	$1\frac{1}{2}$	$1\frac{1}{2}$
$\frac{1}{2}$	++ −	$3, 4, 5 = 4$	8	4
$\frac{3}{4}$	+	6	6	
1	++	$7, 8 = 7\frac{1}{2}$	15	
$1\frac{1}{4}$	+	9	9	
		Smaller rank total =	(−)	$5\frac{1}{2}$

The Wilcoxon's Signed Ranks Table on page 181 shows that, for $n = 9$, a value of R as small as $5\frac{1}{2}$ could be expected by chance if there was no significant difference between the 2 drugs with a probability of 5%. The difference between the drugs is therefore deemed to be probably significant; looking back over the original data shows that, if one drug is superior to the other, it must be the Phenobarbitone which is the better one.

Ex. 9. To illustrate the breaking of ties, consider this example. A research chemist has discovered a new hair bleach, but before marketing the product he wants to know whether it weakens hair or not. He collects 7 lady volunteers, and in each case determines the breaking point of 6 of their hairs before bleaching and again after bleaching. For each volunteer he then works out the average breaking point (to the nearest gram) before and after bleaching. The results were as follows –

Volunteer #	I	II	III	IV	V	VI	VII
Av. breaking point before bleaching	105	105	93	120	111	80	91
Av. breaking point after bleaching	97	95	93	117	108	85	86
Difference	8	10	0	3	3	−5	5

Notice that volunteers IV to VII involve ties in the difference values. Case III has to be excluded from the test because there is no difference between the before and after averages, which makes $n = 6$. Anyway, let us proceed to find the smaller rank total –

Difference values	Tally	Rank values	+ Ranks	− Ranks
3	+ +	$1, 2 = 1\frac{1}{2}$	3	
5	+ −	$3, 4 = 3\frac{1}{2}$	$3\frac{1}{2}$	$3\frac{1}{2}$
8	+	5	5	
10	+	6	6	
Smaller rank total = (−)				$3\frac{1}{2}$

Referring to the Wilcoxon's Signed Ranks Table on page 181 shows that, for $n = 6$, a value of $R = 3\frac{1}{2}$ indicates a probability level of more than 10% (actually P = 19%).

Now watch the effect of breaking the ties. The chemist goes back to his original measurements, and calculates the averages of those measurements involved in ties to a further decimal place. The result is as follows –

Volunteer #	I	II	III	IV	V	VI	VII
Av. breaking point before bleaching	105	105	93·2	120·1	111·4	80·1	91·3
Av. breaking point after bleaching	97	95	93·0	117·1	108·3	84·7	86·0
Difference	8	10	0·2	3·0	3·1	−4·6	5·3

Difference values	Tally	Rank values	+ Ranks	− Ranks
0·2	+	1	1	
3·0	+	2	2	
3·1	+	3	3	
4·6	−	4		4
5·3	+	5	5	
8	+	6	6	
10	+	7	7	
Smaller rank total = (−)				4

Now we find that, for $n = 7$, a value of $R = 4$ has a probability of just over 10% (actually $P = 11\%$, compared with 19% before the ties were broken). The power of the test has thus been increased by breaking the ties. Even so, the probability level still indicates that a significant difference is not proven; in other words, the experiment does not indicate that the new bleach weakens hair. However, the experiment is on quite a small scale, and there is always the possibility that a larger series of tests might change this conclusion and show a significant difference before and after bleaching.

This may suggest to you that the chemist should have made use of all 6 measurements from each volunteer, giving 42 measurements before bleaching to compare with 42 measurements after bleaching. But this would not be legitimate. For the breaking point of any 6 hairs from one person's scalp will show some variation (hence the desirability of using the average rather than single measurements in this experiment), but there would be a much greater degree of variation from one person to another. Hence to list all 42 measurements for comparison would cause a mingling of related with non-related measurements, and the test would no longer be valid. Another way of looking at it is to realize that while the before and after averages of each volunteer are truly matched, there would be no real matching between any one of the 'before' measurements with an 'after' measurement in each group of six.

Ex. 10. My wife wanted to know whether putting cut flowers into a certain chemical solution (we'll call it 'Flower-Life') would prolong their life, so we designed the following experiment. She bought 2 fresh blooms of 25 different kinds of flowers – 2 roses, 2 irises, 2 carnations, and so on, until the florist asked if she was related to Noah. We then put one of each pair in a vase of water, and their partners in a vase containing 'Flower-Life'. Both vases were put side by side in the same room, and the length of life of each flower was noted.

We then had 2 matched samples, so the results could be tested for significance by Wilcoxon's Signed Ranks Test. This revealed a smaller rank total of 50. Is there a statistical difference between 'Flower-Life' and plain water?

Data: $n = 25$; $R = 50$

Calculation: Since n is more than 20, we must calculate z. No doubt you've guessed it, but we shall make easy work of that horrible looking z Formula by doing it in 3 stages.

$$z = \frac{\frac{1}{2}n(n + 1) - 2R}{\sqrt{\frac{1}{6}n(n + 1)(2n + 1)}}$$

First, the top half of the fraction –

$$\frac{1}{2}n(n + 1) - 2R$$
$$= \frac{1}{2}(25)(25 + 1) - (2 \times 50)$$
$$= (25 \times 13) - 100$$
$$= 325 - 100$$
$$= 225$$

Secondly, the bottom half of the fraction –

$$\sqrt{\frac{1}{6}n(n + 1)(2n + 1)}$$
$$= \sqrt{\frac{1}{6}(25)(25 + 1)(50 + 1)}$$
$$= \sqrt{\frac{1}{6} \times 25 \times 26 \times 51}$$
$$= \sqrt{\frac{1}{6} \times 33{,}150}$$
$$= \sqrt{5{,}525}$$
$$= 74 \cdot 35$$

Finally, reconstituting the fraction we get –

$$z = \frac{225}{74 \cdot 35} = \underline{\underline{3 \cdot 03}}$$

Reference to the z Table on page 154 shows this to indicate a probability of almost 0·2%. If there was no difference between water and 'Flower-Life', then, the observed differences could be expected to occur by chance once in about 500 times. We therefore reject this outside possibility, and accept that the difference is statistically significant.

Questions

Q50. A car club decided to run a test to see if an upper cylinder lubricant added to gasoline affected fuel consumption. Feeling that the test should embrace several makes of car, they got 8 different cars to make the same round trip of 100 miles, first on the ordinary gasoline, and then a second time using the upper cylinder lubricant added to the same brand of gasoline. Each time the quantity of fuel used was measured carefully. The results, expressed in miles per gallon, were as follows –

Car #	I	II	III	IV	V	VI	VII	VIII
Plain gasoline	17·1	29·5	23·8	37·3	19·6	24·2	30·0	20·9
With lubricant	14·2	30·3	21·5	36·3	19·6	24·5	26·7	20·6

Is there a significant difference between these 2 sets of results?

Q51. To illustrate how common sense led scientists to adopt good designs for their experiments in the days before statistical methods had been developed properly and applied to the problems of design, Fisher (*The Design of Experiments*, Oliver & Boyd, 1960, p. 27) quoted the case of Charles Darwin who, in 1876, reported a series of experiments in which he had compared the growth of cross-fertilized plants with that of self-fertilized ones. Darwin intelligently used matched pairs of plants, and grew them under exactly the same experimental conditions. However, the interpretation of the results caused him some concern. He wrote –

I long doubted whether it was worth while to give the measurements of each separate plant, but have decided to do so, in order that it may be seen that the superiority of the cross-fertilized plants over the self-fertilized ones does not commonly depend on the presence of two or three extra fine plants on the one side, or of a few very poor plants on the other side. . . . As only a moderate number of cross- and self-fertilized plants were measured, it was of great importance to me to learn how far the averages were trustworthy. . . . I may premise that if we took by chance a dozen or score of men belonging to two nations and measured them, it would I presume be very rash to form any judgement from such small numbers on their average national heights.

Darwin thus intuitively recognized that the results must be judged by the consistency of the superiority of one group over the other, and not merely on the difference between the averages. (Compare the case of the local anaesthetics, p. 56.) He therefore turned for help to his cousin, Galton, who was the foremost statistician of his era, but received the reply, 'I doubt, after making many tests, whether it is possible to derive useful conclusions from these few observations. We ought to have at least 50 plants in each case, in order to be in a position to deduce fair results.' Imagine Darwin's disappointment – his experiments had taken him 11 years, and still he had not produced sufficient evidence!

Thanks to the completeness of Darwin's reporting, however, we are nowadays able to give a statistical verdict to his work. In one experiment, with 15 matched pairs of maize plants, the results were as follows –

Height of cross-fertilized plant (inches)	Height of comparable self-fertilized plant (inches)	Difference (eighths of an inch)
$23\frac{4}{8}$	$17\frac{3}{8}$	49
12	$20\frac{3}{8}$	−67
21	20	8
22	20	16
$19\frac{1}{8}$	$18\frac{3}{8}$	6
$21\frac{4}{8}$	$18\frac{5}{8}$	23
$22\frac{1}{8}$	$18\frac{5}{8}$	28
$20\frac{3}{8}$	$15\frac{2}{8}$	41
$18\frac{2}{8}$	$16\frac{4}{8}$	14
$21\frac{5}{8}$	18	29
$23\frac{2}{8}$	$16\frac{2}{8}$	56
21	18	24
$22\frac{1}{8}$	$12\frac{6}{8}$	75
23	$15\frac{4}{8}$	60
12	18	−48

Fisher then proceeded to test the significance of these results, using the classical form of t Test devised by 'Student' in 1908.

This involves calculating the mean of these differences, and the sum of their squares; from the latter is subtracted the product of the mean of the differences and the sum of the differences; this value is divided by 14 and then by 15, and the square root of this answer is divided into the mean of the differences. This gives the value of $t = 2 \cdot 15$. Referring now to 'Student's' t Table on page 162 shows that for $n = 15$, this value of t indicates a probability of *just under 5%*; the difference between the 2 varieties of plant is therefore probably significant.

Now, when you were reading about distribution-free statistical tests on page 167, you may have got the impression that, because they are often fairly simple to do, they just couldn't be as good as their more complicated alternative tests. I have, in fact, worked through a large number of sets of data using both kinds of test, and my findings are in agreement with those statisticians who maintain that, at a practical level, these simple tests compare very favourably with their 'big brothers'. Try the present case for yourself. Apply Wilcoxon's Signed Ranks Test to Darwin's data. How does the answer compare with that obtained by the t Test?

Q52. On page 117 it was stated that for a given number of observations, independent sample groups were less able to show a statistical difference than paired samples. Prove this statement by treating the data of *Ex. 8* (p. 183) as though the results had been obtained by giving Phenobarbitone to one group of people, and the Nockout pills to a different group, and applying a suitable significance test.

WILCOXON'S STRATIFIED TEST (1946)

Purpose

To compare 2 independent stratified random samples (p. 47) of measurements which have comparable strata, and the same number of measurements in both samples.

It thus serves to compare the effect of 2 treatments when both are tested at various levels (p. 122), or when both are applied to 2 or more independent sample groups (which then constitute strata).

Principle

This extension of Wilcoxon's Sum of Ranks Test (p. 166) is applicable when the samples are grouped or stratified. Suppose it is desired to compare 2 methods of teaching arithmetic (say, the Cuisenaire system versus the conventional method), it is obvious that different standards of examination would be necessary for children of different ages; if children were compared only after their first year of training, a difference may be apparent, but it may not be maintained in subsequent years. It would therefore be desirable to compare the results after say 1, 2, 4, and 6 years of each system. This would create a stratified sample of results obtained with one method of teaching, and a comparable stratified sample of results from the other method of teaching. We would then have to compare the comparable groups (strata) of the samples, for an 80% mark obtained after 1 year's training is hardly to be judged against an 80% mark obtained at a much harder examination after 6 years of training. Then, to find out whether one method is better than the other, in overall way, we would need to be able to add the results of comparing each level.

You might imagine that the strata measurements could simply be added and then compared, but to assess the significance of these totals, we must know something about their dispersion. And you will recall (p. 68) that standard deviations can't be added up as simply as beans. An easy, yet reasonably accurate way around this problem was discovered by Dr Frank Wilcoxon, a year after his Sum of Ranks Test. He described it in the *Journal of Economic Entomology*, 1946, pp. 269–70. In essence, it consists of adding the results of Sum of Ranks Tests applied to each stratum. Then, as we are dealing with rank values instead of the actual measurements, it is not hard to combine the standard deviations concerned.

We begin by assuming that there is no significant difference between the 2 samples, and tentatively attribute any observed difference to chance. If our assumption is correct, the rank totals earned by each sample ought to be about the same; if they differ greatly, our assumption is probably not correct. In fact, this test enables us to determine just what is the probability of our tentative assumption being true.

Data Required

k = number of strata or groups (I, II, III, etc.) in each sample.

n = number of measurements in each stratum of each sample, if this number is the same in each, or

n_I, n_{II}, n_{III}, etc. = number of measurements in stratum I, II, III, etc., of each sample, if the strata are of different sizes.

Plus all the individual measurements in both samples.

Procedure

(1) Prepare a combined data and calculation table with a column on the left for a numbering or description of the various strata or groups, and then 2 wide columns, one for each of the 2 samples or treatments, thus –

Stratum	Sample A Obs. Ranks	Sample B Obs. Ranks
I		
II		

(2) Enter the observed measurements in each stratum of each sample down the left of the second and third columns. It is convenient (but not essential) to list each set of measurements in order, from smallest to largest.

(3) Assign rank values to the observations in both samples combined, *for each stratum separately*, working from the smallest to the largest measurement in each case.

Give average rank values to identical measurements occurring in the same stratum.

(4) Add up whichever is the smaller of the 2 sets of total sample rank values. Call this smaller rank total ' R '.

Interpretation

The smaller the value of R, the less the chance of our tentative negative assumption being correct.

(a) *With 2 to 7 strata or groups*, and an equal number of measurements (up to 7) in each stratum of each sample, the critical values of the smaller rank total (R) are given in Wilcoxon's Table opposite. The values of R shown in the body of this Table are those which could be expected by chance if there was no significant difference between the 2 samples or treatments with a probability of 5% or 1%. If your calculated value of R is as small as shown, accept that the 2 samples or treatments are probably or practically certainly different, according to the level reached.

(b) *With 8 or more strata or groups*, or in any case having unequal numbers of measurements in the various strata, calculate the significance of your value of R by determining z thus –

(i) With k strata or groups, each containing n measurements in each stratum of each sample –

$$z = \frac{kn(2n + 1) - 2R}{\sqrt{\dfrac{kn^2(2n + 1)}{3}}}$$

(ii) When the number of measurements is not the same in all strata, the above formula becomes –

$$z = \frac{n_I(2n_I + 1) + n_{II}(2n_{II} + 1), \text{etc.} - 2R}{\sqrt{\dfrac{n_I^2(2n_I + 1) + n_{II}^2(2n_{II} + 1), \text{etc.}}{3}}}$$

The significance of the value of z so obtained is given in the z Table on page 154.

For comments on the significance of different probability levels, see page 140.

Incidentally, the above formulae are merely expanded versions

R TABLE FOR WILCOXON'S STRATIFIED TEST

No. of groups or strata k	No. of pairs in each group n	Values of R indicating	
		$P = 5\%$	$P = 1\%$
2	2	6	—
	3	15	13
	4	26	24
	5	42	38
	6	61	55
	7	83	77
3	2	11	9
	3	24	21
	4	42	39
	5	66	61
	6	96	89
	7	131	123
4	2	15	13
	3	33	30
	4	58	54
	5	91	85
	6	131	124
	7	179	170
5	2	19	17
	3	42	39
	4	75	70
	5	116	110
	6	168	159
	7	228	218
6	2	24	22
	3	52	49
	4	91	86
	5	142	135
	6	204	195
	7	277	266
7	2	28	26
	3	62	58
	4	108	102
	5	168	160
	6	241	231
	7	327	314

Adapted from F. Wilcoxon, *Biometrics Journal*, 1947, Vol. 3, p. 122.

of the z formula used for Wilcoxon's Sum of Ranks Test (p. 169), as shown by the fact that if there was only one stratum, they would reduce to the Sum of Ranks formula.

Examples

Ex. 11. In his article describing this test, Wilcoxon gave the following example. Two different DDT preparations were tested on flour beetles, to see if one killed significantly more beetles than the other. Each test was done 4 times, and each DDT preparation was tested in 3 strengths. The percentage of beetles killed is shown in the combined data and calculation table below. Is there a significant difference between Preparation A and Preparation B?

Data and Calculation: In the following table, the results are presented in order of size, not in the order in which they occurred in the experiments.

Concentration of DDT	Preparation A % killed	Ranks	Preparation B % killed	Ranks
25 mg %	18	1	34	4
	26	2	42	5
	30	3	53	7
	50	6	63	8
50 mg %	33	1	60	5
	42	2	62	6
	44	$3\frac{1}{2}$	66	7
	44	$3\frac{1}{2}$	80	8
100 mg %	44	1	74	5
	50	2	77	6
	56	3	84	7
	64	4	92	8
Smaller rank total =	32			(—)

There are 3 strata, so $k = 3$. Each sample has 4 measurements in each stratum, so $n = 4$. Each stratum is ranked separately from 1 to 8, and the sum of the rank values for Preparation A

is seen to be the smaller total. Referring to Wilcoxon's Table on page 193 shows that for $k = 3$ and $n = 4$, a value of $R = 39$ has a probability of 1%. Our value of $R = 32$, so the probability of there being no significant difference between the 2 preparations is less than 1%. We therefore conclude that a significant difference between the preparations does, in fact, exist.

Notice that it would have been possible to test this data by averaging the results of each stratum for each preparation, and applying Wilcoxon's Signed Ranks Test, for the overall results in each stratum could be considered satisfactorily matched (even though the individual results were not matched, a situation exactly the same as that of the measurements of the breaking point of 6 hairs, discussed on page 185). However, only 3 pairs of measurements would then be obtained, and Wilcoxon's Signed Ranks Test does not reach $P = 5\%$ until there are at least 6 pairs of measurements. The present test, by utilizing more information, therefore proves to be more powerful.

Ex. 12. Suppose we want to compare 2 treatments for curing acne (pimples). Suppose, too, that for practical reasons we are obliged to use a presenting sample of patients (p. 49). We might then decide to alternate the 2 treatments strictly according to the order in which the patients arrive (*A*, *B*, *A*, *B*, and so on). Let us agree to measure the cure in terms of weeks to reach 90% improvement (this may prove more satisfactory than awaiting 100% cure, for some patients may not be completely cured by either treatment, and many patients might not report back for review when they are completely cured).

This design would ordinarily call for Wilcoxon's Sum of Ranks Test, but there is one more thing to be considered: severity of the disease. For it could happen that a disproportionate number of mild cases might end up, purely by chance, in one of the treatment groups, which could bias the results in favour of this group, even if there was no difference between the 2 treatments. It would clearly be better to compare the 2 treatments on comparable cases, and this can be done by stratifying the samples. Suppose we decide to group all patients into one or other of 4 categories – mild, moderate, severe, and very severe. Then all the mild cases would be given the 2 treatments alternatively, and likewise with the other groups.

Given the results tabulated below (in order of size, not of their actual occurrence), is the evidence sufficient to say that one treatment is better than the other?

Category	Treatment A Weeks	Treatment A Ranks	Treatment B Weeks	Treatment B Ranks
(I) Mild	2	1½	2	1½
	3	3	4	4
(II) Moderate	3	1	4	2
	5	3	6	4½
	6	4½	7	6
	10	8	9	7
(III) Severe	6	1	9	3
	8	2	14	5½
	11	4	14	5½
(IV) Very severe	8	1	12	4
	10	2	14	5
	11	3	15	6
Smaller rank total =		34		(—)

Notice that, on account of the manner of sampling, it so happens that there are different numbers of observations in the various categories. We shall therefore have to calculate z to assess the results.

We shall need to use the second z formula, viz. –

$$z = \frac{n_I(2n_I + 1) + n_{II}(2n_{II} + 1), \text{etc.} - 2R}{\sqrt{\dfrac{n_I{}^2(2n_I + 1) + n_{II}{}^2(2n_{II} + 1), \text{etc.}}{3}}}$$

This formula is not as fearsome as it looks, especially if we break it into sections, as follows.

$$n_I = 2, \quad \text{so } n_I(2n_I + 1) = 2 \times 5 = 10$$
$$n_{II} = 4, \quad \text{so } n_{II}(2n_{II} + 1) = 4 \times 9 = 36$$
$$n_{III} = 3, \quad \text{so } n_{III}(2n_{III} + 1) = 3 \times 7 = 21$$
$$n_{IV} = 3, \quad \text{so } n_{IV}(2n_{IV} + 1) = 3 \times 7 = 21$$

The sum of these, which is the first part of the fraction, is $10 + 36 + 21 + 21 = 88$.

Next, $R = 34$, so $2R = 68$.

Finally, $n_I{}^2(2n_I + 1) = 4 \times 5 = 20$

$$n_{II}{}^2(2n_{II} + 1) = 16 \times 9 = 144$$
$$n_{III}{}^2(2n_{III} + 1) = 9 \times 7 = 63$$
$$n_{IV}{}^2(2n_{IV} + 1) = 9 \times 7 = 63$$

The sum of these $= 20 + 144 + 63 + 63 = 290$.

That fearsome-looking z formula is now reduced to this –

$$z = \frac{88 - 68}{\sqrt{\frac{1}{3} \times 290}} = \frac{20}{\sqrt{96 \cdot 7}} = \frac{20}{9 \cdot 83} = \underline{\underline{2 \cdot 03}}$$

Referring this value to the z Table on page 154 shows it to indicate a probability of less than 5%. This means that the observed differences between the 2 treatments are probably significant.

Using rank values instead of the actual measurements naturally simplifies the calculations, but ranks only show differences and not the size of the differences, and this tends to reduce the sensitivity of the test. Thus in the present example, the precise probability indicated by this value of z is 4·2%, whereas the more sensitive (and much more laborious) test of Analysis of Variance yields a probability of 2·6%. Even so, both answers indicate 'probably significant'.

Incidentally, you may have noticed that Treatment A seems to show its superiority chiefly in the more severe cases. This suggests that interaction may be present. Wilcoxon's Stratified Test cannot assess the significance of possible interactions; only Analysis of Variance can do this (in the present example, the interaction is not proven to be significant, $P > 5\%$).

Questions

Q53. That master of escape, Houdini, used to challenge his audiences to an escape competition, using standard police handcuffs. One evening his challenge was taken up by a man who called himself Mr Anon (because he had recently escaped from standard police handcuffs, and had no desire to be recaptured). The competition entailed each competitor escaping 3 times from 2 different models of handcuffs, the winner to be the one with the quickest average time. Here are the results (the escape times are quoted in seconds) –

Type	Houdini	Mr Anon
Model I	10·9 11·3 10·2	10·3 10·6 12·2
Model II	13·8 15·1 14·3	16·3 15·2 15·8
Totals Averages	75·6 12·6	80·4 13·4

So Houdini, with an average of 12·6 seconds, won the competition. But is the difference statistically significant?

Q54. When Joe got back from Bolivia, the first thing he did was to test the 'magic powder' that the natives there had claimed would prevent materials from shrinking when they were being laundered. He got 9 different sorts of cloth, cut them into strips, measured them, laundered them with the 'magic powder', and then measured them again. He did this twice with each type of cloth, and then repeated the whole process (using fresh cloths, of course) without the 'magic powder', to serve as his controls. Here are the results; is there a statistically significant difference between them?

Material	With 'Magic Powder' % Shrinkage		With ordinary detergent % Shrinkage	
	1st test	2nd test	1st test	2nd test
A	0	1	3	3
B	2	5	4	6
C	6	4	5	7
D	10	8	7	11
E	4	1	3	2
F	1	2	1	4
G	6	5	9	9
H	0	2	2	3
I	4	7	3	5

SPEARMAN'S CORRELATION TEST (1904)

Purpose

To test for correlation between 2 measurable characteristics. The measurable characteristics may exist either –

(a) *in each individual* of a sample group, e.g. weight and height, or
(b) *at the same time*, e.g. the temperature in England and the wearing of overcoats in Germany.

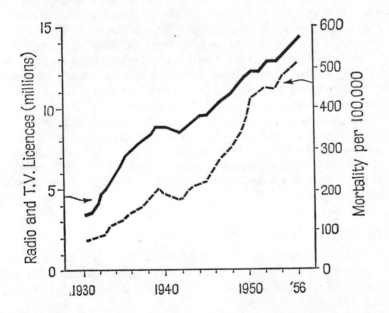

Fig. 11. There is a direct correlation between the number of radio and television licences and the death rate from coronary artery disease, but this does not prove that one causes the other. (From Dr J. Yudkin, *The Lancet*, July 1957.)

Correlation is the mutual relationship between 2 numerical variables. *Direct correlation* exists when a rise or fall in the value of one characteristic is associated with a corresponding rise or fall in the value of the other. Fig. 11 shows a fairly high degree of direct correlation between the number of radio and television licences in England and the death rate from coronary artery disease between 1930 and 1956. On the other hand, when the temperature in England falls, the number of overcoats worn in Germany increases, and vice versa, so these variables stand in *inverse correlation* with each other.

Of course, one must always be on guard against the fallacy of assuming that correlation between 2 things implies that one of them *causes* the other. The drop in temperature during the English winter does not cause the increased wearing of overcoats in Germany; both are caused by a third factor – the northern winter. At most, the demonstration that any 2 factors are correlated (beyond the bounds of chance) can only suggest that it may be worth while designing an investigation into the possible causal relationship between them. The correlation in Fig. 11 is a case in point. It would be absurd to think that radio and television caused people to die from heart attacks, just as it would be to imagine that deaths from coronary disease caused relatives of the deceased to rush out and attend to that long-forgotten licence. Rather, such correlation points to a trend of the times, and was, in fact, part of an investigation into the effect of diet on the coronary artery disease. But watch out, for even if the amount of fat in the English diet was also directly correlated with the coronary artery mortality rate, it wouldn't prove that a causal relationship existed between them, would it?

Principle

This test for correlation was the first statistical test to use ranks. It was worked out by a London psychologist, Professor Charles Spearman (*Amer. Journ. of Psychol.*, 1904, pp. 72–101). It works like this.

Each of the 2 sets of measurements is given its own set of rank values, from smallest to largest. Now, if the sets of measurements are perfectly and directly correlated, there will be no difference between the ranking order of the 2 sets, so if we subtracted each rank value in the one set from its partner in the other set, the total of the differences would be zero. On the other hand, if the 2 sets are perfectly but inversely correlated, the difference between their rank values will be maximal for the number of observations concerned. Thus if there are 4 pairs of observations, the largest possible total of the differences is given when the ranks are in perfect inverse correlation, like this –

A	B	Difference
1	4	3
2	3	1
3	2	1
4	1	3
Total = 8		

Any other arrangement will give a total of less than 8. Prove it for yourself.

Between these extremes of perfect correlation there will be a range of rank differences likely to occur if the 2 sets of measurements are not correlated. But we must remember that, in practice, the influence of chance is ever-present, so that things which are highly correlated will sometimes yield samples of measurements which, by chance, show little or no correlation; contrariwise things which are not correlated will sometimes give samples which do show more or less correlation.

In testing for correlation, then, we start by *assuming that there is no significant correlation* between the 2 sets of measurements, and then determine the probability of getting, by chance, a difference between the rank values as small, or as large, as that which is observed. If the probability is remote, we reject our original tentative negative assumption, and accept instead that the sets are correlated. If the probability is one which could occur by chance more than once in 20 trials (i.e. $P > 5\%$), it is considered insufficient to refute our assumption of no-correlation.

Since we convert each measurement into a rank value, this test can also be applied to proportions or percentages or even to those types of characteristic (such as taste) which cannot be measured in units but which can at least be ranked in order of merit or preference.

Data Required

n = number of pairs of measurements and/or ranked observations. This must be at least 5 before a 5% probability level can be reached.

Plus all the individual measurements or rank values of the 2 characteristics, A and B.

Procedure

(1) Prepare a table for the calculation with the following headings –

Data values		Rank values		Difference	d^2
A	B	A	B	d	

(2) In the first column write down all the observed values of A, in any convenient order, and beside them put the corresponding values of B.

(3) In the next column assign rank values, first to the A measurements, and then (quite separately) to the B measurements. Work from the smallest to the largest data values in each set.

If 2 or more observations in the one set of measurements have the same value, break the tie if possible by taking the measurements concerned to an additional decimal place. If this can't be done, give an average rank value to the tied observations as described on page 168.

(4) Subtract to find the difference (d) between the rank values of each pair of observations.

(5) Square each difference value (d^2), using the Table of Squares at the back of this book if necessary.

(6) Add up all the squares of the differences; call this total 'D^2'.

(7) If there are any tied observations, a correction factor (T) must be added to the value of D^2. This is necessary because each set of ties involving x observations falsely lowers the value of D^2

by an amount equal to $\dfrac{x^3 - x}{12}$. The correction is made as follows:

Let t_2 = number of ties involving 2 observations,

t_3 = number of ties involving 3 observations,

t_4 = number of ties involving 4 observations, and so on.

Then T = the sum of all the values of $\left(\dfrac{x^3 - x}{12}\right)t_x$, that is –

$T = \tfrac{1}{2}t_2 + 2t_3 + 5t_4 + 10t_5 + 17{\cdot}5t_6 + 28t_7$ and so on.

(8) Finally, add $D^2 + T$.

Interpretation

Remember, we started with the tentative negative assumption that there is no real correlation between the 2 sets of characteristics, and tentatively attribute any apparent correlation merely to Lady Chance. We will now find out the probability of this assumption being correct.

(a) *If n = 5 to 30*, refer the value of $D^2 + T$ to the Spearman's Correlation Test Table on page 204, which shows the probability of $D^2 + T$ being as small, or as large, as observed if there is no significant correlation between the 2 sets of measurements.

(b) *If n > 30*, calculate z from this formula –

$$z = \sqrt{n - 1} \cdot \left| 1 - \frac{D^2 + T}{\tfrac{1}{6}(n^3 - n)} \right|$$

To save you work, values of $\tfrac{1}{6}(n^3 - n)$ for $n = 31$ to 70 are given in the special Table on page 205.

The value of $\dfrac{D^2 + T}{\tfrac{1}{6}(n^3 - n)}$ is less than 1 if the correlation tends to be direct; it is more than 1 with inverse correlation.

The critical values of z were given in the z Table on page 154.

The significance of the critical probability levels was discussed on page 140.

$D^2 + T$ TABLE FOR SPEARMAN'S CORRELATION TEST

No. of pairs n	Direct correlation			Inverse correlation		
	P=10%	P=5%	P=1%	P=10%	P=5%	P=1%
5	3	1	—	37	39	—
6	8	5	1	62	65	69
7	17	12	5	95	100	107
8	31	23	11	137	145	157
9	49	38	21	191	202	219
10	73	59	35	257	271	295
11	104	85	54	336	355	386
12	144	121	84	428	451	488
13	191	163	115	537	565	613
14	247	213	154	663	697	756
15	313	272	201	807	848	919
16	390	342	257	970	1,018	1,103
17	479	423	322	1,153	1,209	1,310
18	581	515	398	1,357	1,423	1,540
19	697	621	484	1,583	1,659	1,796
20	827	740	583	1,833	1,920	2,077
21	972	873	695	2,108	2,207	2,385
22	1,134	1,022	820	2,408	2,520	2,722
23	1,313	1,188	960	2,735	2,860	3,088
24	1,509	1,370	1,115	3,091	3,230	3,485
25	1,725	1,570	1,286	3,475	3,630	3,914
26	1,961	1,790	1,475	3,889	4,060	4,375
27	2,217	2,028	1,681	4,335	4,524	4,871
28	2,495	2,288	1,906	4,813	5,020	5,402
29	2,795	2,569	2,150	5,325	5,551	5,970
30	3,119	2,872	2,414	5,871	6,118	6,576

Adapted from D. B. Owen, *Handbook of Statistical Tables*, #13.2 (Addison–Wesley, 1962), and for $n > 11$ calculated from t-distribution.

TABLE OF VALUES OF $\frac{1}{6}(n^3 - n)$ FOR $n = 31$ to 70

n	$\frac{1}{6}(n^3 - n)$	n	$\frac{1}{6}(n^3 - n)$
31	4,960	51	22,100
32	5,456	52	23,430
33	5,984	53	24,800
34	6,545	54	26,240
35	7,140	55	27,720
36	7,770	56	29,260
37	8,436	57	30,860
38	9,139	58	32,510
39	9,880	59	34,220
40	10,660	60	35,990
41	11,480	61	37,820
42	12,340	62	39,710
43	13,240	63	41,660
44	14,190	64	43,680
45	15,180	65	45,760
46	16,220	66	47,900
47	17,300	67	50,120
48	18,420	68	52,390
49	19,600	69	54,740
50	20,820	70	57,160

Examples

Ex. 13. Television came to Australia in 1957. Over the next few years the number of new radio and television licences issued were as follows –

Year	New radio licences (thousands)	New television licences (thousands)
1957	171	74
1958	178	224
1959	251	300
1960	160	404
1961	155	323

From Commonwealth Bureau of Census & Statistics, Pocket Compendium, 1964.

Do these figures indicate that the coming of television has been associated with a significant alteration in the number of new radio licences? In other words, is there a significant correlation (direct or inverse) between the 2 sets of observations?

Calculation

Data values		Rank values		Difference	d^2
A	*B*	*A*	*B*	*d*	
171	74	3	1	2	4
178	224	4	2	2	4
251	300	5	3	2	4
160	404	2	5	3	9
155	323	1	4	3	9
$n = 5$					$D^2 = 30$

There are no tied ranks, so $T = 0$.

We now make the assumption that the 2 characteristics are unrelated, and test this by referring our calculated $D^2 + T = 30$ to the Spearman's Correlation Test Table on page 204. Here we find that for $n = 5$, the value of $D^2 + T$ would have to be as small as 1, or as large as 39, before our assumption would be improbable, so we conclude that correlation is not proven on the evidence presented.

Ex. 14. Mr Smith was asked to judge the beauty contest. There were 8 lovely young ladies in the contest, and the results were as follows –

Contestant	Place
Amelia, aged 17	1
Betsy, aged 16	=2
Carolyn, aged 18	=2
Daisy, aged 20	4
Eve, aged 18	5
Freda, aged 18	6
Georgina, aged 20	7
Helen, aged 23	8

Their ages are also quoted, because I suspect that Mr Smith shows a bias towards youth. Let us see if this apparent relationship between age and place is likely to be a mere chance correlation, or not. In other words, if Mr Smith has shown no tendency to favour the young contestants, what is the probability of getting the observed results merely by chance?

Calculation

Data values		Rank values		Difference	d^2
A	B	A	B	d	
1	17	1	2	1	1
$=2$	16	$2\frac{1}{2}$	1	$1\frac{1}{2}$	$2\frac{1}{4}$
$=2$	18	$2\frac{1}{2}$	4	$1\frac{1}{2}$	$2\frac{1}{4}$
4	20	4	$6\frac{1}{2}$	$2\frac{1}{2}$	$6\frac{1}{4}$
5	18	5	4	1	1
6	18	6	4	2	4
7	20	7	$6\frac{1}{2}$	$\frac{1}{2}$	$\frac{1}{4}$
8	23	8	8	0	—
$n = 8$					$D^2 = 17$

The present example shows the use of Spearman's Correlation Test when one of the characteristics (beauty) is not measured in units, but is merely ranked.

Now, the calculation Table shows there are 3 sets of ties –

(1) Betsy and Carolyn tied for 2nd place,
(2) Daisy and Georgina are both aged 20,
(3) Carolyn, Eve, and Freda are all aged 18.

We thus have 2 sets of ties involving 2 observations each (so $t_2 = 2$), and 1 set involving 3 tied observations (so $t_3 = 1$). The only parts of the T Formula relevant in the present case are therefore –

$$T = \tfrac{1}{2}t_2 + 2t_3$$
$$= (\tfrac{1}{2} \times 2) + (2 \times 1)$$
$$= 1 + 2 = 3$$
$$\therefore D^2 + T = 17 + 3 = 20$$

We now refer this value of 20 to the Spearman's Correlation Test Table on page 204, and find that for $n = 8$, this value could be expected by chance with a probability of less than 5%, so the direct correlation is probably significant.

Ex. 15. Professor Z. was well-known for his unusual investigations. One time he decided to investigate whether or not there was any correlation between children's spelling ability and the size of their feet. He went to a school, chose 37 children at random, measured the length of each child's feet, and gave them all a spelling test consisting of 100 words of various degrees of difficulty. The results are shown in the data columns of the table opposite; a quick glance reveals a tendency for those children with the largest feet to have scored the highest marks in the spelling test, but we must apply Spearman's Correlation Test to see if the apparent correlation is significant or just a chance effect.

Calculation: The observations concerning the children's foot size have been listed in the Table opposite from smallest to largest, so ranking these values is easy. Several ties are present; they are bracketed so you can pick them out easily. These tied values are assigned average rank values (e.g. $6\frac{3}{4}$ in. shares 2nd and 3rd place, so each is given a rank value of $2\frac{1}{2}$). The ranking of the spelling results called for a tally chart like that used for Wilcoxon's Sum of Ranks Test (p. 173). The difference between the rank values of each pair (d), the square of each difference (d^2), and the sum of the squares of the differences (D^2), are all shown in the calculation table.

Next, note the ties. How many ties involve 2 observations? There are 6 in A, and 4 in B, so $t_2 = 10$. There are no ties involving 3 observations, so $t_3 = 0$. There are 3 ties involving 4 observations in A, and 1 in B, so $t_4 = 4$. Finally, there is 1 set of ties involving 5 observations, in A, so $t_5 = 1$. Accordingly –

$$T = \tfrac{1}{2}t_2 + 2t_3 + 5t_4 + 10t_5$$
$$= (\tfrac{1}{2} \times 10) + (2 \times 0) + (5 \times 4) + (10 \times 1)$$
$$= 5 + 0 + 20 + 10$$
$$= 35$$

$$\therefore D^2 + T = 1{,}446 + 35 = 1{,}481$$

Now $n = 37$, which is beyond the limit of our ready reference Table, so we must calculate z –

$$z = \sqrt{n - 1} \cdot \left| 1 - \frac{D^2 + T}{\tfrac{1}{6}(n^3 - n)} \right|$$

Calculation of Example 15

Data values A (inches)	B (spelt correctly)	Rank values A	B	d	d²
6½	16	1	4	3	9
6¾	28	2½	6½	4	16
6¾	46	2½	12	9½	90¼
7	14	4	2	2	4
7¼	41	5	9	4	16
7½	10	7½	1	6½	42¼
7½	56	7½	17	9½	90¼
7½	43	7½	11	3½	12¼
7½	15	7½	3	4½	20¼
7¾	21	10	5	5	25
8	50	12½	15	2½	6¼
8	28	12½	6½	6	36
8	57	12½	18	5½	30¼
8	65	12½	24	11½	132¼
8¼	42	15	10	5	25
8½	36	16	8	8	64
8¾	71	17	28½	11½	132¼
9	47	18	13½	4½	20¼
9¼	47	19½	13½	6	36
9¼	66	19½	25	5½	30¼
9½	71	23	28½	5½	30¼
9½	71	23	28½	5½	30¼
9½	61	23	21	2	4
9½	55	23	16	7	49
9½	86	23	33	10	100
9¾	62	26½	22	4½	20¼
9¾	78	26½	31½	5	25
10	71	28½	28½	0	—
10	59	28½	19	9½	90¼
10¼	60	30½	20	10½	110¼
10¼	63	30½	23	7½	56¼
10½	68	33½	26	7½	56¼
10½	98	33½	36½	3	9
10½	88	33½	34	½	¼
10½	91	33½	35	1½	2¼
10¾	98	36½	36½	0	—
10¾	78	36½	31½	5	25

$n = 37$ $\qquad\qquad D^2 = 1446$

The Table on page 205 shows that for $n = 37$, the value of $\frac{1}{6}(n^3 - n) = 8{,}436$.

$$\therefore z = \sqrt{37 - 1} \cdot \left| 1 - \frac{1{,}481}{8{,}436} \right|$$
$$= \sqrt{36} \cdot \left| 1 - 0{\cdot}176 \right|$$
$$= 6 \times 0{\cdot}824$$
$$= \underline{\underline{4{\cdot}94}}$$

Referring this value to the z Table on page 154 shows that if there was no correlation between foot size and spelling ability, the probability of getting the observed results by chance is much less than $0{\cdot}2\%$. Under these circumstances we must reject our null hypothesis, and accept that a statistically very significant degree of correlation does exist.

I sincerely hope that no one, on this evidence, will think that big feet *cause* good spelling! For the correlation was brought about by the influence of age – the best spellers tend to be the oldest children, and the oldest children tend to have the biggest feet.

Questions

Q55. The yield of a certain chemical process was noticed to vary from day to day, without apparent reason. A number of investigations were unsuccessful in tracing the cause, so it was decided to find out exactly what effect different temperatures had on the process. The following results show the yield at various temperatures; is there a significant evidence to establish correlation between temperature and yield?

Temperature (°C)	Yield (lb)
15	66
18	69
21	69
24	70
27	64
30	73
33	75

Q56. The number of letters posted each year in Australia is compared in the following table with the number of post offices handling the mail. Notice that the amount of mail has increased during the period observed, whereas the number of post offices has decreased. Is there a statistically significant degree of inverse correlation, or might these results be reasonably attributed to chance? For practice, confirm your answer by calculating z too.

Year	No. of letters (millions)	No. of Post Offices
1951	1,486	8,315
1952	1,448	8,315
1953	1,473	8,261
1954	1,570	8,244
1955	1,619	8,234
1956	1,705	8,222

From Commonwealth Bureau of Census & Statistics, Pocket Compendium, 1962.

KRUSKAL AND WALLIS' TEST (1952)

Purpose

To compare 3 or more unmatched random samples of measurements.

Principle

This test was devised jointly by two American statisticians, Professors W. H. Kruskal and W. A. Wallis (*Journ. Amer. Statist. Assoc.*, 1952, pp. 583–618).

Functionally, this test extends the range of Wilcoxon's Sum of Ranks Test (p. 166) to cases where there are more than 2 sets of measurements. As in that test, we start by making the tentative assumption that there is no significant difference between the samples, as if they all really came from the same parent group. Any differences between them are thus looked on (tentatively) as being the result of chance variation from one sample to the next. Then, as in Wilcoxon's Sum of Ranks Test, we pool all the measurements from the different samples, and assign them rank values, from the smallest measurement in the combined pool to the largest one.

Now, if our assumption of no significant difference is correct, extracting the rank values assigned to each sample group from the combined pool should reveal that the rank totals for each sample are about the same as each other (if the number of observations in each sample group is the same). If each of these sample rank totals is then squared, and the squares added together, the answer will be minimal when the rank totals of each sample are identical, and proportionately greater than minimal if any inequality is present. For example, if there are 3 samples each containing 3 measurements, and each sample has a rank total of 15, the sum of the squares of the rank totals will be $15^2 + 15^2 + 15^2 = 675$. But if the rank totals were unequal, say 13, 15, and 17, the sum of their squares will be larger, viz. $13^2 + 15^2 + 17^2 = 683$.

In this test the probability of getting any particular sum of squares, under the hypothesis of no significant difference between the samples, is determined by converting the sum of the squares into a value called χ^2 (chi-squared), about which we shall have a lot more to say later on (p. 269). For the moment it will suffice to know that the significance of the value of χ^2 is given by reference to special Tables provided here. These Tables tell us the probability of getting a sum of squares of the rank totals as large as that observed, if there is really no significant difference between the samples.

Before proceeding, a very important point must be mentioned here. Given a set of 3 or more samples, you may often be tempted to compare only the best with the worst. To do so can be very misleading. For if you take a series of samples from the one parent group, the more samples you take, the greater will become the difference between the best and the worst of the samples (in the same way as the range increases with increasing sample size, p. 53). In other words, a significant difference will often become apparent between the extreme samples, even when all the samples have genuinely been derived from the same source. The safest thing to do is to look upon any such apparently significant differences purely as suggestions for further experimentation, in which the particular pair of circumstances can be deliberately compared. A guide as to whether this is worth doing can be obtained by considering how many possible different pairs you could have chosen from the set of samples, which is given by –

$$\frac{k \times (k-1)}{2}$$

– where k = the number of samples. Having chosen but one of these possible pairs, it is then only fair to tighten our minimal probability requirement from 5% down to –

$$\frac{5\%}{\text{Number of possible pairs}}$$

For instance, suppose we choose 2 samples having the highest and lowest averages out of a set of 6 samples. The number of possible different pairs that might have been chosen from 6 samples is –

$$\frac{6 \times (6-1)}{2} = \frac{6 \times 5}{2} = 15$$

Before accepting that the difference between our 2 specially chosen samples is statistically probably significant, the difference would have to be such that Wilcoxon's Sum of Ranks Test applied to them gave a probability level, not of 5% but of –

$$\frac{5\%}{15} = 0.33\%$$

If such a level was attained, it would be roughly equivalent to getting a 5% probability level in an ordinary test of 2 samples, so it would be worth putting the 2 specified circumstances to a fresh, separate experimental test.

In order to avoid the trouble and expense of having to continue such experiments, significance tests have recently been devised which permit the direct comparison of any or all of the individual samples or treatments in a multiple comparison experiment. This is made possible by incorporating a suitable correction factor to allow for the multiplicity of comparisons. As overall tests the simpler of these newcomers tend to be less powerful than Kruskal and Wallis' Test; the latter should therefore be carried out in the first place, and if a significant difference is demonstrated, it is safe to presume that at least the same level of significance must apply to the difference between the best and worst samples in the whole set. However, if you then want to know whether the best sample is significantly different from the second best (without having to do a further experiment to compare them specifically), one of these '*selected comparison tests*' will prove more powerful than the relatively crude method of applying Wilcoxon's Sum of Ranks Test to the relevant 2 samples and interpreting the result with a modified critical probability level (according to the total number of possible pairs, as described above).

The same kind of situation may be encountered when comparing 2 or more new products or treatments with a control (or comparing a new product with several established ones). Kruskal and Wallis' Test may well give the answer you seek if the control turns out to be the worst (or best) sample of the set, but if the control happens to be one of the intermediate samples, a selected comparison test will be needed to compare the control with the others. The internal correction factor is slightly different for these cases, because the number of possible comparisons is less; with k sets of measurements, one of which is the control, there are $k - 1$ possible pairs. A brief review of the theory underlying these selected comparison tests is given in F. Wilcoxon and R.

Wilcox, *Some Rapid Approximate Statistical Procedures*, 1964,
pp. 9–12 (published and distributed by Lederle Laboratories,
Pearl River, New York).

Data Required

n_A, n_B, n_C, etc. = number of measurements in Sample A,
Sample B, Sample C, etc.

N = total number of measurements in all the samples (so
that $N = n_A + n_B + n_C$, etc.)

Plus all the individual measurements in all the samples.

Procedure

(1) Prepare a calculation table with the following headings –

Data values	Tally	Rank values	A ranks	B ranks	C ranks

(2) Look over the measurements in all the samples and find
the smallest and the largest measurement among them all.

(3) In the first column write down an ordered list of data values,
starting with the smallest measurement and ending with the
largest.

(4) In the second column enter the tally. Record each measurement of Sample A by entering an A beside the appropriate data
value, and then put a B for each of the measurements in Sample
B, and then enter a C for each of Sample C's measurements, and
so on for each sample being compared.

(5) Assign a rank value to each observation. Do this from the
smallest measurement to the largest, though the direction of
ranking makes no difference to this test (even if the samples contain different numbers of observations).

Break any ties between measurements if possible by working
to an additional decimal place, failing which give tied measurements an average rank value (as described on p. 168).

(6) Now transfer the rank values belonging to each sample into separate columns.

(7) Add up the rank values of each sample; call these totals R_A, R_B, R_C, etc. As a check on your calculation so far, note that the sum of these totals equals $\frac{1}{2}N(N + 1)$.

(8) Finally, calculate the value of χ^2 from this formula –

$$\chi^2 = \frac{12}{N^2 + N}\left(\frac{R_A{}^2}{n_A} + \frac{R_B{}^2}{n_B} + \frac{R_C{}^2}{n_C}, \text{etc.}\right) - 3(N + 1)$$

Interpretation

The higher the value of χ^2, the greater the likelihood that the observed differences are not just from chance but are due to genuine differences in the parent groups from which the samples have been drawn.

(a) *With up to 5 measurements in each sample*, refer the value of χ^2 to Kruskal and Wallis' Table opposite, which shows the values that χ^2 must reach to indicate a significant difference between the samples at the 5% and 1% levels. (Samples of smaller size than those shown in this Table cannot reach the 5% level with this test; see Lord's Range Test, p. 366.)

(b) *With more than 5 measurements in each sample, or with more than 3 samples of measurements*, refer your answer to the ordinary χ^2 Table on page 276. You will need to know the 'degrees of freedom' in order to use the correct row of this Table – the degrees of freedom for this test are 1 less than the number of samples (k) being compared, so –

$$\text{D.O.F.} = k - 1$$

The significance of these probability levels was discussed on page 140.

Note: If there are an appreciable number of ties and you get an answer indicating a probability slightly greater than 5% or 1%, it may be worth while applying a special *correction for ties*, as described on page 385.

χ^2 TABLE FOR KRUSKAL AND WALLIS' TEST

No. in each group			Minimal values of K. and W.'s χ^2 indicating		
n_A	n_B	n_C	P = 10%	P = 5%	P = 1%
1	2	5	4·2	5·0	—
1	3	3	4·6	5·1	—
1	3	4	4·0	5·2	—
1	3	5	4·0	4·9	6·5
1	4	4	4·1	4·9	6·67
1	4	5	4·0	4·9	6·9
1	5	5	4·1	5·0	7·1
2	2	3	4·5	4·7	—
2	2	4	4·5	5·1	—
2	2	5	4·3	5·1	6·4
2	3	3	4·6	5·2	6·3
2	3	4	4·5	5·4	6·35
2	3	5	4·5	5·2	6·82
2	4	4	4·5	5·3	6·9
2	4	5	4·5	5·3	7·12
2	5	5	4·5	5·3	7·3
3	3	3	4·6	5·6	6·5
3	3	4	4·7	5·7	6·75
3	3	5	4·5	5·6	7·0
3	4	4	4·5	5·6	7·14
3	4	5	4·5	5·6	7·44
3	5	5	4·5	5·6	7·55
4	4	4	4·6	5·7	7·6
4	4	5	4·6	5·6	7·75
4	5	5	4·5	5·6	7·8
5	5	5	4·6	5·7	7·98

Adapted from W. H. Kruskal and W. A. Wallis, *Journ. Amer. Statist. Assoc.*, 1952, pp. 614–17 and 1953, p. 910.

Example

Ex. 16. In their article announcing this test, Kruskal and Wallis gave the following example.

Three machines were making bottle caps. Their output on days selected at random were –

Machine	Daily output				
A	340	345	330	342	338
B	339	333	344		
C	347	343	349	355	

Is there a significant difference between the output of these 3 machines?

Calculation: The values of R_A, R_B, and R_C are calculated in the Table below.

Data values	Tally			Rank values	A ranks	B ranks	C ranks
330	A			1	1		
333		B		2		2	
338	A			3	3		
339		B		4		4	
340	A			5	5		
342	A			6	6		
343			C	7			7
344		B		8		8	
345	A			9	9		
347			C	10			10
349			C	11			11
355			C	12			12
				$N = 12$	$R_A = 24$	$R_B = 14$	$R_C = 40$
					$n_A = 5$	$n_B = 3$	$n_C = 4$

$$\chi^2 = \frac{12}{N^2 + N}\left(\frac{R_A{}^2}{n_A} + \frac{R_B{}^2}{n_B} + \frac{R_C{}^2}{n_C}\right) - 3(N + 1)$$

$$= \frac{12}{12^2 + 12}\left(\frac{24^2}{5} + \frac{14^2}{3} + \frac{40^2}{4}\right) - 3(12 + 1)$$

$$= \frac{12}{156}\left(\frac{576}{5} + \frac{196}{3} + \frac{1,600}{4}\right) - (3 \times 13)$$

$$= \frac{1}{13}(115 \cdot 2 + 65 \cdot 33 + 400) - 39$$

$$= \left(\frac{1}{13} \times 580 \cdot 53\right) - 39$$

$$= 44 \cdot 66 - 39$$

$$= \underline{\underline{5 \cdot 66}}$$

All 3 samples contain 5 or less measurements, so we refer this value of χ^2 to Kruskal and Wallis' Table on page 217. In the present instance, $n_A = 5$, $n_B = 3$, and $n_C = 4$, but for interpreting the results it does not matter which is which – the main thing is that we have 3 samples sized 3, 4, and 5. In the Table you will find the various sample sizes listed in simple numerical order, so the line to use in the present case is the one which shows $n_A = 3$, $n_B = 4$, and $n_C = 5$. This row shows that our calculated value of χ^2, which was 5·66, has a probability of occurring (if there really is no significant difference between the samples) of just less than 5%. We conclude therefore that the observed difference between the 3 machines is probably significant.

You may be interested to know how this result compares with that obtained by applying the more complicated technique of Fisher's Analysis of Variance to the same data. The specific probability with Kruskal and Wallis' Test is 4·9%; Analysis of Variance gives an answer of 5·1%. The correspondence is good.

Selected Comparisons

In experiments which compare 3 or more *unmatched* samples of measurements, the difference between any 2 of the samples can be tested for significance quite easily provided that all samples contain the same number of measurements (that is, $n_A = n_B = n_C$, etc.).

The test is done by calculating the value, K, from the following formula –

$$K = \frac{d - 0 \cdot 8}{n \cdot \sqrt{n}}$$

where d = difference between the rank totals of the 2 samples being compared (these totals being those in the tabular part of Kruskal and Wallis' Test),

and n = number of measurements in each sample.

The critical values of K are given in the Table below, the left side of which is used when the experiment has been designed to compare any or all of the samples with one another, while the right side is reserved for comparing a number of samples individually with a control. If your calculated K equals or exceeds the values in the body of the Table, it indicates that a significant difference exists between the particular samples being compared, at the probability level shown at the top of the Table.

It is quite legitimate to apply this test to any or all pairs of samples in a multiple comparison experiment.

Performing the test will be demonstrated in Q57 (below).

K TABLE FOR SELECTED COMPARISONS
with unmatched samples

Total No. of samples in experiment k	When comparing any pair of samples		When comparing any sample with a control	
	Values of K indicating			
	P = 5%	P = 1%	P = 5%	P = 1%
3	2·89	3·60	2·72	3·45
4	4·22	5·12	3·86	4·80
5	5·60	6·69	5·00	6·16
6	7·01	8·30	6·17	7·53
7	8·46	9·92	7·37	8·94
8	9·94	11·58	8·55	10·33
9	11·43	13·25	9·77	11·77
10	12·97	14·95	11·01	13·19

Derived from data in F. Wilcoxon and R. Wilcox, *Some Rapid Approximate Statistical Procedures*, Tables 3 and 4 (Lederle Laboratories, 1964).

If the number of measurements is not the same in all samples, individual pairs of samples can still be compared by applying

Wilcoxon's Sum of Ranks Test and reducing the acceptable probability level according to the total number of possible pairs (p. 213).

However, in any experiment comparing 3 or more samples, there is no objection to testing any or all pairs of samples with Wilcoxon's Sum of Ranks Test, without adjusting the probability answer, provided that it was definitely planned to make these individual comparisons *before* seeing any of the results. If there is the slightest doubt about this intention, one must use the safer procedures described above.

Question

Q57. The 3 prima donnas were having a chat over a morning glass of champagne. 'My dears,' remarked Fifi, 'did you hear about my performance at La Scala last month? The audience quite adored me – called me back 36 times.'

Adelina spat back, 'But darling, I got 39 curtain calls at the Met last Saturday night.' By the time a truce was declared (because Maria's poodle had an appointment with the hair-dresser), the scores at the last 4 appearances of each singer were found to be –

Singer	Number of curtain calls			
Fifi	36	22	19	16
Adelina	39	14	20	18
Maria	21	32	28	22

Could one of the singers fairly claim to have won the battle of the curtain calls, or could the observed differences be reasonably attributed to chance?

For practice, try a selected comparison test on Fifi and Maria.

FRIEDMAN'S TEST (1937)

Purpose

To compare 3 or more random samples of measurements which are matched. The matching may be achieved by –

(*a*) carefully replicating members of each sample group, e.g. by dividing homogeneous substances or specimens into portions and allotting one portion to each sample group for different treatments, or

(*b*) using 1 sample group for different treatments, different methods of testing, different observers, or on different occasions (as in paired comparisons, p. 120).

Principle

The usual tentative assumption is made that there is no significant difference between the various samples, and the probability of getting the observed differences as a result of chance is then calculated.

This is done by ranking each set of matched measurements. Then, if our tentative negative assumption is correct, there should be little difference between the total rank values scored by each sample. The significance of any disparity can then be tested in a way which is similar in principle to that described for Kruskal and Wallis' Test (p. 212). The calculation, which was devised by an American economist, Milton Friedman (*Journ. Amer. Statist. Assoc.*, 1937, pp. 675–701), gives a χ^2 value from which the probability of our tentative negative assumption being correct can be told.

Data Required

k = number of samples or treatments (*A*, *B*, *C*, etc.) being compared.

n = number of measurements in each sample.

Plus all the individual measurements in all the samples.

Procedure

(1) Prepare a calculation table with a column for each of the k samples or treatments. Each column must be wide enough to accommodate a double entry – the value of the observations and the rank value that will be assigned to each observation. It is

desirable, too, to have a column at the left of the calculation table for a numbering or description of the matched test conditions, thus –

Conditions	Sample A Obs. Rank	Sample B Obs. Rank	Sample C Obs. Rank

The n measurements in each sample or treatment will then extend down each column, but note carefully that the matching measurements of the various samples run across the table in the rows.

(2) Enter the measurements observed in Sample A in the appropriate column of the calculation table, in any convenient order. Then enter those of Sample B, being careful to align them so that the results obtained under the matched conditions correspond with the Sample A results. Do likewise with the other sample or samples.

(3) Assign rank values to the matched measurements, proceeding across each row. Each row is ranked independently of the other rows, so that there will be n sets of ranks, each set extending from 1 to k.

It is easiest to rank from the smallest to the largest measurement in each row, but it does not affect the result if you choose to work in the opposite direction as long as you are consistent throughout the test.

If 2 or more measurements across a row have the same value, break the tie if possible by working to an additional decimal place, but if this can't be done, give tied measurements an average rank value as described on page 168.

(4) Add the rank values in each column, to get the score of each sample. As a check on your calculation so far, note that the sum of these rank totals equals $\frac{1}{2}nk(k + 1)$.

(5) Square each rank total, using the Table of Squares at the back of this book.

(6) Add the squares of the rank totals; call this total 'R'.

(7) Finally, calculate the value of χ^2 from the following formula –

$$\chi^2 = \frac{12R}{nk(k + 1)} - 3n(k + 1)$$

χ^2 TABLE FOR FRIEDMAN'S TEST

No. of columns k	No. of rows n	Minimal χ^2 indicating P = 5%	P = 1%	No. of columns k	No. of rows n	Minimal χ^2 indicating P = 5%	P = 1%
3	3	5·8	—	5	3	8·4	10·1
	4	6·4	7·8		4	8·8	10·9
	5	6·2	8·3		5	9·0	11·4
	6	6·4	8·7		6	9·1	11·7
	7	6·1	8·7		8	9·2	12·1
	8	6·2	9·0		10	9·2	12·4
	9	6·2	8·7		15	9·3	12·7
	10	6·1	9·0		20	9·4	12·8
	11	6·2	9·0		100	9·5	13·2
	12	6·2	8·9	6	3	9·9	11·7
	13	6·0	9·0		4	10·2	12·6
	14	6·1	9·0		5	10·4	13·1
	15	6·2	8·9		6	10·5	13·4
	20	6·0	8·9		8	10·7	13·9
	100	6·0	9·1		10	10·8	14·1
4	2	6·0	—		15	10·9	14·4
	3	7·1	8·3		20	10·9	14·6
	4	7·6	9·4		100	11·0	15·0
	5	7·7	9·9	7	3	11·2	13·3
	6	7·5	10·2		4	11·6	14·2
	7	7·6	10·3		5	11·8	14·7
	8	7·6	10·4		6	12·0	15·1
	10	7·7	10·5		8	12·1	15·5
	15	7·7	10·8		10	12·2	15·8
	20	7·7	10·9		15	12·4	16·1
	100	7·8	11·3		20	12·4	16·3
					100	12·6	16·7

Adapted from D. B. Owen, *Handbook of Statistical Tables*, #14.1 (Addison-Wesley, 1962), and M. Friedman, *Ann. Math. Statist.*, 1940, p. 89.

Interpretation

The higher the value of χ^2, the greater the likelihood that there is a significant difference between the samples.

(a) *With 3 to 7 samples* (that is, $k = 3$ to 7), refer the calculated value of χ^2 to Friedman's χ^2 Table above. This shows the values which χ^2 must reach to indicate the critical probability levels.

(b) *With 8 or more samples* (that is, $k = 8$ or more), refer your answer to the ordinary χ^2 Table on page 276, using the row with $k - 1$ degrees of freedom.

The significance of these probability levels was discussed on page 140.

Note: If there are a lot of ties, you can strengthen the test by applying a *correction for ties*, in the way described on page 385 but with $n(k^3 - k)$ replacing $N(N^2 - 1)$ as divisor in the T-formula.

Example

Ex. 17. Andy, Bill, Cyril, and Derek were 4 friends who met for a round of golf every 2 months. After 6 such games, Andy declared that a handicap system was called for, in order to even the scores. 'After all, Bill,' he said, 'you've won 3 of the games, and come second twice, which is a lot more than we've been able to do.'

Here are their actual scores. Could the differences be reasonably attributed to chance variation on the assumption that there is no significant difference between the players?

Data and Calculation

Game	Andy Obs.	Andy Rank	Bill Obs.	Bill Rank	Cyril Obs.	Cyril Rank	Derek Obs.	Derek Rank
I	80	2	77	1	81	3	82	4
II	80	1	81	2	83	3	85	4
III	85	3	82	1	84	2	87	4
IV	90	4	86	2	85	1	87	3
V	85	3	80	1	86	4	81	2
VI	81	2	82	$3\frac{1}{2}$	82	$3\frac{1}{2}$	79	1
Rank totals		15		$10\frac{1}{2}$		$16\frac{1}{2}$		18
Ranks squared		225		$110\frac{1}{4}$		$272\frac{1}{4}$		324

We are comparing the scores of the 4 players, *A*, *B*, *C*, and *D*, so $k = 4$. Friedman's Test is applicable here, because the measurements in each of the 4 samples have been truly matched by being obtained under the same test conditions (this would apply even if the 6 games were played on 6 different golf courses).

Each sample contains 6 observations, so $n = 6$.

The sum of the squares of the rank totals is –

$$R = 225 + 110\tfrac{1}{4} + 272\tfrac{1}{4} + 324 = 931 \cdot 5$$

We can now fill in Friedman's Formula, thus –

$$\chi^2 = \frac{12R}{nk(k+1)} - 3n(k+1)$$

$$= \frac{12 \times 931 \cdot 5}{(6 \times 4)(4+1)} - (3 \times 6)(4+1)$$

$$= \frac{11{,}178}{24 \times 5} - (18 \times 5)$$

$$= \frac{11{,}178}{120} - 90$$

$$= 93 \cdot 15 - 90$$

$$= \underline{\underline{3 \cdot 15}}$$

Since the number of samples being compared (k) is 4, we refer this answer to Friedman's χ^2 Table on page 224. There we find that for $k = 4$ and $n = 6$, a value of χ^2 of 3·15 indicates a probability of more than 5%. This means that if our tentative assumption that there is no significant difference between the 4 players is correct, the observed results could have arisen by chance more than once in 20 such trials. This is altogether too likely to allow us to reject our tentative assumption, so we conclude that a significant difference is not proven. Andy's claim for a handicapping system is thus not warranted on the present evidence.

Now, in the above example we have compared the rank totals scored by the 4 players. Suppose instead we had wanted to find out whether golf scores are affected by seasonal influences. Our golfers played 6 rounds through the course of a year, so we have enough data available to assess the effects of season. But to do this we would have to repeat the analysis, turning the table on its side, with the games forming the column headings, and the players forming the test conditions down the left of the table. In this arrangement the 6 games become 6 samples that we are going to compare, and the players simply provide the matched information needed for this comparison. We still rank the measurements across each row separately; k becomes 6, and n becomes 4. Here is the rearranged data table; it shows the im-

portance of setting up your table correctly, with the things you
want to compare forming the columns.

Player	Game number					
	I	II	III	IV	V	VI
Andy	80	80	85	90	85	81
Bill	77	81	82	86	80	82
Cyril	81	83	84	85	86	82
Derek	82	85	87	87	81	79

The beauty of Friedman's Test is that it allows this 2-way
analysis of matched measurements. As such, it is applicable to
certain factorial designs (p. 120). For the 4 players we could substitute 4 varieties of wheat, and instead of the 6 games, we could
use 6 different fertilizing treatments. Or we could test 6 cars with
4 different brands of gasoline. In addition, by pooling the results
of say the first 3 games, and comparing these with the pooled
results of the second set of 3 games, you will see that this Test
can also be extended to investigating 3 or more factors simultaneously. The only shortcoming is that Friedman's Test will not
detect or assess interactions between factors (p. 123) – only
Fisher's Analysis of Variance will do that.

To even the score, however, Friedman's Test can do one thing
that Fisher's Analysis of Variance can't do, namely handle data
which is only capable of being ranked in order of merit or
preference. We shall meet an example of this in *Q60*.

Selected Comparisons

In testing any particular pair of *matched* samples in a multiple
comparison experiment, it is to be noted that the remarks on
page 213 also apply here. If Friedman's Test indicates a significant
difference between the whole set of samples, you can then proceed to make individual comparisons by calculating *F* as follows –

$$F = \frac{d}{\sqrt{n}}$$

where d = difference between the rank totals of the 2 samples
being compared (taken from the tabular part of
Friedman's Test),

and n = the number of measurements in each sample.

The significance of F is given in the accompanying Table, which is used in exactly the same way as its companion on page 220.

Examples of how to do this test are given in the questions below.

F TABLE FOR SELECTED COMPARISONS

with matched samples

Total No. of samples in experiment k	When comparing any pair of samples		When comparing any sample with a control	
	Values of *F* indicating			
	P = 5%	P = 1%	P = 5%	P = 1%
3	3·32	4·12	3·12	3·94
4	4·70	5·68	4·30	5·34
5	6·10	7·28	5·46	6·70
6	7·54	8·90	6·64	8·10
7	9·00	10·52	7·86	9·50
8	10·50	12·22	9·04	10·92
9	12·02	13·90	10·26	12·36
10	13·54	15·62	11·52	13·78

Derived from data in F. Wilcoxon and R. Wilcox, *Some Rapid Approximate Statistical Procedures*, Tables 5 and 6 (Lederle Laboratories, 1964).

Questions

Q58. Determine whether or not there is a significant seasonal influence on the 4 friends' golf scores, using the data Table on page 227. Try the correction for ties. Then see if there is a significant difference between the results of Games I and III.

Q59. Reverting to the first part of *Ex. 17*, in which a significant difference between the 4 players was not proven, Andy was still not satisfied. 'That's all very well when you lump us all in together,' he said, 'but surely it isn't fair to put Derek (who came 4th in 3 games) in the same class as Bill (who came 1st in 3 games). I still reckon we ought to handicap Bill.'

What would you reply?

Q60. To see how Friedman's Test works when the data is only available in rank values, try this one. The Italiano family had a little sweet shop near a school. They wanted to know which of 3 makes of sausage rolls the kids would like best. There was a slight difference in price between them, and Mama Italiano was all in favour of the cheapest variety – 'They'll like 'em just as well as the dearest kind,' she said. But Papa was more scientific. He gave 8 kids one of each type of sausage roll to try, and asked which one they liked best, which second best, and which least. Here are the little gourmets' opinions; is there a significant difference between their preferences?

Tester	8¢ roll	10¢ roll	11¢ roll
Johnny	3	2	1
Gordon	2	3	1
Juliet	2	3	1
Rebecca	3	1	2
Lotte	3	1½	1½
Alex	2	3	1
Charlie	3	1	2
Luigi	1	3	2

POISSON'S TEST (1837)

Purpose

To compare –

(*a*) the number of *isolated occurrences* (such as accidents) in a random sample of a certain size or duration, and the expected number for such a sample as indicated by a large set of observations (at least 10 times the sample size or duration), provided that the number in the sample is *40 or less* (if more, use the *zI* Test, p. 245), or

(*b*) the proportions of a random *binomial sample*, and an average derived from a much larger set of observations, provided that the average proportion of the smaller class is *less than 10%*, and the number in the smaller class of the sample is *40 or less* (if more than 40, use the *zI* Test, p. 245). Remember, a binomial sample is one made up of 2 mutually exclusive classes, such as men and women (p. 29).

Principle

This test is a straightforward application of Poisson's Probability Formula (p. 32).

You will recall that Poisson evolved this formula by considering what happened to the Binomial Formula when the proportions of one class of a binomial group became very small. It should therefore be no surprise to find that although Poisson's Formula applies primarily to isolated occurrences, it also provides a good approximation for binomial situations when the average proportion of the smaller class is less than 10%.

The correspondence between the Poisson and binomial probabilities is very close when the smaller class average is only 1% or 2%, but even at 9% the results are generally in agreement to within 1% at the 5% probability level, and to within $\frac{1}{2}$% at the 1% probability level. As an example, if the smaller class average is 9%, and 6 instances are observed in the smaller class when 13 are expected, the true binomial two-sided probability is 4·07%, while Poisson's Formula indicates a probability of 5·09%. Had the observed number been 5, the binomial probability would have been 1·36%, compared with a Poisson answer of 1·15%. And remember, the agreement becomes even better as the average proportion of the smaller class approaches 0%.

The use of Poisson's Test for these binomial situations offers simplicity, combined with sufficient accuracy for most practical purposes, plus safety. The latter is due to the fact that the Poisson answer is nearly always a shade weaker than the true binomial probability, so there is very little risk of getting a 'significant' answer when you shouldn't. However, this does mean that you will occasionally get an answer of 'significant difference not proven' when a statistician, using the Binomial Formula or special binomial tables, would arrive at an answer of 'difference probably significant'.

To apply Poisson's Test, we calculate the expected number of occurrences or instances for a sample of the size concerned, according to the known average. The tentative negative assumption is then made that any difference between this expected number (E) and the number actually present (x) in the sample is not a sign of a significant difference, but has merely come about as a result of chance. The probability of this being the case is then found to see if our tentative assumption is tenable or not.

Provided that the number observed (x) in the sample is not more than 40, the probability can be found without further calculation, simply by referring to the special Table provided on page 233, which has been derived from Poisson's Formula.

Data Required

With isolated occurrences –

x = number of occurrences in the sample.
n = sample size, i.e. duration or dimension.
P_x = average number of occurrences per unit time or size in the large set of observations.

With binomial cases –

x = number of instances in the smaller class in the sample.
n = sample size, i.e. total number of observations making up the sample.
P_x = average proportion of the smaller class in the large set of observations. This must be less than 10%.

Procedure

(1) Calculate the expected number (E) of occurrences or instances for a sample of the size being investigated, thus –

$$E = P_x \times n$$

(2) Refer this value of E to the Table for Poisson's Test (p. 233). First find the line to use according to the observed value of x in the sample, then decide whether the left- or the right-hand section of the Table is appropriate, according to whether x is larger or·smaller than E. The values in the body of the Table are the critical levels that E must reach to indicate the probability levels shown at the top of each column.

Interpretation

The probability levels indicated by the Table for Poisson's Test are the probabilities of there being no significant difference between the sample and the much larger set of observations.

The significance of these levels was discussed on page 140.

Examples

A number of examples will be given in order to show the versatility of Poisson's Test.

Ex. 18. It is known that a certain kind of machine requires adjustment on an average of once a month. In one particular month, it gets out of adjustment 3 times. Is this likely to be due to bad luck (i.e. chance), or does it incriminate some factor other than chance (such as careless operating of the machine)?

Data: $x = 3$ (the number of breakdowns in the sample)

       $n = 1$ month (the sample duration)

       $P_x = 1$ (the average number of breakdowns per month)

Calculation: This is a typical example of isolated occurrences. We can state how many times the occurrence did occur, but not the number of times it did not occur.

The expected number (E) in the sample is obviously 1. The observed number (x) in the sample is 3. What is the probability of E being as small as 1 when x is 3? We turn to the E Table for Poisson's Test opposite and, using the left-hand section of the Table (because x is larger than E), we find the line for $x = 3$. Then we run across to the right to find where a value of $E = 1$ would be placed. It is between 5% and 10%.

This means that the observed number of occurrences could be expected to arise by chance in more than 5% occasions, so a significant difference from the average is not proven.

E TABLE FOR POISSON'S TEST

No. in sample *x*	Probability when *x* > *E*				Probability when *x* < *E*			
	P = 10%	P = 5%	P = 1%	P = 0·2%	P = 10%	P = 5%	P = 1%	P = 0·2%
0	—	—	—	—	2·8	3·6	5·1	6·6
1	0·105	0·051	0·010	0·002	4·75	5·55	7·15	9·1
2	0·53	0·355	0·149	0·065	6·25	7·1	9·15	11·1
3	1·10	0·82	0·435	0·243	7·75	8·6	10·65	13·1
4	1·74	1·37	0·82	0·52	8·8	10·1	12·6	14·7
5	2·35	1·97	1·28	0·87	10·3	11·55	14·1	16·5
6	2·75	2·61	1·78	1·27	11·75	13·05	15·5	18·0
7	3·55	3·15	2·35	1·72	12·8	14·1	17·0	19·5
8	4·3	3·5	2·91	2·20	14·3	15·7	18·5	21·5
9	4·7	4·45	3·5	2·72	15·6	17·5	20·0	22·5
10	5·75	5·1	4·1	3·25	17·0	18·5	21·5	24·5
11	6·2	5·5	4·7	3·8	18·0	19·5	23·0	25·5
12	7·25	6·6	5·05	4·4	19·5	21·0	24·0	27·0
13	7·7	7·0	5·8	5·0	20·5	22·5	25·5	28·5
14	8·7	8·05	6·65	5·6	22·0	23·5	27·0	30·0
15	9·5	8·5	7·05	6·15	23·0	25·0	28·0	31·5
16	10·25	9·55	7·7	6·5	24·5	26·0	29·5	32·5
17	11·25	10·0	8·65	7·1	25·5	27·5	31·0	34·0
18	11·7	11·1	9·05	8·0	26·5	28·5	32·0	35·5
19	12·75	11·5	10·0	8·6	28·0	29·5	33·5	37·0
20	13·45	12·55	10·6	9·0	29·0	31·0	35·0	38·0
21	14·25	13·0	11·15	9·9	30·5	32·5	36·0	39·5
22	15·1	14·05	12·1	10·6	31·5	33·5	37·5	41·0
23	15·9	14·9	12·55	11·05	32·5	34·5	38·5	42·0
24	16·5	15·5	13·6	12·0	34·0	35·5	40·0	43·5
25	17·5	16·2	14·0	12·6	35·0	37·0	41·0	45·0
26	18·0	17·0	15·0	13·1	36·0	38·5	42·5	46·0
27	19·0	18·0	15·5	14·15	37·0	39·5	43·5	47·5
28	20·0	19·0	16·0	14·6	38·5	40·5	45·0	48·5
29	21·0	19·5	17·0	15·1	39·5	41·5	46·0	50·0
30	21·5	20·0	18·0	16·0	40·5	43·0	47·5	51·0
31	22·5	21·0	18·5	16·5	42·0	44·0	48·5	52·5
32	23·5	22·0	19·5	17·0	43·0	45·5	50·0	54·0
33	24·0	23·0	20·0	18·0	44·0	46·5	51·0	55·0
34	25·0	23·5	21·0	19·0	45·5	47·5	52·0	56·5
35	26·0	24·5	22·0	19·5	46·5	48·5	53·5	57·5
36	27·0	25·0	22·5	20·0	47·5	50·0	54·5	58·5
37	27·5	26·0	23·5	21·0	48·5	51·0	56·0	60·0
38	28·5	27·0	24·0	22·0	50·0	52·5	57·0	61·5
39	29·5	28·0	25·0	22·5	51·0	53·5	58·0	62·5
40	30·0	29·0	26·0	23·0	52·0	54·5	59·5	64·0

Adapted from General Electric Company's *Tables of the Individual and Cumulative Terms of Poisson Distribution* (1962, copyright by D. Van Nostrand Company Inc., Princeton, N.J.), and 'Chartwell' Poisson Probability Graph Sheets, Nos. 5591/A & B.

Ex. 19. There were 150 road accidents in 5 years on a highway between *A* and *B*. In an attempt to reduce this dreadful toll, the speed limit was reduced. The volume of traffic remained the same, but in the first 3 months after this new law was introduced, there were only 4 accidents. Is this a significant reduction, or rather, is this a significant difference from average (p. 143)?

Data: $x = 4$ (accidents in the sample)

$n = 3$ months (the sample duration)

$P_x = 150$ per 60 months (the average proportion in the parent group) $= \frac{150}{60}$ accidents per month

Calculation: $E = P_x \times n$

$= \frac{150}{60} \times 3 = 7.5$

If $x = 4$, what is the probability of E being as large as 7·5? The right-hand section of the Poisson Table is appropriate here, and shows the probability to be greater than 10%. The reduction in accident rate may therefore be reasonably ascribed to chance. A significant difference from average is not proven.

Ex. 20. Continuing *Ex. 19*, suppose that in the next 3 months there are again only 4 accidents. Assuming the trial conditions to be unchanged, does this additional information affect the statistical verdict?

Data: $x = 4 + 4 = 8$ accidents

$n = 3 + 3 = 6$ months

$P_x = 150$ per 60 months

Calculation: $E = \frac{150}{60} \times 6 = 15$

The Poisson Table shows that for $x = 8$, a value of $E = 15$ indicates a probability of between 5% and 10%. A significant reduction in accidents is still not proven, but the trend suggests that it may be worth following the observations for a longer period. Perhaps in the next 3 months the results will pass beyond the 5% level.

Ex. 21. Over a period of years, a manufacturer of television sets found that a certain valve, the brand of which we shall call *A*, became faulty during the guarantee period in an average of 2% sets.

In order to reduce production costs, a change was made to a cheaper brand of valve (*B*). However, the first 150 sets equipped with this new valve provided 7 instances where this cheaper valve broke down during the guarantee period. Does this mean that the brand *B* valves are really inferior to the original brand *A* ones, or might this outcome simply be due to bad luck (i.e. chance)?

Data: $x = 7$ (the number of instances in the smaller, faulty class in the sample)

$n = 150$ (the sample size)

$P_x = 2\%$ (the average proportion of the smaller, faulty class in the large parent group)

Calculation: This is a binomial example, there being 2 classes of valves, faulty and non-faulty. Poisson's Test is applicable because the average proportion of the smaller class is less than 10% and the observed number in the smaller class is not more than 40. The procedure follows the same lines as for isolated occurrences.

$$E = P_x \times n$$
$$= \tfrac{2}{100} \times 150$$
$$= 3\cdot0$$

This means that if brand B valves were as good as brand A, one might have expected only 3 valves to become faulty in this batch of 150 television sets. Instead, there were 7.

To assess whether this difference is significant or not, we turn to the E Table for Poisson's Test, armed with these values, $x = 7$, and $E = 3\cdot0$. There, on the line for $x = 7$, we see that a value of $E = 3\cdot0$ falls between the 5% and 1% levels. Our value of E thus has a probability of less than 5%; the difference is therefore probably significant.

As a matter of interest, the precise Poisson probability for this data is 3·35%, a figure which compares favourably with the true binomial probability for the data, which is 3·20%. The approximation is very close when P_x is 2%.

Ex. 22. If 99% of a certain publisher's books are ordinarily bound perfectly, would the finding of 3 imperfectly bound books in an order of 60 books be significant of a lowered quality, or might it just be bad luck?

Data: $x = 3$

$n = 60$

$P_x = 100\% - 99\% = 1\%$ (remember, P_x is the proportion of the smaller class, which in this case consists of the faulty books)

Calculation: $E = \tfrac{1}{100} \times 60 = 0\cdot6$

Reference to the Poisson Table shows that when $x = 3$, the probability of getting a value of E as small as 0·6 is less than 5%. This indicates that the finding of 3 faulty books in this sized order is probably significant of a real lowering of the quality.

Questions

Q61. New drugs are always tested for side-effects before being released for general use. Suppose a new drug was tested on 250 people, and was found to cause an allergic skin rash in 1 of them.

Now if you prescribed this drug for 25 patients, and found that 2 of them got an allergic rash from the drug, is it likely that this result is due to chance, or does it suggest that the preliminary trial was inadequate?

Q62. Over a period of some years, it has been found that accidents occur in a particular factory on the average of once a fortnight. Is the occurrence of 6 accidents in 6 weeks significant of some factor other than chance?

Q63. In 10 years, 320 flying saucers were observed over Bazookaland. The following year there were only 17 such reports. Is this a significant difference?

Q64. With standard treatment, a certain type of snake-bite proves fatal in 7% cases. A new treatment is tried on 50 patients, all of whom recover. Is this sufficient evidence to establish that the new treatment is superior to the old?

BINOMIAL TEST (1713)

Purpose

This test compares –

(*a*) the proportions of a random *binomial sample*, and an average derived from a much larger set of observations, when the average proportion of the smaller class is *10% or more*, provided that when the observed number in the smaller class of the sample is less than expected, it does *not exceed 4*, or if more than expected, does *not exceed 20*. A binomial sample is one made up of 2 mutually exclusive classes, such as faulty and non-faulty (p. 29), or

(*b*) 2 random samples of *isolated occurrences* (such as accidents), when the samples have different durations or come from parent sources of different sizes, provided that the smaller sample duration or smaller parent source represents *10% or more* of the combined pair, and provided that, in the sample with the smaller duration or coming from the smaller parent, if the observed number is less than expected, it does *not exceed 4*, or if more than expected, it does *not exceed 20*.

In either situation –

If proportion is 50%, use 50% Probability Test (p. 254).
If proportion is less than 10%, use Poisson's Test (p. 230).
If the numbers exceed 4 or 20, use the *zI* Test (p. 245).

Principle

This test is the practical application of the Binomial Formula, which was discovered by Jacques Bernoulli and published in 1713 (p. 30). This formula tells the probability of getting proportions other than average in a sample from a binomial parent group.

The assumption is therefore made that any discrepancy between the sample proportions and that of the parent group is simply due to chance, and the Binomial Formula will then tell us just what the likelihood is of this being the case.

However, except for very small samples, computing the Binomial Formula involves a great deal of work. This seriously limited its application in practical affairs until recent years, when these massive calculations were performed by electronic computing machines, and the results published in king-size volumes. Fortunately, it is possible to extract certain critical figures from these tables, and to present them in a simple, compact format. For dealing with proportions other than those shown in the Tables on pages 240–1, acceptable results can be obtained by simple linear interpolation.

When comparing 2 samples of isolated occurrences with each other, we start with the tentative negative assumption that there is no significant difference between them, after making due allowance for any difference in the duration of the samples or in the dimensions of the parent groups from which the samples have arisen. The expected numbers in each sample are then determined by their relative duration, or by the parent dimensions, and any observed difference from such expectations is tentatively attributed to chance. The probability of this being the case is then obtained by looking on each sample as a class in a binomial sample, the proportions of which are known from the parent source.

Data Required

With a binomial sample –

x = number of instances in that class of the sample corresponding to the smaller class of the parent group.

n = total number of observations making up the sample.

P_x = the average proportion of the smaller class in the parent group. This must be 10% or more.

With 2 samples of isolated occurrences –

x = number of occurrences in the sample with the smaller duration, or coming from the smaller parent source.

n = total number of occurrences in the 2 samples.

P_x = the proportion of the x sample duration, or its parent dimensions, to that of the combined pair.

Procedure

(a) *If P_x is 10%, 15%, 20%, 25%, 30%, 35%, 40%, or 45% –*

No calculation is required. The answer is directly obtainable from the n Tables for the Binomial Test (pp. 240–1). First decide whether the observed number (x) is less than or more than expected from the average (you can calculate $E = P_x \times n$ if necessary), in order to know whether to use the upper or the lower part of the Tables.

Then find the line corresponding to the observed value of x, run across to the appropriate column of P_x, and there you will find the critical values of the sample size, n. (Note that these are not Tables of E, as in the case of the Table for Poisson's Test, p. 233.)

(b) *If P_x is some other proportion between 10% and 50% –*

Calculate the critical values of n from the adjacent values in the Binomial Tables, by simple proportions. The formula for doing this is as follows –

$$n = n^- - \frac{(P_x - P_x^-)(n^- - n^+)}{5}$$

where $n = $ critical intermediate value of n,

$n^- = $ critical value of n for the proportion just smaller than the intermediate proportion,

$n^+ = $ critical value of n for the proportion just larger than the intermediate proportion,

$P_x = $ the intermediate proportion (expressed as a percentage figure),

$P_x^- = $ the proportion just smaller than the intermediate proportion (also expressed as a percentage figure).

This technique gives results which are usually correct to within 1% at the 5% probability level, and to within $\frac{1}{2}$% at the 1% probability level. If anything, the results tend to be slightly on the weak side (hence we are not led to believe a difference is significant when it is not), because the exact probabilities of the values of n in the Binomial Tables are generally slightly smaller than indicated at the tops of the columns. This is unavoidable when handling whole numbers; for instance with $P_x = 25\%$, and $x = 15$, the exact two-sided probability of $n = 32$ is 0.72%, while the next number up, $n = 33$, has a probability of 1.37%. As in-between values of n are not possible, it is therefore necessary to show $n = 32$ as the critical value for 1%.

Nevertheless, this degree of accuracy is ample for most purposes, but statisticians' tables can be consulted if greater precision is required.

n TABLES FOR BINOMIAL TEST

No. in sample x	$P_x = 10\%$		$P_x = 15\%$		$P_x = 20\%$		$P_x = 25\%$	
	Probability when $x < E$							
	P = 5%	P = 1%	P = 5%	P = 1%	P = 5%	P = 1%	P = 5%	P = 1%
0	32	48	22	33	17	23	12	19
1	52	68	35	46	24	33	20	27
2	67	88	44	56	32	43	26	33
3	82	103	55	69	39	50	32	39
4	97	118	65	79	47	58	38	45
	Probability when $x > E$							
2	3	2	2	—	2	—	—	—
3	8	5	6	3	4	3	3	—
4	14	9	10	6	7	5	6	4
5	20	14	14	10	10	7	9	6
6	27	19	18	13	14	10	11	9
7	31	25	21	17	16	13	13	11
8	37	31	26	21	20	16	17	13
9	47	37	31	25	23	20	19	16
10	51	43	34	29	27	21	22	18
11	60	47	41	32	31	25	25	21
12	66	54	44	38	34	29	28	24
13	74	63	51	42	38	32	31	26
14	81	67	54	45	41	35	35	29
15	89	74	61	52	45	39	37	32
16	96	83	64	55	50	42	41	35
17	106	87	72	60	53	47	43	38
18	111	97	74	66	58	48	47	40
19	121	102	81	69	61	54	50	44
20	126	110	87	76	66	57	53	46

Interpretation

The Binomial Tables indicate the probability of there being no significant difference between the observed and expected numbers being tested. If this probability is remote, the difference is held to be statistically significant; otherwise a difference is not proven.

The significance of the 5% and 1% probability levels was discussed on page 140.

Examples

Ex. 23. If 15% of white Australian people are allergic to penicillin, would the finding of only 1 such instance in a random

n TABLES FOR BINOMIAL TEST

No. in sample *x*	$P_x = 30\%$		$P_x = 35\%$		$P_x = 40\%$		$P_x = 45\%$		$P_x = 50\%$	
	Probability when $x < E$									
	P = 5%	P = 1%	P = 5%	P = 1%	P = 5%	P = 1%	P = 5%	P = 1%	P = 5%	P = 1%
0	10	14	9	13	7	10	6	8	6	8
1	17	21	13	17	12	15	9	13	9	12
2	22	26	17	23	15	19	13	17	12	15
3	27	33	22	27	19	23	16	20	15	18
4	30	38	26	31	22	26	19	24	17	21
	Probability when $x > E$									
2	—	—	—	—	—	—	—	—	—	—
3	3	—	3	—	—	—	—	—	—	—
4	5	4	4	—	4	—	4	—	—	—
5	7	5	6	5	6	—	5	—	—	—
6	9	7	8	7	7	6	(6)	6	6	—
7	11	9	10	8	9	8	8	7	7	—
8	14	12	12	10	11	9	9	8	9	8
9	16	13	14	12	13	11	12	10	10	9
10	19	15	16	14	14	13	13	12	12	10
11	21	18	18	16	17	14	15	13	13	12
12	24	20	21	17	18	16	17	15	15	13
13	26	22	23	20	21	18	18	16	17	15
14	29	25	25	22	22	19	20	18	18	16
15	31	27	28	24	25	22	22	19	20	18
16	34	30	29	26	27	23	24	22	21	19
17	37	32	32	28	28	25	26	23	23	21
18	39	35	34	30	31	27	27	25	25	22
19	42	37	36	33	33	29	29	26	26	24
20	45	40	39	34	34	31	31	28	28	26

Adapted from *Tables of the Cumulative Binomial Probabilities*, prepared and published by the USA Ordnance Corps, 1952.

sample of 50 Australian aboriginees represent a significant difference (perhaps of racial origin)?

Data: $x = 1$
$n = 50$
$P_x = 15\%$

Calculation: Obviously the value of x is less than expected (E would be 7·5), so we use the upper part of the Binomial Table on page 240. There we find the line for $x = 1$, and run across to the column for $P_x = 15\%$. There we find that $n = 46$ is the critical 1% value of n. Our value exceeds this, so the probability of there being no significant difference between the 2 sets of observations is less than 1%. This implies that the observed difference is statistically significant.

Ex. 24. Over a period of some years, a car manufacturing firm finds that 18% of their cars develop body squeaks within the guarantee period. A new design of body is then introduced, and of the first 20 new cars to reach the end of the guarantee period, none develops squeaks. Is this a significant reduction?

Data: $x = 0$

$n = 20$

$P_x = 18\%$

Calculation: The Binomial Tables provided in this book do not have a column for $P_x = 18\%$, so we must find the critical values of n by interpolating between the values for $P_x = 15\%$ and $P_x = 20\%$, using the formula –

$$n = n^- - \frac{(P_x - P_x^-)(n^- - n^+)}{5}$$

First consider the 5% level. Reference to the Table on page 240 shows that for $x = 0$ and $P_x = 15\%$, the critical value of $n^- = 22$. Likewise, for $x = 0$ and $P_x = 20\%$, the 5% critical value of $n^+ = 17$. The required value of n for $P_x = 18\%$ must therefore lie between 22 and 17. Just what its value is will be given by applying the above formula –

$$5\% \ n = 22 - \frac{(18 - 15)(22 - 17)}{5}$$

$$= 22 - \frac{3 \times 5}{5}$$

$$= 19 \cdot 0$$

In the same way we determine the critical 1% level of n –

$$1\% \ n = 33 - \frac{(18 - 15)(33 - 23)}{5}$$

$$= 33 - \frac{3 \times 10}{5}$$

$$= 27 \cdot 0$$

We can now draw up a little table for ourselves –

x	$P_x = 18\%$ $P = 5\%$ $\quad$ $P = 1\%$
0	19·0 $\qquad$ 27·0

Our data gave us $n = 20$. Where does this fit in the above little table? Between the two, indicating a probability of less than 5%. The observed difference between the old and the new body design is thus probably significant.

You may be interested to know that the exact probability for this data is 2·60%, so our approximation of 'less than 5%' is quite correct.

Ex. 25. A firm owns 2 factories, *A* and *B*. Factory *A* has 4 presses of a certain kind, while *B* has 6 of them. During the same observation period, *A*'s presses break down 7 times, while *B*'s do so only 3 times. Is this sufficient evidence for management to complain about the way *A*'s presses are being used, or might this difference be reasonably attributed to chance?

Data: $x = 7$ (breakdowns in the sample having the smaller parent source, viz. factory *A* with its 4 presses)

$n = 7 + 3 = 10$ (breakdowns in both samples)

$P_x = \dfrac{4}{4 + 6} = \dfrac{4}{10} = 40\%$ (which is the proportion of the x sample's parent dimensions to that of the combined pair)

Calculation: The value of x is distinctly higher than expected, so the lower part of the Binomial Table on page 241 is appropriate. There we find that for $x = 7$, in the column for $P_x = 40\%$, a value of $n = 10$ represents a probability of more than 5%. A significant difference is therefore not proven; the difference may well be due to chance.

Questions

Q65. Suppose that a large survey of sporting injuries in schools revealed that 10% of students who played sport sustained some sporting injury (other than minor bruises) in the course of a year. A similar survey on a random sample of 100 university

SIGNIFICANCE TESTS

students showed 17 who sustained some such injury. Would this be adequate evidence of increased danger in university sports?

Q66. It has long been known that Hodgkin's disease is twice as common in men as in women. A doctor, reporting on a new treatment for this disease, presented 12 patients, 7 of whom were women. Would you accept that his sample group was random?

Q67. The well-known Rock-'n-Roll singer, Johnny Blowhard, claimed 15 girls had completely swooned away at his last 3 concerts. His competitor, Dizzie Hotso, could only remember his most recent concert; only 1 female 'victim', hardly worth mentioning really. But is the difference statistically significant? (Assume the concerts were equally attended. After answering the question, work out the expected numbers on the assumption that there is no real difference.)

zI TEST (1733 and later)

Purpose

The *z* Test for Instances (hence '*zI*') applies to situations which are beyond the limits of the Poisson and Binomial Tables. Thus it is used for comparing –

(*a*) the number of *isolated occurrences* (such as accidents) in a random sample, and an average, when the observed number in the sample is *more than 40* (if less, use Poisson's Test, p. 230), or

(*b*) 2 random samples of *isolated occurrences*, when the samples have different durations or come from parent sources of different sizes, provided that the smaller sample duration or smaller parent source is *10% or more* of the combined pair, and provided that when the number of occurrences in the sample with the smaller duration or coming from the smaller parent is less than expected, it *exceeds 4*, or *exceeds 20* when the observed number is greater than expected. (With smaller observed numbers, use Binomial Test, p. 237), or

(*c*) the proportions of a random *binomial sample* (i.e. with 2 classes, such as men and women), and an average derived from a much larger set of observations, when –

(i) the average proportion of the smaller class is *less than 10%*, and the observed number in the smaller class of the sample is *more than 40* (if less, use Poisson's Test, p. 230), or

(ii) the average proportion of the smaller class is *10% or more*, and the observed number in this class of the sample is less than expected but is *more than 4*, or is more than expected and is *more than 20*. (With smaller numbers, use Binomial Test, p. 237. If the average is 50%, save time by using the 50% Probability Test, p. 254.)

Principle

This test is essentially an adaptation of the *zM* Test (p. 152) for use with numbers of instances instead of measurements.

This adaptation is possible because, as pointed out on pages 35–6, the Normal Probability Formula provides a close approximation to the Binomial and Poisson's Formula, at least for the situations specified above. Under these circumstances, this approximation can be trusted to be correct to within about 1%

at the 5% probability level, and to within $\frac{1}{2}$% at the 1% probability level. This degree of correspondence is ample for most practical purposes, and we are thereby spared a great deal of arithmetic.

When comparing a sample and a known average, we start with the tentative assumption that there is no significant difference between them, and proceed to calculate the probability that the sample did, in fact, come from this parent source. The difference between the observed number and the expected number (as indicated by the average) leads to a value of z, the 'standardized difference' from the mean, the critical probability levels of which are provided in a Table (p. 248). If the difference between the sample and the average can reasonably be put down to chance variation (to which all samples are prone), our assumption is not contradicted. On the other hand, if the difference is only remotely likely to have happened by chance, our assumption is almost certainly wrong, so we accept instead that the difference is significant.

When comparing 2 samples of isolated occurrences with each other, we make the tentative assumption that there is no significant difference between them. Under these conditions, the expected number in each sample would be determined simply by the relative duration of the 2 samples or by the relative size of their parent sources. The probability of the observed difference from expectations can then be calculated as if the 2 samples were the 2 classes of a single binomial sample, the average proportions of which were known. As above, this is done by calculating the value of z.

Incidentally, when the zI Test is used with the conventional correction factor of -0.5, it gives answers which are identical with those obtained from a 2-cell χ^2 Test with Yates' correction. However, when the observed number in the smaller group is less than expected, a considerable improvement in the accuracy of the answers (compared with the true Poisson or true binomial probabilities) is achieved with the new correction factors described herein. As a typical instance, if a random binomial sample consisted of 24 in one class and 50 in the other, when the expected proportion of the smaller class was 42%, the exact two-sided binomial probability is 10.0%. The answer given by the improved zI Test is 10.5%; a 2-cell χ^2 Test with Yates' correction gives a probability of 12.1%.

With any approximate test, it is very desirable that you should have some idea of its accuracy under a variety of conditions. The zI Test is essentially an approximate test, so in the examples below, the exact probabilities will also be quoted to let you judge the approximation for yourself.

Data Required

For comparing a sample of *isolated occurrences* and an average –

> x = number of occurrences in the sample.
> n = duration or dimension of the sample.
> P_x = average number of occurrences per unit time or size in the large set of observations.
> P_y = 1. This is the virtual proportion of 'non-occurrences', reflecting the fact that the actual occurrences only take up a minute fraction of the available time or space.

For comparing *2 samples of isolated occurrences* with each other –

> x = number of occurrences in the sample which has the smaller duration, or is derived from the smaller parent group.
> y = number of occurrences in the other sample.
> n = total number of occurrences in the 2 samples (thus $n = x + y$).
> P_x = fractional proportion of the x sample duration, or parent source, to the whole.
> $P_y = 1 - P_x$

For comparing a *binomial sample* and a large parent group –

> x = number of instances in the smaller class of the sample.
> y = number of instances in the larger class of the sample.
> n = total number of cases in the sample ($n = x + y$).
> P_x = fractional proportion of the x (smaller) class in the large parent group.
> $P_y = 1 - P_x$. This is the proportion of the y (larger) class in the parent group.

Procedure

(1) Calculate the expected number (E) of occurrences or instances for a sample of the size observed using this formula –

$$E = P_x \times n$$

(2) Then calculate z from this formula –

$$z = \frac{|E - x| - c}{\sqrt{E \times P_y}}$$

– where c is a correction factor, usually granted a value of 0·5 regardless of the circumstances. However, for getting the correct two-sided probabilities needed for significance tests, the results are appreciably more accurate if c is assigned a value as follows –

c TABLE

For comparing	Value of c	
	If $x < E$	If $x > E$
A sample of isolated occurrences and an average	0	0·5
2 samples of isolated occurrences	0·2	0·5
A binomial sample and an average	0·2	0·5

Interpretation

The answer to the above calculation, z, is the difference between the 2 values being compared, expressed in terms of standard deviations. The larger z is, the less the likelihood of our no-significant-difference assumption being correct.

The critical values of z are as follows –

z TABLE

Probability of no significant difference between E and x			
P = 10%	P = 5%	P = 1%	P = 0·2%
$z = 1·64$	1·96	2·58	3·09

Adapted from E. S. Pearson and H. O. Hartley, *Biometrika Tables for Statisticians*, Vol. 1, Table 4 (CUP, 1966).

The significance of these probability levels was discussed on page 140.

Examples

Ex. 26. Suppose it has been found that the average number of telephones going out-of-order in a town is 7 per day. If the number of telephones remains the same, would the finding of 67 telephones out-of-order in 1 week be likely to be due to chance, or should some physical cause be sought?

Data: $x = 67$
$n = 7$ days
$P_x = 7$ per day
$P_y = 1$ (because we are comparing a sample of isolated occurrences and an average)

Calculation: Don't fall for the temptation to reduce the sample number (x) to a daily figure of $67 \div 7 = 9\frac{1}{2}$ for comparison with the daily average of 7. No, what we want is the expected number (E) for a sample of this size, according to the average rate. Thus –

$$E = P_x \times n$$
$$= 7 \times 7$$
$$= 49$$

Our Poisson's Test is inapplicable because x is more than 40, so we proceed to calculate z. Since x is larger than E, c is 0·5.

$$z = \frac{|E - x| - 0\cdot5}{\sqrt{E \times P_y}}$$
$$= \frac{|49 - 67| - 0\cdot5}{\sqrt{49 \times 1}}$$
$$= \frac{17\cdot5}{7} = \underline{\underline{2\cdot50}}$$

Referring to the z Table opposite, we see that this almost reaches the 1% level of significance. The observed difference could therefore be expected to occur by chance almost once in 100 times, which is sufficiently remote that it would certainly seem desirable to seek a physical cause for the increase, possibly water in some underground connections.

Concerning the accuracy of this test, you may be interested to know that the exact Poisson probability for this data is 1·48%, whereas the precise probability indicated by this value of z is 1·24%.

Ex. 27. Mr Ling Fu thought up a wonderful idea for improving the blooms of his favourite orchid. He tried it out, but was disappointed to find that, instead of the usual 70 blooms, he only got 54. Could this be chance, or does it signify that his 'wonderful idea' was in fact inhibiting the development of the blooms?

Data: $x = 54$
$n = 1$ year
$P_x = 70$ per year
$P_y = 1$

Calculation: E is obviously 70. Since x is less than E, and we are dealing with a sample of isolated occurrences and an average, c is zero. Therefore –

$$z = \frac{|E - x|}{\sqrt{E \times P_y}}$$

$$= \frac{70 - 54}{\sqrt{70 \times 1}}$$

$$= \frac{16}{\sqrt{70}}$$

$$= \frac{16}{8 \cdot 37} \quad \text{(from Table of Square Roots)}$$

$$= 1 \cdot 91$$

Referring this answer to the z Table (p. 248) shows the probability of this being a chance occurrence is slightly more than 5%. A significant difference between the observed and expected number of blooms is therefore not proven.

The exact Poisson probability of this data is 5·56%. The specific probability of our calculated value of z is 5·60%. The probability that would be obtained if we had subtracted the usually-recommended correction factor of 0·5 turns out to be 6·40%.

Ex. 28. In a consignment of 10,000 army boots, 1,100 were substandard, which was an incidence of 11%. A great fuss was made, and the manufacturer promised to do better in future. The next batch of 150 boots contained 8 faulty boots. Is this significantly different from the previous average?

Data: $x = 8$
$n = 150$
$P_x = 11\% = \frac{11}{100} = 0\cdot11$
$P_y = 1 - 0\cdot11 = 0\cdot89$

Calculation: $E = P_x \times n = 0\cdot11 \times 150 = 16\cdot5$

Since we are comparing a binomial sample and an average, and x is less than E, the correction factor (c) to use is 0·2.
Therefore –

$$z = \frac{|E - x| - 0·2}{\sqrt{E \times P_y}}$$

$$= \frac{|16·5 - 8| - 0·2}{\sqrt{16·5 \times 0·89}}$$

$$= \frac{8·5 - 0·2}{\sqrt{14·7}}$$

$$= \frac{8·3}{3·83} \quad \text{(from Table of Square Roots)}$$

$$= \underline{\underline{2·17}}$$

The z Table (p. 248) shows this to have a probability of less than 5%. The observed difference is therefore probably significant; the manufacturer appears to have kept his promise.

Binomial Tables show the exact probability for this data to be 1·96%. The specific probability of our calculated value of z is 3·04%; this is 1·08% too high, which is unusually high for this test. However, had we subtracted the conventional correction factor of 0·5, the answer would have been a probability of 3·68% which is 1·72% too high.

Ex. 29. The news item read: 'SAFETY HELMETS HAVE REDUCED MOTOR-CYCLE FATALITIES. Since the beginning of 1961, it has been compulsory in Victoria for all motor-cycle and motor-scooter riders to wear safety helmets. A report concerning the effectiveness of this measure has just been released. In the 2 years before the introduction of this law, 65 motor-cyclists were killed in accidents on Victorian roads, whereas in the 2 years after this, only 31 were killed. After studying all relevant factors, it was concluded that this reduction in fatalities was directly attributable to the compulsory wearing of safety helmets.'

This certainly sounds convincing, doesn't it? But can we be sure that it wasn't just a matter of chance? Or maybe the reduction merely resulted from a smaller number of motor-cyclists on the roads. So I went to the original report for full details. Here are the essential figures; do they indicate a statistically significant reduction in fatalities, or not?

Years	Av. No. of registered motor-cycles in Victoria	No. of motor-cyclists killed in road accidents in Victoria
1959–60 (before legislation)	21,704	65
1961–62 (after legislation)	17,021	31
Totals =	38,725	96

(From L. Foldvary and J. C. Lane, *Journal of Australian Road Research*, September 1964, pp. 15–16.)

Data: $x = 31$
$n = 96$
$P_x = \dfrac{17,021}{38,725} = 0{\cdot}4395$
$P_y = 1 - 0{\cdot}4395 = 0{\cdot}5605$

Calculation: If, as we tentatively assume, there is no significant difference between the 2 samples, the distribution of the 96 fatalities ought to be in proportion to the number of motor-cycles on the roads. Therefore, the expected number (E) in the second (smaller) sample would be –

$$E = P_x \times n = 0{\cdot}4395 \times 96 = 42{\cdot}2$$

Next, since our problem is one of comparing 2 samples of isolated occurrences, and $x < E$, the Table of Correction Factors on p. 248 shows $c = 0{\cdot}2$.
Therefore –

$$z = \frac{\mid 42{\cdot}2 - 31 \mid - 0{\cdot}2}{\sqrt{42{\cdot}2 \times 0{\cdot}5605}}$$

$$= \frac{11{\cdot}2 - 0{\cdot}2}{\sqrt{23{\cdot}7}}$$

$$= \frac{11{\cdot}0}{4{\cdot}87}$$

$$= \underline{\underline{2{\cdot}26}}$$

Turning now to the z Table on page 248, we see that this value of z indicates that the probability of our no-significant-difference assumption being true is less than 5%. It is to be concluded, then, that the wearing of safety helmets probably has reduced the risk of death.

The exact probability of this value of z is 2·38%. The true binomial probability is 2·33%, so once again we see that the zI Test has given a satisfactory answer.

Questions

Q68. Pablo's hobby was fishing. At last he found a superb spot, and week after week he brought home to his family and friends an average of 80 fish. Then one week he only caught 43 fish. Might this be reasonably attributed to bad luck, or is it more likely that his supply was drying up?

Q69. 'Daddy, why does this die always seem to show a six when I specially don't want it to?' asked my younger daughter as she was once more obliged to slide down a snake in a game of Snakes and Ladders. I examined the die, and found that the dots on its faces had been made by drilling superficial holes and partly filling these holes with paint. This makes the face with the six dots lighter than the others, which can cause this face to land uppermost more often than it should. So I decided to test the die for bias, and resolved to get rid of it if the difference from expectation reached the 5% probability level. (It's generally a good idea to make a decision as to the probability level you'll accept before doing any experiment.) I then threw the die 108 times, and got 25 sixes. Did we keep the die?

Q70. A police report states that in the past month there have been 25 burglaries in the northern half of a city, in contrast with the southern half in which there have been 25 burglaries in the past 3 months. Is this difference statistically significant?

50% PROBABILITY TEST (1713)

Purpose

This test will compare any 2 numbers of things which are expected to differ from one another only by chance.

It will therefore compare –

(*a*) the 2 classes of a random *binomial sample* (p. 29), when the expected number in each class is the same, or

(*b*) 2 random samples of *isolated occurrences* (p. 31), when the expected number in each sample is the same, or

(*c*) 2 sets of *matched observations* (p. 118), when the observations are classified into categories (as distinct from matched observations of measurements) to see if there is a significant difference between the 2 sets. The matched observations may be made either on matched pairs or on a single random sample group.

Principle

This test is merely an extended version of the Binomial Test (p. 237) for cases in which the known or expected average is 50%. A 50% Binomial Table is provided which gives immediate answers for cases in which the smaller number of the pair being compared is up to 40. Beyond that point the zI Test (p. 245) could be used, and with the conventional correction factor of 0·5 it gives results which are correct to within 0·1% of the true binomial probability. However, answers which are identical to those of the zI Test can be obtained with the 50% χ^2 Test, which is even easier to do. The nature of χ^2 will be described later (p. 269); for the moment it is sufficient to know that it is a kind of cousin to the z of the zI Test. Like the latter, χ^2 is based on the Normal Probability Formula.

With a *binomial sample*, this significance test works as follows. Consider a large parent group made up of equal numbers of 2 classes of things, say males and females. If a sample group is chosen at random from this parent group, we would expect to find roughly equal numbers of each class in the sample. Of course, some inequality would be the rule rather than the exception, and would be due to chance. We know, too, that small variations from equality would occur commonly, whereas large variations would occur uncommonly.

On page 30, using the number of heads and tails in a set of coin tosses for our example, we saw that the probability of chance causing any particular degree of inequality in such cases can be calculated exactly by means of the Binomial Formula of Jacques Bernoulli. If the sample contains 100 or more observations, the Normal Probability Formula provides a perfectly adequate substitute which is much easier to compute.

We start, then, by assuming that the sample has been drawn from a binomial parent group whose proportions are 50%, and ascribe any disproportion in the sample group to chance. We then determine the probability of this being the case, using the Binomial Formula or the Normal Probability Formula according to the sample size. If this indicates that our tentative assumption is very unlikely to be true, we prefer to accept that the sample did not derive from such a parent group; otherwise, we provisionally accept our original assumption as being not disproven.

This test can also be applied to comparing 2 samples of *isolated occurrences* if the expected number in each sample is the same. This is the case when both samples have the same duration and are drawn from parent groups of the same size. Thus we could expect an equal number of accidents to occur in 2 cities of the same size in the same period of observation. Any difference from equality can then be tentatively put down to chance. We deny this hypothesis if the probability turns out to be remote. We find this probability by treating the number of occurrences in one sample as if they formed one class of a binomial sample (say, heads), and the occurrences in the other sample as if they formed the other class of the binomial sample (i.e. tails). Then, by assuming that this binomial sample has been drawn from a parent group of 50:50 proportions, we can then test any discrepancy from equality between the 2 sample groups in the same way as we would deal with coin tossing. The conventional way of comparing 2 samples of isolated occurrences is to use a 2-cell χ^2 Test with Yates' correction, but this 50% Probability Test gives identical answers with large samples, and more accurate answers with small samples, and is even easier to do into the bargain.

The differences between *matched categorized observations* form a third group to which this 50% Probability Test applies. The matching may be achieved by –

(*a*) pairing members of 2 sample groups, carefully, or symmetrically, or by splitting (see p. 118), or (*b*) using 1 sample group for different treatments, different methods of testing, different observers, or on different occasions (see p. 120).

The virtue of matched observations is that the degree of variability that normally exists between independent groups is greatly reduced, which enables significant differences between the 2 test conditions to be detected much more readily. We saw this with measurements in the petrol–mileage experiment (p. 116), and it only remains to be said that exactly the same advantages obtain with matched observations which fall into classes.

To examine 2 sets of matched categorized observations for the presence of a significant difference, it is necessary to have the results in 3 classes, viz. –

(*a*) those cases showing no difference between the 2 test conditions,
(*b*) those showing a difference in one direction, and
(*c*) those showing a difference in the other direction.

For example, to find out which is the better of 2 ointments for treating dermatitis, we could take a random sample of patients suffering from dermatitis on symmetrical areas of their bodies, and get them to apply one ointment to one side, and the other ointment to the other side. After a period of such treatment, the relative effectiveness of the 2 ointments could then be noted as 'Prefers *A*', or 'Prefers *B*', or 'No Difference' (i.e. both equally good, bad, or indifferent). We now make the usual assumption that there is no significant difference between the 2 sets of results; any observed difference is provisionally ascribed to chance. If this is really the case, any patients preferring *A* ought to be balanced by about the same number preferring *B*, and any inequality between these numbers can be tested by the 50% Probability Test.

This raises an interesting point. If we test only those who show a difference between the 2 test conditions, it would seem as though we were ignoring the information provided by the group who show no difference. Not only would this be wasteful, but surely if a large number of cases were in this category, it ought to reflect a lessened degree of difference between the 2 test conditions. In fact, such worrying thoughts are merely an illusion. These cases are not wasted, for the larger this no-difference group, the smaller will be the group of remaining cases, and this has the effect of making it harder for any inequality among these remaining cases to register as a significant difference. The reverse applies if the no-difference group is small. You will get this idea more clearly if you imagine that we are testing a taped stack of coins for bias, and this particular coin stack can land heads, tails, or balanced on its edge. The more times it lands on its edge, the less times there are left for either heads or tails to predominate. If the coin stack was so thick that it landed on its edge 980 times out of 1,000 tosses, one would expect the remaining 20 results to consist of about 10 heads and 10 tails. But *if* these remaining results were all heads, the result would be statistically significant in that such an outcome would be unlikely to be due to chance; in so far as a bias is demonstrated, it is decidedly a bias towards heads, but the slightness of the bias would be indicated by the fact that it would take as many as 400 tosses to show statistical significance at the 1% level. And it would take many more than 400 tosses if the situation was not uniformly biased in one direction. So don't think that the no-difference group is not exerting its due effect, even though we ignore its size in the actual calculation. (You will recall that we did exactly the same thing in the Wilcoxon's Signed Ranks Test, p. 180).

Data Required

  With a binomial sample –

  x = number of cases in the smaller class of the sample.
  y = number of cases in the larger class of the sample.
  $n = x + y$ = sample size.

  With 2 samples of isolated occurrences –

  x = number of occurrences in the smaller sample.
  y = number of occurrences in the larger sample.
  $n = x + y$ = number of occurrences in the 2 samples.

  With 2 sets of matched observations –

  x = number of cases in the smaller of the 2 groups showing a difference between the 2 test conditions.
  y = number of cases in the larger of the 2 groups showing a difference between the 2 test conditions.
  $n = x + y$ = total number of cases showing a difference between the 2 test conditions (*not* the sample size).

Note: In all cases, n must be at least 6 before a 5% probability level can be reached.

Procedure

(a) *If x = 40 or less –*

No calculation is required. Simply refer to the *n* Table for the 50% Probability Test opposite. Find the line corresponding to the observed value of *x*, and then run across the Table to place your value of *n*.

The numbers in the body of this Table are the critical values of *n* for various values of *x*. The Table is really only an extension of the $P_x = 50\%$ part of the Binomial Table (p. 241). The probabilities indicated are, of course, two-sided (p. 143).

(b) *If x = more than 40 –*

Calculate the value of χ^2 using the following formula –

$$\chi^2 = \frac{(|\,x - y\,| - 1)^2}{n}$$

This simple formula is algebraically identical with that used in a 2-cell χ^2 Test with Yates' correction, when the known or expected average is 50%. Its enunciation and use is due to Professor Quinn McNemar of Stanford University (*Psychometrika Journal*, 1947, pp. 153–7).

Interpretation

The higher the value of χ^2, the less the likelihood of our no-significant-difference assumption being correct.

The critical values of χ^2 for this test are as follows –

50% χ^2 TABLE

Probability of no significant difference between x and y			
P = 10%	P = 5%	P = 1%	P = 0·2%
$\chi^2 = 2\cdot71$	3·84	6·63	9·55

Adapted from E. S. Pearson and H. O. Hartley, *Biometrika Tables for Statisticians*, Vol. 1, Table 8 (CUP, 1966).

The significance of these probability levels was discussed on page 140.

n TABLE FOR 50% PROBABILITY TEST

x	P = 10%	P = 5%	P = 1%	P = 0·2%
0	5	6	8	12
1	8	9	12	14
2	11	12	15	18
3	13	15	18	21
4	16	17	21	24
5	18	20	24	27
6	21	23	26	30
7	23	25	29	33
8	26	28	32	35
9	28	30	34	38
10	30	33	37	41
11	33	35	39	43
12	35	37	42	46
13	37	40	44	49
14	40	42	47	51
15	42	44	49	54
16	44	47	52	56
17	47	49	54	59
18	49	51	57	61
19	51	54	59	64
20	53	56	61	66
21	56	58	64	69
22	58	61	66	71
23	60	63	69	74
24	62	65	71	76
25	65	67	73	79
26	67	70	76	81
27	69	72	78	83
28	71	74	80	86
29	74	77	83	88
30	76	79	85	91
31	78	81	87	93
32	80	83	90	95
33	82	86	92	98
34	85	88	94	100
35	87	90	97	102
36	89	92	99	105
37	91	94	101	107
38	93	97	104	110
39	96	99	106	112
40	98	101	108	114

Adapted from *Tables of the Cumulative Binomial Probabilities*, prepared and published by the USA Ordnance Corps, 1952.

Examples

Ex. 30. A coin is tossed 20 times, and shows 14 heads and 6 tails. Do these figures suggest that the coin or tossing technique is biased, or could such a departure from average be expected with fair frequency purely as a result of chance?

Data: x, the smaller class $= 6$
y, the larger class $= 14$
$n = x + y = 6 + 14 = 20$

Calculation: Since x does not exceed 40, no calculation is needed. Just refer to the 50% Probability Table (p. 259), and you will see that for a value of $x = 6$, a value of $n = 20$ indicates a probability of more than 10%. In other words, if on 10 occasions you tossed an unbiased coin 20 times, you could expect to get a result differing from the average by this extent (i.e. either 14 heads and 6 tails, or 6 heads and 14 tails) in at least 1 of those trials. A significant difference from average (or if you prefer, a significant difference between the number of heads and tails) is not proven; the coin is not proven to be biased.

Ex. 31. In 1 year, 13 people are accidentally shot while duck shooting, so a new law is introduced in an attempt to reduce this dreadful state of affairs. In the next year there are 5 such accidents. Assuming an equal number of duck shooters in both years, do these figures indicate that the new law is effective, or might the reduction be merely a result of chance?

Data: x, the smaller sample number $= 5$
y, the larger sample number $= 13$
$n = x + y = 5 + 13 = 18$

Calculation: Since x does not exceed 40, we can get the answer immediately from the n Table for the 50% Probability Test (p. 259). This shows that for $x = 5$, a value of $n = 18$ corresponds to a probability of 10%. The difference between the 2 accident rates is therefore judged as not proven to be statistically significant.

Ex. 32. As an example of matched observations made on a single sample group, 75 people were invited to taste 2 kinds of wine, and to name which they preferred. It was found that 25 preferred one kind, and 50 preferred the other. Is this result to be attributed to chance, or does it suggest a significant difference in popularity between the 2 wines?

Data: x, the smaller number $= 25$
  y, the larger number $= 50$
  $n = x + y = 25 + 50 = 75$

Calculation: Reference to the 50% Probability Table shows that for $x = 25$, a value of $n = 75$ has a probability of less than 1%. It is therefore almost certain that the results were not due to chance, but indicate a real preference for the second wine.

Ex. 33. At a discussion recently, an engineer maintained that Australia's own car, the Holden, was now so popular that fully 50% of the cars on Australian roads were Holdens. To check this statement, I asked my daughters to take a census of cars passing our home, with the following result –

Holdens	not-Holdens	Total
253	320	573

Assuming that our sample is equivalent to a random sample, are these figures consistent with the engineer's statement, i.e do they differ significantly from a 50:50 proportion or not?

Data: x, the smaller class $= 253$
  y, the larger class $= 320$
  $n = x + y = 253 + 320 = 573$

Calculation: Since x is more than 40, we calculate χ^2, thus –

$$\chi^2 = \frac{(|x - y| - 1)^2}{n}$$

$$= \frac{(|253 - 320| - 1)^2}{573}$$

$$= \frac{(67 - 1)^2}{573}$$

$$= \frac{66^2}{573}$$

$$= \frac{4{,}356}{573} \quad \text{(from Table of Squares)}$$

$$= \underline{\underline{7 \cdot 60}}$$

Referring this value of χ^2 to the 50% χ^2 Table on page 258 shows that the probability of there being no significant difference between the number of Holdens and of not-Holdens is less than 1%. The difference is therefore held to be significant, and the engineer's statement must be considered somewhat exaggerated.

Ex. 34. Two makes of radar equipment are used with equal frequency at an airport. Over a 12-year period, Type A has developed faults on an average of 6 times a year, and Type B on an average of 8·5 times a year. Do these results indicate that Type A is significantly superior to Type B in this respect, or might the observed differences be reasonably ascribed to chance?

Data: x, the smaller sample $= 12 \times 6 = 72$ (We must use the actual numbers, not just the averages, for this test.)

y, the larger sample $= 12 \times 8\cdot5 = 102$

$n = x + y = 72 + 102 = 174$

Calculation: Since x is more than 40, we calculate χ^2, thus –

$$\chi^2 = \frac{(|\,72 - 102\,| - 1)^2}{174}$$

$$= \frac{(30 - 1)^2}{174}$$

$$= \frac{29^2}{174}$$

$$= \frac{841}{174} \quad \text{(from Table of Squares)}$$

$$= \underline{\underline{4\cdot83}}$$

The 50% χ^2 Table on page 258 shows this value has a probability of occurring by chance of less than 5%, so the difference is probably significant.

Ex. 35. The research design called paired comparison (p. 115) is so important that it will bear a more extended illustration. Suppose a doctor wants to compare a new tranquilliser tablet with an established one. The amount of variation between one anxious patient and the next is apt to be considerable, so the experimental design is tightened up by testing both the new and the old remedy on each patient in a random or presenting sample group (rather than treating one group with one drug, and another independent group with the other drug). Some patients, decided by a Table of Random Numbers, are given the new tablet first, while others start on the old tablet; this is necessary to equalize any tendency for the second course of tablets to get a flying start as a consequence of any benefit derived from the first course of tablets.

It is virtually impossible to measure the degree of improvement in a clinical trial of this kind. The only results that can be obtained and trusted are the patients' straightforward comments of 'felt better', 'felt worse', or 'felt about the same as before'. Our question being one of preference between 2 remedies, we should therefore classify our results as follows –

Result	No. of patients
Prefers A	3
Prefers B	11
No difference	4
Total =	18

Data: x, the smaller group showing a preference $= 3$
y, the larger group showing a preference $= 11$
$n = x + y = 3 + 11 = 14$

Note that we exclude the 4 patients who found no difference between the remedies, for the reason discussed on page 256.

Calculation: If there is no significant difference between the 2 tranquillizers, one would expect that of those who do show a preference for one or other of them, that about half would prefer one, and half would prefer the other. But knowing that chance variations can occur, as with tosses of a coin, we need to find out, 'What is the chance of getting 3 and 11 in a group of 14 when the expectation is 7 and 7?'

Since x is only 3, we can get the answer directly from the 50% Probability Table (p. 259). There we find that $n = 14$ indicates a probability of more than 5%, so a significant difference is not proven by the data on hand.

Comment: You may not be satisfied with this answer. Perhaps most of the patients who preferred Tablet *B* were very enthusiastic in their praise of it, while those preferring Tablet *A* tended to be rather lukewarm about the issue. What do you do? Continue the investigation. It may simply be a matter of collecting more evidence.

If, when the sample size has been doubled, the proportion of patients favouring Tablet *B* remains the same as before, the results would be –

Result	No. of patients
Prefers *A*	6
Prefers *B*	22
No difference	8
Total =	36

Of the 28 patients who show a preference for one or the other remedy, the smaller group now contains 6. The *n* Table for the 50% Probability Test shows that for $x = 6$, a value of $n = 28$ has a probability of less than 1%. The results now show that a significant difference is present.

In some cases it is reasonable to assign a points score to each remedy according to the degree of benefit produced. Thus no benefit would score 0, slight benefit 1, moderate benefit 2, and great benefit 3. The relative magnitude as well as the direction of differences then enables a more powerful statistical test – Wilcoxon's Signed Ranks Test (p. 179) – to be applied. However, considerable care must be exercised in doing this, for it may lead to a false degree of accuracy. There will always be trouble classifying those patients who report, 'Some days I feel fine, but other days I'm as nervy as ever.' Then there is the problem of judging patients' responses consistently when the trial extends over a period of several months. Again, it may lead to results which look more accurate than they really are when half a dozen doctors co-operate and pool their results, for some doctors may judge the results more stringently than others. And last, but not least, the scoring scale must be designed with intervals spaced as uniformly as possible, for the nature of Wilcoxon's Signed Ranks Test is such that it is really only valid when the measurement units belong to an equi-intervalled scale (p. 52).

However, let us see how this works before going any further. Here are the results of the original 18 patients, awarded points as described above –

Patient No.	Response to A	Response to B
I	1	3
II	0	3
III	2	1
IV	0	2
V	3	3
VI	3	1
VII	0	3
VIII	2	1
IX	0	0
X	0	1
XI	1	2
XII	2	3
XIII	1	3
XIV	2	2
XV	0	2
XVI	0	3
XVII	0	1
XVIII	1	1

Q71. Apply Wilcoxon's Signed Ranks Test to this data. How does the result compare with that obtained by the 50% Probability Test?

Ex. 36. There were 200 entries in the Annual Art Competition last year. The judges were the Mayor and the Police Commissioner. In order to simplify matters, they agreed to select a small group of finalists from the whole group, and to this end they independently classified each painting into one or other of 3 categories – 'very good', 'average', and 'terrible'. Their results thus formed 2 sets of matched observations on a single sample group.

The tabulation of such results calls for a special kind of table. The structure deserves close attention. Look –

		Mayor's Judgements			Row totals
		V.G.	Av.	Terr.	
Police	V.G.	⑩	9	1	20
Commissioner's	Av.	5	⑩⑤	25	135
Judgements	Terr.	0	5	㊵	45
Column totals		15	119	66	200

You will notice that $10 + 105 + 40 = 155$ of the entries were given the same rating by both judges. In a like manner we can extract the fact that $5 + 0 + 5 = 10$ entries were given a higher rating by the Mayor than by the Police Commissioner, while $9 + 1 + 25 = 35$ entries were rated higher by the Police Commissioner.

Now the question is, is there a significant difference between the standards of the 2 judges, or is it reasonably likely that the observed differences could simply be due to chance?

Data: $x = 10$
$\qquad y = 35$
$\qquad n = x + y = 45$

Calculation: No further calculation is needed, because x does not exceed 40. The 50% Probability Table shows that for $x = 10$, a value of $n = 45$ has a probability of less than 0·2%. It can therefore be concluded that the observed difference is not likely to be due to chance, so it must represent a real difference in the judges' standards.

This same method can also be used on tables with more than 3 categories.

Questions

Q72. Lady Poppingate silenced the dinner party conversation with her question, 'I've heard, young man, that you have clairvoyant powers. You see through closed doors and things, don't you?' The shy, bearded fellow admitted that this was so.

'Tommy-rot! Impossible!' said Sir Alfred, 'though I must admit I'd like to see you try.' So they arranged a simple experiment. They took 2 packs (decks) of ordinary playing cards, shuffled them, and spread them out in neat rows, face down, on the library floor, and asked the young man to write down the order of red and black cards, aided only by his sixth sense. (They decided on this design to avoid any ambiguities, such as might arise if he was asked to name the cards and claimed that '7 of spades' was near enough if the correct answer was '8 of spades'.)

The young man got 67 out of the 104 answers correct. Is it likely that he was just guessing? What do you think of the design of this experiment?

Q73. Catsear and Bark are veterinary surgeons, each with the same sized practice. In 5 years, Catsear sustains occupational injuries 6 times. In the same period, Bark gets bitten 12 times. Should Bark get advice from his colleague on how to reduce the number of his injuries, or might he simply have been having a run of bad luck?

Q74. Potkins and Smith are the only contenders for a certain seat in Parliament. A week before the election, Potkins has a survey carried out on 200 people chosen at random in the electorate, in order to assess his chance of winning the seat. The results were as follows –

> Pro-Potkins 86
> Pro-Smith 100
> Undecided 14

Assuming that Potkins may not get any of the 'undecided' voters' votes, is the outlook for his success dismal or not?

Q75. Eric always uses an exposure meter to determine his camera settings when taking colour photographs. Jack never does; he just relies on his visual judgement. After arguing the relative merits of both methods, they decide to put the matter to an actual test. Each will take 36 colour photos of a variety of subjects with his own technique, and the results will be compared when the films have been processed. To keep the test fair, they take matching pictures, with each subject photographed by each at the same time, in the same light, from the same angle, and so on.

Jack got 27 perfect exposures, while Eric got 33. 'There you are,' said Eric, 'that proves my method is better than yours.'

Well, does it really prove it? Here is the analysis of their results in tabular form –

Eric's exposures	Jack's exposures	
	Perfect	Wrong
Perfect	25	8
Wrong	2	1

Q76. Suppose that 1,000 people attend a film featuring the movie star, Brigitte Bardot. As they enter the theatre, each person is asked what they think of her acting ability (on the evidence of her previous performances). After the film each person is asked whether they still hold the same opinion of her ability or not. Would the following figures indicate that this film increased a significant number of peoples' opinions of her acting ability, or might they just be the result of chance variation?

Opinion unchanged	Opinion changed	
	to good	to bad
700	200	100

χ² TEST (1900)

Purpose

This test compares any 3 or more groups or classes of things. It will thus compare –

(*a*) a set of 3 or more random samples of *isolated occurrences* (such as accidents), or

(*b*) the numbers in each class of a random sample composed of *3 or more mutually exclusive classes* (such as the outcome of a sample of die throws), or

(*c*) 2 or more random samples, each composed of *3 or more classes*, or

(*d*) 3 or more random *binomial samples* (each composed of 2 classes, such as faulty and non-faulty), or

(*e*) *matched observations* made on a single random sample group, for evidence of association between 2 qualities (such as brains and beauty), when at least 1 of these qualities is divided into 3 or more categories (such as beautiful, average, and ugly).

In all cases, the *expected number* should be 5 or more in each class or category.

Principle

The symbol χ^2 ('chi-squared' – pronounced ki-squared) stands for the sum of the squares of all the standardized differences (z) from the expected frequencies in a group of categorized observations. The value of χ^2 for any given situation is calculated from the formula –

$$\chi^2 = \text{the sum of all values of } \frac{(O - E)^2}{E}$$

where $O =$ the observed frequencies, and $E =$ the expected frequencies.

Through its relationship to z, this test is seen to be a derivative of the Normal Probability Formula (p. 34), the usefulness of which is thus extended still further afield. However, the Normal Probability Formula applies strictly only to continuous variables (such as measurements), and when applied to things which can only be whole numbers (as in the zI Test and the χ^2 Test), the answers are apt to differ slightly from the true probabilities. To get the true probabilities, however, requires quite complicated calculations, whereas the χ^2 Test is simple to do, and gives answers which are correct to within about 1% at the 5% probability level, and to within $\frac{1}{2}\%$ at the 1% probability level. This degree of accuracy is perfectly adequate for all practical purposes.

The theoretical calculations underlying the χ^2 Test were first worked out in 1875 by a German physicist called Helmert. However, it was not until 1900 that an English statistician, Professor Karl Pearson, showed how these calculations could be put to use as a significance test (*Phil. Mag.*, 1900, Series 5, **50**, pp. 157–72).

When the χ^2 Test is used for assessing the significance of an observed *difference* between independent, *unmatched samples*, the usual tentative assumption is made that there is no significant difference between them, and the probability of this being the case is then calculated to find out if this assumption is reasonable or not. This involves comparing the observed numbers with the expected numbers in each class or category; the expected numbers are worked out by simple proportions, pooling the results of the various samples in accordance with our original hypothesis that there is no significant difference between the samples.

The way of estimating the expected numbers will be understood best by an example. A group of 818 people who were exposed to cholera in Calcutta in 1894–6 was studied. Of this group, 279 had been inoculated with Haffkine's anti-cholera vaccine, while the remaining 539 had not been so inoculated. The study group was thus divisible into 2 independent sample groups, the inoculated and the non-inoculated. The number of persons who contracted cholera in each sample group was then noted, so that the 818 persons were finally classified into 4 categories – the inoculated persons who caught cholera, the inoculated ones who didn't catch it, the non-inoculated ones who did, and the non-inoculated ones who didn't. In all cases like this, the data is correctly displayed in what is called a *contingency table for unmatched samples*, thus –

	Infected with cholera	Not infected	Row totals
Inoculated	3	276	279
Non-inoculated	66	473	539
Column totals	69	749	818

(From M. Greenwood and G. U. Yule, *Proc. Royal Soc. Med.*, Epidemiology Section, 1915, p. 122.)

As you can see, 69 of the 818 people became infected with cholera. Now, if the inoculation against cholera was useless, the cholera-rate in the inoculated and non-inoculated groups would be the same (apart from any chance variations). Knowing that the proportion of inoculated persons in the whole study group is $\frac{279}{818}$, we can now proceed to say that if there is no significant difference between the inoculated and non-inoculated groups, the inoculated group ought to have $\frac{279}{818}$ of the observed cholera cases. This proportionate share, $\frac{279}{818}$ of 69, works out to be 23·5; the remainder, $69 - 23\cdot5 = 45\cdot5$, ought to be the number of cases of cholera in the non-inoculated group. In fact, only 3 occurred in the first category (where 23·5 were expected), and 66 occurred in the second category (where 45·5 were expected). In exactly the same way, we say that of the 749 persons who did not get cholera, one would expect $\frac{279}{818}$ of $749 = 255\cdot5$ to be in the inoculated group, and the remainder, $749 - 255\cdot5 = 493\cdot5$, to be in the non-inoculated group. These proportionate calculations give us the expected numbers in each category, and these expected numbers are then used for determining the value of χ^2.

The above contingency Table is called a 2×2 table, for there are 2 columns and 2 rows of categories. It serves nicely to demonstrate how expected numbers are calculated, even though the ordinary χ^2 Test applies only to 3×1 tables (i.e. with 3 columns and 1 row) or larger.

The χ^2 Test can also be applied to assessing the part played by chance in producing apparent *association* between 2 qualities or characteristics, when at least 1 of these qualities is divided into 3 or more categories. In such cases, we are necessarily dealing with *matched observations* made on a single sample group. For instance, to test for a possible association between brains and beauty in the females of our species, we could examine a random sample of females concerning both of these features. To keep the discussion simple, suppose it is decided to classify the ladies into 3 categories of intelligence (high, average, or low), and into 2 categories of beauty (attractive or plain). As always, the results of matched observations must be put into a table which shows one characteristic spread across the top of the table, and the other characteristic down the left of the table. Such a *contingency table for matched observations* looks like this –

		Intelligence			Row totals
		High	Average	Low	
Beauty	Attractive	a	b	c	$a+b+c$
	Plain	d	e	f	$d+e+f$
Column totals		$a+d$	$b+e$	$c+f$	$a+b+c$ $+d+e+f=N$

In this 3×2 table, each of the 6 compartments, or cells, a to f, would contain the number of ladies observed to have each particular combination of the characteristics being examined. Thus, cell a would contain those who were both highly intelligent and attractive; cell b would contain the number who were of average intelligence and were attractive, and so on. (You will recall that we have met this kind of table before, on page 266, when we were testing for a difference between matched observations.)

To test the significance of an apparent association between categorized things, the tentative negative assumption is that they are quite unrelated, and the probability of getting the observed results by chance is determined by calculating χ^2, first estimating the expected number in each cell of the table using the same method as described above for unmatched samples.

Notice that association between categorized things is the counterpart of correlation between measurements, for which Spearman's Test is used (p. 199). As in the case of correlation, the demonstration that an observed association is statistically significant (and not just likely to be due to chance) must never be accepted as proof that one feature causes the other, for a third feature may be the cause of both of them. Association tells us when things tend to go together; proof of cause demands the production of the effect by the causal agent under strict experimental conditions.

Data Required

  a, b, c, etc. = the numbers of instances in each category or class.

If the data is quoted as percentages, convert to actual numbers of things for this test.

Procedure

(1) Prepare a data table, complete with row totals (R), column totals (C), and grand total (N).

For unmatched, independent samples, the structure of the data table will be like this –

	I	Classes II	III	Row totals
Sample A	a	b	c	$a+b+c$
Sample B	d	e	f	$d+e+f$
Column totals	$a+d$	$b+e$	$c+f$	N

For matched observations made on 1 sample group, the structure of the data table will be like this –

		Characteristic A ++ + −		Row totals
Characteristic B $\{$	$+$	a	b	c	$a+b+c$
	$-$	d	e	f	$d+e+f$
Column totals		$a+d$	$b+e$	$c+f$	N

(2) Now prepare a table for calculating χ^2, with the following headings –

O	$\dfrac{R \times C}{N} = E$	$\lvert O - E \rvert = D$	D^2	$\dfrac{D^2}{E}$

(3) In the first column of the calculation table, labelled O, list the observed numbers in each cell of the data table (a, b, c, etc.). Do this systematically, working across each row in turn.

(4) In the next column, calculate the expected number (E) for each cell, as follows –

(a) When a set of N observations is being compared with some known or expected proportions, if the expected proportion in any cell is P, the expected number (E) in that cell will be –

$$E = P \times N$$

(b) When 2 or more categorized sets of observations are being compared with one another, the expected number in each cell is given (for reasons described on p. 271) by –

$$E = \frac{\text{Row total}}{\text{Grand total}} \times \text{Column total} = \frac{R \times C}{N}$$

Remember, the χ^2 Test tends to be inaccurate unless all the cells have an expected number of 5 or more. If E is less than 5 in any cell, pool the results of 2 or more adjacent cells to raise the value of E to the necessary level.

It is a good idea to check your arithmetic at this stage by adding all the calculated values of E; this sum must always equal the grand total, N.

(5) By subtraction, find the difference (D) between each observed and expected value.

(6) Square each value of D, using the Tables of Squares at the back of this book.

(7) Divide each value of D^2 by its corresponding value of E.

(8) Add up all the answers in the last column. This total is χ^2.

Interpretation

If our tentative assumption of no significant difference or association is correct, each value of $| O - E |$ will tend to be small, so χ^2 will also tend to be small. Contrariwise, the higher the value of χ^2, the less the likelihood of our assumption being true.

The critical values of χ^2 vary with the number of *degrees of freedom* (D.O.F.) in the data being tested. The degrees of freedom refer to the number of observations which are free to vary according to restrictions inherent in the data. For instance, if a group of 40 articles is divided into 2, say one group of 15 and another consisting of the remainder, there is only 1 degree of freedom, because the size of the remainder-group is automatically determined by the fixing of the size of the first group. But if the group of 40 articles was divided into 3 groups, one of which contained 15, while the partition of the remainder was left to chance, there would be 2 degrees of freedom.

The degrees of freedom for the χ^2 Test can be found quite easily as follows –

(*a*) When a single set of numbers is being compared with known or expected proportions, the data table will consist of 1 row divided into c classes or groups. For such $c \times 1$ tables –

$$\text{D.O.F.} = c - 1$$

(*b*) When 2 or more sets of numbers are being compared with each other, the data table will consist of c columns and r rows. For such $c \times r$ tables –

$$\text{D.O.F.} = (c - 1)(r - 1)$$

The critical values of χ^2 for up to 100 degrees of freedom are given in the χ^2 Table on page 276. To obtain critical values of χ^2 for degrees of freedom other than those shown in the range 30 to 100, use simple proportions. Thus, for 35 degrees of freedom, the critical values of χ^2 are midway between those listed for 30 and 40 degrees of freedom.

This χ^2 Table tells the probability of the observations yielding a value of χ^2 as high as calculated, if there is no significant difference or association between them. The interpretation of these probability levels was discussed on page 140.

χ^2 TABLE

Degrees of freedom	P = 10%	P = 5%	P = 1%	P = 0·1%
2	4·61	5·99	9·21	13·82
3	6·25	7·81	11·34	16·27
4	7·78	9·49	13·28	18·47
5	9·24	11·07	15·09	20·52
6	10·64	12·59	16·81	22·46
7	12·02	14·07	18·48	24·32
8	13·36	15·51	20·09	26·12
9	14·68	16·92	21·67	27·88
10	15·99	18·31	23·21	29·59
11	17·28	19·68	24·73	31·26
12	18·55	21·03	26·22	32·91
13	19·81	22·36	27·69	34·53
14	21·06	23·68	29·14	36·12
15	22·31	25·00	30·58	37·70
16	23·54	26·30	32·00	39·25
17	24·77	27·59	33·41	40·79
18	25·99	28·87	34·81	42·31
19	27·20	30·14	36·19	43·82
20	28·41	31·41	37·57	45·31
21	29·62	32·67	38·93	46·80
22	30·81	33·92	40·29	48·27
23	32·01	35·17	41·64	49·73
24	33·20	36·42	42·98	51·18
25	34·38	37·65	44·31	52·62
26	35·56	38·89	45·64	54·05
27	36·74	40·11	46·96	55·48
28	37·92	41·34	48·28	56·89
29	39·09	42·56	49·59	58·30
30	40·26	43·77	50·89	59·70
40	51·81	55·76	63·69	73·40
50	63·17	67·50	76·15	86·66
60	74·40	79·08	88·38	99·61
70	85·53	90·53	100·42	112·32
80	96·58	101·88	112·33	124·84
90	107·56	113·14	124·12	137·21
100	118·50	124·34	135·81	149·45

Adapted from E. S. Pearson and H. O. Hartley, *Biometrika Tables for Statisticians*, Vol. 1, Table 8 (CUP, 1966).

Examples

Ex. 37. On page 138 was mentioned the case of the accident rate in 3 shifts at a factory. In a month there are 7 accidents in shift *A*, 7 in shift *B*, and 1 in shift *C*. Is there a significant difference between these rates?

Data:

No. of Accidents in Shift			Total
A	*B*	*C*	(*N*)
7	7	1	15

Calculation: In this example we are comparing 3 samples of isolated occurrences. Since the sample duration was the same for each, if there was no significant difference between them we would expect an equal number of accidents in each shift; there being 3 shifts, we would therefore expect $\frac{1}{3}$ of the total number of accidents in each shift. $P = \frac{1}{3}$. We proceed with the tabular calculation, as shown below.

O	$P \times N = E$	$\lvert O - E \rvert = D$	D^2	$\dfrac{D^2}{E}$
7	$\frac{1}{3} \times 15 = 5\cdot00$	$2\cdot00$	$4\cdot00$	$\dfrac{4\cdot00}{5\cdot00} = 0\cdot80$
7	$\frac{1}{3} \times 15 = 5\cdot00$	$2\cdot00$	$4\cdot00$	$\dfrac{4\cdot00}{5\cdot00} = 0\cdot80$
1	$\frac{1}{3} \times 15 = 5\cdot00$	$4\cdot00$	$16\cdot00$	$\dfrac{16\cdot00}{5\cdot00} = 3\cdot20$
D.O.F. $= c - 1 = 2$				$\chi^2 = 4\cdot80$

To find the probability level of this value of χ^2, we turn to the χ^2 Table opposite, and there on the top line (which is the line relevant for 2 degrees of freedom) we see that a value of $4\cdot80$ indicates that the observed differences between the shifts could be expected by chance on more than 5% occasions. A significant difference is therefore not proven.

Ex. 38. To exemplify the comparing of numbers of instances in a multi-class random sample, with a set of known proportions of a parent group, consider the following case. I wanted to know if the ages of my patients were a representative cross-section of the age-groups in the community. My secretary took a systematic sample (p. 49) of 100 history cards from my files, and tallied the patients' ages into 6 classes. We then worked out the proportion of each of these classes in the Australian community from the 1960 Census (in the Commonwealth Bureau's 1962 Pocket Compendium). Our results are given in the table below; is there a significant difference between the age distribution of my patients and that of the community at large?

| | Age groups (years) | | | | | | Total |
	0–14	15–29	30–44	45–59	60–74	75–	
My patients	19	32	22	19	6	2	100
Australian proportions	30%	20%	21%	16%	10%	3%	100%

Calculation: If we were testing the above data to see if there was a significant difference between the numbers in each class in the sample, we would tentatively assume no significant difference, and compare each observed value with an expected value of $\frac{1}{6}$ of 100. However, this is not the case, for we are interested in comparing the observed number in each class with an expected number determined by the community proportions. (The comparison would otherwise be very unfair on elderly patients, who are relatively infrequent in the community.)

O	$P \times N = E$	$\lvert O - E \rvert = D$	D^2	$\dfrac{D^2}{E}$
19	30% of 100 = 30·00	11·00	121·00	$\dfrac{121}{30} = 4\cdot03$
32	20% of 100 = 20·00	12·00	144·00	$\dfrac{144}{20} = 7\cdot20$
22	21% of 100 = 21·00	1·00	1·00	$\dfrac{1}{21} = 0\cdot05$
19	16% of 100 = 16·00	3·00	9·00	$\dfrac{9}{16} = 0\cdot56$
$\left.\begin{matrix} 6 \\ 2 \end{matrix}\right\}8$	$\left.\begin{matrix} 10\% \text{ of } 100 = 10\cdot00 \\ 3\% \text{ of } 100 = 3\cdot00 \end{matrix}\right\}13\cdot00$	5·00	25·00	$\dfrac{25}{13} = 1\cdot92$
D.O.F. $= c - 1 = 4$				$\chi^2 = 13\cdot76$

Note that the expected number in the last class is less than 5, so we have pooled the observed numbers and the expected numbers in the last 2 classes in order to make $E = 5$ or more in all classes. This decreases the number of classes to 5, and therefore there are 4 degrees of freedom.

The value of χ^2 for this data is seen to be 13·76. The χ^2 Table (p. 276) shows that with 4 degrees of freedom, this value of χ^2 indicates a probability of there being no significant difference between the observed and expected numbers of slightly less than 1%. A significant difference is therefore present.

Ex. 39. The comparing of 2 independent samples, each divided into 3 classes, will now be illustrated. Monsieur Alphonse is director of the Continental Mannequin Academy. He is desirous of a raise in salary, so is telling his Board of Management how his results are better than those of their competitor, Mrs Batty's Establishment for Young Models. 'At the Institute's last examinations,' he goes on, 'we got 10 honours, 45 passes, and 5 failures, whereas Mrs Batty's Establishment only got 4 honours, 35 passes, and had 11 failures.'

Do his figures prove his point, or might they be reasonably ascribed to chance?

Data: We draw up a data table, and add up the row totals (R), the column totals (C), and the grand total (N), thus –

| | Exam. results | | | R |
	Honours	Passes	Failures	
Alphonse	10	45	5	60
Batty	4	35	11	50
C	14	80	16	$110 = N$

Calculation:

O	$\dfrac{R \times C}{N} = E$	$\begin{array}{c}\|O - E\|\\= D\end{array}$	D^2	$\dfrac{D^2}{E}$
10	$\dfrac{60 \times 14}{110} = 7\cdot64$	2·36	5·57	$\dfrac{5\cdot57}{7\cdot64} = 0\cdot73$
45	$\dfrac{60 \times 80}{110} = 43\cdot64$	1·36	1·85	$\dfrac{1\cdot85}{43\cdot64} = 0\cdot04$
5	$\dfrac{60 \times 16}{110} = 8\cdot73$	3·73	13·91	$\dfrac{13\cdot91}{8\cdot73} = 1\cdot59$
4	$\dfrac{50 \times 14}{110} = 6\cdot36$	2·36	5·57	$\dfrac{5\cdot57}{6\cdot36} = 0\cdot88$
35	$\dfrac{50 \times 80}{110} = 36\cdot36$	1·36	1·85	$\dfrac{1\cdot85}{36\cdot36} = 0\cdot05$
11	$\dfrac{50 \times 16}{110} = 7\cdot27$	3·73	13·91	$\dfrac{13\cdot91}{7\cdot27} = 1\cdot91$
D.O.F. $= (c - 1)(r - 1) = 2 \times 1 = 2$				$\chi^2 = 5\cdot20$

Our tentative negative assumption in this case is that there is no significant difference between Monsieur Alphonse's and Mrs Batty's results, so we calculate the expected number in each category by means of the overall totals. The expected number in all cells is more than 5 so our test will be valid. You may care to check that the sum of the E's equals 110·00, as it should.

The value of χ^2 works out to be 5·20, and the degrees of freedom are 2. Reference to the χ^2 Table (p. 276) shows this to have a probability of more than 5%. A significant difference between the 2 samples is not proven. Monsieur Alphonse did not get his raise!

The calculation would have been exactly the same if we had been comparing 3 binomial samples. Thus the honour students would have formed one sample, the pass students a second sample, and the failed students a third sample. All students went to one or other of 2 schools, hence each is a binomial sample. Is there a significant difference in the proportion of students attending each school in the 3 samples? The calculation and answer would be identical with that already worked out above.

Ex. 40. To find out whether there is a significant association between the sensitivity of the skin to sunlight and the colour of a person's eyes, we could take a random sample of, say, 100 people, and test them all with a standard dose of ultra-violet rays, noting both their reaction to this test and their eye colour. Both of these features are noted for each person in the trial, so the observations are matched. The results could then be tabulated in an association table like this –

		Ultra-violet reaction			Row totals
		++	+	−	
Eye colour	Blue	19	27	4	50
	Grey	6	6	3	15
	Green	1	2	2	5
	Brown	1	13	16	30
Column totals		27	48	25	100

Now it is obvious that there are not enough grey- and green-eyed people in the sample to give expected numbers of 5 in each cell; it will not disturb the purpose of the experiment if we combine the grey- and green-eyed categories, as follows –

		Ultra-violet reaction			R
		++	+	−	
Eye colour	Blue	19	27	4	50
	Grey or green	7	8	5	20
	Brown	1	13	16	30
	C	27	48	25	$100 = N$

We have thus converted a 3 × 4 table into a 3 × 3 one, and as will be shown, the expected numbers in all cells are now sufficient for the χ^2 Test.

Calculation:

O	$\dfrac{R \times C}{N} = E$	$\begin{aligned}\lvert O - E \rvert\\= D\end{aligned}$	D^2	$\dfrac{D^2}{E}$
19	$\dfrac{50 \times 27}{100} = 13\cdot50$	$5\cdot50$	$30\cdot25$	$\dfrac{30\cdot25}{13\cdot50} = 2\cdot24$
27	$\dfrac{50 \times 48}{100} = 24\cdot00$	$3\cdot00$	$9\cdot00$	$\dfrac{9\cdot00}{24\cdot00} = 0\cdot38$
4	$\dfrac{50 \times 25}{100} = 12\cdot50$	$8\cdot50$	$72\cdot25$	$\dfrac{72\cdot25}{12\cdot50} = 5\cdot78$
7	$\dfrac{20 \times 27}{100} = 5\cdot40$	$1\cdot60$	$2\cdot56$	$\dfrac{2\cdot56}{5\cdot40} = 0\cdot47$
8	$\dfrac{20 \times 48}{100} = 9\cdot60$	$1\cdot60$	$2\cdot56$	$\dfrac{2\cdot56}{9\cdot60} = 0\cdot27$
5	$\dfrac{20 \times 25}{100} = 5\cdot00$	0	—	—
1	$\dfrac{30 \times 27}{100} = 8\cdot10$	$7\cdot10$	$50\cdot41$	$\dfrac{50\cdot41}{8\cdot10} = 6\cdot22$
13	$\dfrac{30 \times 48}{100} = 14\cdot40$	$1\cdot40$	$1\cdot96$	$\dfrac{1\cdot96}{14\cdot40} = 0\cdot14$
16	$\dfrac{30 \times 25}{100} = 7\cdot50$	$8\cdot50$	$72\cdot25$	$\dfrac{72\cdot25}{7\cdot50} = 9\cdot63$
Total $E = 100\cdot00$				$\chi^2 = 25\cdot13$
D.O.F. $= (3 - 1)(3 - 1) = 4$				

Under the assumption that there is no association between eye colour and ultra-violet-ray skin sensitivity, this value of χ^2 (for 4 degrees of freedom) shows that the observed figures could arise by chance less than once in 1,000 times. We therefore reject our tentative negative assumption, and accept instead that the association observed is statistically highly significant. This confirms our suspicion that blue-eyed persons tend to sunburn more readily than brown-eyed ones.

It is to be noted, however, that the χ^2 Test measures the probability of an association being due to chance. It does not measure the degree of the association, nor does it imply a practical significance (p. 148), nor indicate causation (p. 273).

Ex. 41. You may have wondered why the χ^2 Table goes to 100 degrees of freedom. The main use of these upper reaches is for combining independent estimates of χ^2. For, remarkably enough, you can simply add up any number of values of χ^2, each obtained independently of one another, and interpret the result after similarly adding up all the degrees of freedom concerned. This has the overall effect of increasing the size of the samples and, as on page 264, this may convert an answer of 'significant difference not proven' into one of 'the difference is significant'. Alternatively, it may cause a doubtful result, say P between 5% and 10%, to swing more distinctly towards 'a difference is not proven' by increasing the probability to more than 10%.

For an illustration of this technique, let us go back to 1894, when the early anti-cholera vaccine was being tested in India (p. 270). Five small reports were available, each of which yielded a value of χ^2 with 1 degree of freedom as follows –

Source	χ^2	Probability
East Lancashire Regiment	2·04	15·3%
British Troops at Cawnpore	1·83	17·7%
British Troops at Dinapore	1·60	20·8%
Gya Jail	5·90	1·5%
Durbhanga Jail	3·18	7·5%

(From M. Greenwood and G. U. Yule, *Proc. Royal Soc. Med.*, Epidemiology Section, 1915, pp. 122–3.)

The results are so varied, what are we to believe? Well, let us combine the results. Is there a significant difference between the combined groups of inoculated and non-inoculated persons, or not?

Calculation: Adding the 5 values of χ^2 in the data Table above gives a total $\chi^2 = 14\cdot55$. As each of the above tests were based on 1 degree of freedom, the 5 tests together are to be judged with 5 degrees of freedom.

Referring to the χ^2 Table (p. 276), we find that this value of χ^2, on the line for 5 degrees of freedom, indicates a probability of less than 5% (in fact, not far off 1%). The observed difference between the cholera-rate in the inoculated and the non-inoculated is therefore probably significant. Further observations, such as those at Calcutta (p. 271), proved the difference to be highly significant.

A word of warning before you charge off and start testing piles of data in this way. You surely realize that the data you test must not be hand-picked. The bad results must be tested along with the good. But a source of error can be present despite your best intentions, for it often happens that authors of unsuccessful experiments fail to publish their results. Suppose, for example, you have conducted an experiment on 20 of your friends to find out if any of them have powers of mental telepathy, and fail to find any such evidence, the chances are high that your experiment will never be reported; but if your experiment showed that one of your subjects had remarkable powers of this kind, sooner or later this would quite likely end up in print. In a case like this, it would therefore be invalid to add up the χ^2 scores of published results, for it might easily produce a spuriously high value of χ^2.

Questions

Q77. There are 3 equally popular skiing slopes at a certain ski resort. During a 2 month period, there are 6 accidents on slope *A*, 11 on *B*, and 16 on *C*.

(*a*) Does this prove that *C* is significantly more dangerous than the other two slopes?

(*b*) After what has been said on page 213, would it be allowable to compare only *A* and *C*?

Q78. The laboratory assistant seemed to be rounding his weighings to a final digit of 0 or 5 with undue frequency. The boss pointed out that in his last 60 weighings, there were 12 ending in 0, and 8 ending in 5. Reprimand the assistant if such a result has a chance probability of less than 5%. (Don't forget to take into account the weighings ending in digits other than 0 and 5, and be careful when working out the expected frequency of the 3 groups so formed.)

Q79. If the number of convictions obtained against drunken drivers in 3 cities during the same period was as given in the Table below, could it be said that the courts in any one of these cities were significantly more severe than the others, or might the observed differences be reasonably likely to occur by chance?

| | Accused drunken drivers | |
	Convicted	Acquitted
City A	15	85
City B	2	48
City C	83	767

YATES' χ^2 TEST (1934)

Purpose

This modification of the χ^2 Test is used on 2 × 2 contingency tables (p. 271) for the purpose of comparing –

(*a*) 2 random *binomial samples* (each divided into 2 classes, such as cured and not-cured), for evidence of difference between the samples, or

(*b*) *matched observations* made on a single random sample group, for evidence of association between 2 qualities (such as brains and beauty), when both qualities are divided into 2 classes (such as present or absent).

In either case, the *expected number* in each cell or class should be 5 or more, and the *total number* (*N*) of things or instances observed should be more than 50. (With smaller numbers, use Fisher's Test, p. 292).

If the observed number in any cell is 1 or 2, check on pp. 297–8 to see if Fisher's Test is applicable, and if so, use that Test.

Principle

Given a contingency table with certain marginal totals, the way to calculate the exact probability of getting, by chance, any particular set of values inside the table was discovered by Fisher in 1934 (p. 292). This finding was immediately put to use by Fisher's colleague, Dr Frank Yates, for the purpose of testing the accuracy of the χ^2 Test (*Journ. Royal Statist. Soc.*, Suppl. 1, 1934, pp. 217–35). It was pointed out on page 270 that χ^2 is really appropriate for dealing with things, like measurements, which vary on an uninterrupted scale, with the result that applying it to data which can only take on whole number values (such as numbers of people) causes some degree of inaccuracy. Now, by comparing the results of the χ^2 Test with the true probabilities, Yates found that the correspondence between the answers was good in the case of large contingency tables (3 × 1 or more), but was definitely bad in the case of 2 × 2 tables. Yates was able to suggest how to counteract this inaccuracy, by using a small but important correction factor, which splits the difference between whole numbers. Then, as long as the numbers are not too small, the χ^2 results prove quite satisfactory for all practical purposes.

This correction factor has also been incorporated into a number of other significance tests (50% χ^2 Test, zI Test, etc.).

As in the ordinary χ^2 Test, the tentative assumption is made that there is no significant difference (or association, as the case may be) between the sets of observations, and Yates' χ^2 Test then tells us the probability of this being so. If this probability is remote, we deny our original assumption and accept that the observed difference (or association) is significant.

Data Required

a, b, c, d = the numbers of instances in each of the 4 categories of the sample or pair of samples.

N = the total number of things or instances observed. This must be more than 50.

Procedure

(1) Prepare a data table, as in the ordinary χ^2 Test (p. 273). The structure of the table will depend on whether we are comparing 2 unmatched samples or 2 sets of matched observations.

(2) Next, draw up a table for calculating Yates' χ^2, with the following headings –

O	$\dfrac{R \times C}{N} = E$	$\lvert O - E \rvert = D$	$D - 0.5 = Y$	Y^2	$\dfrac{Y^2}{E}$

(3) Calculate χ^2 in the manner described on page 274, but make Yates' correction by subtracting 0.5 from each value of D before squaring it. If your arithmetic is correct, all 4 values of Y will be the same. Adding up all the answers in the last column yields the value of χ^2.

Provided that you are absolutely sure that your calculation of the first value of E is correct, the other 3 values of E can be found simply by subtraction from row and column totals.

Interpretation

All 2×2 contingency tables have 1 degree of freedom, as told by the formula $(c - 1)(r - 1)$.

The following Table gives the critical values of χ^2 for 1 degree of freedom –

χ^2 TABLE FOR YATES' TEST

P = 10%	P = 5%	P = 1%	P = 0·2%
$\chi^2 = 2\cdot71$	3·84	6·63	9·55

Adapted from E. S. Pearson and H. O. Hartley, *Biometrika Tables for Statisticians*, Vol. 1, Table 8 (CUP, 1966).

Low values of χ^2 indicate that our tentative assumption of no significant difference or association cannot be denied on the evidence presented. High values of χ^2 indicate that the probability of our assumption being correct is remote, so we accept that the difference or association is significant (see p. 140).

Examples

Ex. 42. Suppose you have just read a report concerning a new treatment for 'Boodelitis Terrifica'. The new treatment (*A*) has cured 150 out of 170 patients, a cure rate of 88·2%. Now you happen to have your own results handy; with your treatment (*B*) 130 out of your last 170 patients have recovered, which is a cure rate of 76·5%. Is this sufficient evidence to warrant changing to the new treatment, or might the observed difference simply be a matter of chance?

Data: We are concerned with comparing 2 independent binomial samples, which we shall presume to be equivalent to random samples.

	Number of patients		R
	Cured	Died	
With treatment A	150	20	170
With treatment B	130	40	170
C	280	60	340 = N

Calculation: Yates' χ^2 Test is applicable because E is more than 5 in all cells, and N is more than 50.

The calculation is shown in the calculation Table below.

O	$\dfrac{R \times C}{N} = E$	$\begin{array}{c}\|O - E\| \\ = D\end{array}$	$\begin{array}{c}D - 0 \cdot 5 \\ = Y\end{array}$	Y^2	$\dfrac{Y^2}{E}$
150	$\dfrac{170 \times 280}{340} = 140 \cdot 0$	10·0	9·5	90·25	$\dfrac{90 \cdot 25}{140} = 0 \cdot 645$
20	$\dfrac{170 \times 60}{340} = 30.0$	10·0	9·5	90·25	$\dfrac{90 \cdot 25}{30} = 3 \cdot 01$
130	$\dfrac{170 \times 280}{340} = 140 \cdot 0$	10·0	9·5	90·25	$\dfrac{90 \cdot 25}{140} = 0 \cdot 645$
40	$\dfrac{170 \times 60}{340} = 30 \cdot 0$	10·0	9·5	90·25	$\dfrac{90 \cdot 25}{30} = 3 \cdot 01$
D.O.F. $= (2 - 1)(2 - 1) = 1$					$\chi^2 = 7 \cdot 31$

Having found and confirmed that the first value of E is 140·0, we could have determined the second E by subtracting 170 − 140·0 = 30·0; the third E is likewise 280 − 140·0 = 140·0, and the last E is 170 − 140·0 = 30·0 (or 60 − 30·0 = 30·0).

Referring our answer of 7·31 to the Yates' χ^2 Table on page 287 shows that if there was no difference between the 2 treatments, the observed results could occur by chance less than once in 100 times. Under these circumstances, we conclude that a significant difference does exist, so we should change to the new treatment.

Ex. 43. Does the playing of billiards tend to cause backaches? Such an investigation could be started by comparing a random sample of billiard players and another of non-players to see if one group suffered from more backaches than the other. But this might require large samples to show any difference. Suppose, then, we look for a possible association between billiard playing and backaches. We would then want to know if people who play a lot of billiards get more backaches than those who play an occasional game. If a significant association is demonstrated, it will not prove causation, but at least it will suggest that further investigation is desirable, whereas if there is no significant association to be found, we can probably forget the whole idea.

With this aim, we set off for the club. With the help of some penny-tossing, we asked every second billiard player in sight

how many games a week he averaged, and whether he was prone
to backaches or not. Here are the results –

	Av. No. of games per week					
	$\frac{1}{2}$–1	2–3	4–6	7–14	15 or more	Totals
Backache { +	2	0	2	9	2	15
Backache { –	2	8	6	5	4	25
Totals	4	8	8	14	6	40

A χ^2 Test cannot be applied to this data, because the expected
number in many of the cells is less than 5. So we must re-group
the data, and shall see if there is any association between backaches and playing less than a game a day versus playing one or
more games a day. The data table then becomes –

	Plays		R
	+	++	
Backache { +	4	11	15
Backache { –	16	9	25
C	20	20	$40 = N$

This table shows the borderline between association and difference to have disappeared, for we could equally well view it as a
search for *association* between 2 attributes observed on a single
sample of 40 club members, or as a search for *difference* between
the incidence of backaches among a sample of 20 occasional
players compared with that of a sample of 20 frequent players.
The test is the same either way, so don't worry about it.

By combining groups we have built up the expected numbers
to greater than 5 in all cells (as will be demonstrated shortly),
and have managed to do this without spoiling the aim of the investigation. However, the total number of cases observed is only
40; at this level Fisher's Test (p. 292) is more accurate and easier
to do than Yates' χ^2 Test, but for the purpose of demonstration
let us proceed with Yates' Test, and compare its result with the
exact probability as given by Fisher's Formula.

Data: Of 20 people who played billiards occasionally, 4 were prone to backaches. Of 20 who played a lot of billiards, 11 were prone to backaches. See 2 × 2 contingency Table on page 289.

Calculation:

O	$\dfrac{R \times C}{N} = E$	$\begin{array}{c} \|O - E\| \\ = D \end{array}$	$\begin{array}{c} D - 0.5 \\ = Y \end{array}$	Y^2	$\dfrac{Y^2}{E}$
4	$\dfrac{15 \times 20}{40} = 7{\cdot}50$	3·50	3·00	9·00	$\dfrac{9{\cdot}0}{7{\cdot}5} = 1{\cdot}20$
11	$15 - 7{\cdot}5 = 7{\cdot}50$	3·50	3·00	9·00	$\dfrac{9{\cdot}0}{7{\cdot}5} = 1{\cdot}20$
16	$20 - 7{\cdot}5 = 12{\cdot}50$	3·50	3·00	9·00	$\dfrac{9{\cdot}0}{12{\cdot}5} = 0{\cdot}72$
9	$25 - 12{\cdot}5 = 12{\cdot}50$	3·50	3·00	9·00	$\dfrac{9{\cdot}0}{12{\cdot}5} = 0{\cdot}72$
D.O.F. = 1					$\chi^2 = 3{\cdot}84$

Reference to the χ^2 Table for Yates' Test (p. 287) shows this value of χ^2 to have a probability of exactly 5%. We conclude therefore that the observed association between frequent games of billiards and backaches is probably not just a matter of chance association, but of real association. (Billiard players, relax, this data is fictitious.)

The exact two-sided probability for this data, as determined by Fisher's hypergeometric formula (p. 292), is 4·8%. This is fairly typical of the accuracy of Yates' χ^2 Test. Without Yates' correction factor, the value of χ^2 would have worked out to be 5·23, which has a precise probability of 2·2%. We see that Yates' correction is well worth while.

Questions

Q80. The advertisement claimed, 'Homes with their own swimming pool produce most of our champion swimmers.' A fellow named Thomas decided to check on this, before buying a pool for his children. He went to a swimming competition and asked 100 competitors (chosen quite haphazardly, of course) whether they have a pool at home and whether they have ever won a championship swimming race.

He found 30 champions, 10 of whom had a pool at home, and 70 non-champions, of whom there were also 10 who had a pool at home. In other words, 33% of the champions had a pool, as against 14% of the non-champions. 'There you are, my dear,' he told his wife, 'that proves the ad. was right.' Does it?

Q81. Go back and test the data of the Calcutta cholera investigation, page 271.

(*a*) Was the inoculation effective in reducing the incidence of cholera in this trial to a statistically significant degree?

(*b*) Combine your answer with that obtained in the 5 earlier trials, page 283. Does this change your conclusion?

Q82. In Moroney's *Facts From Figures* (Penguin, 1962), the author reported the following survey.

'In order to investigate whether hair colour was associated with culture among the females of the species, I did a little experiment. At a Bach concert I counted the number of blondes (real and artificial) and the number of brunettes. In the interests of science, I went to a Bebop session and did the same. The results were as shown in the Table. The question is: Would it be fair of me, on the strength of this evidence, to state that blondes prefer Bebop to Bach?' (It will suffice in this instance if you work to 1 decimal place throughout.)

	Blonde	Brunette
Bach	7	143
Bebop	14	108

FISHER'S TEST (1934)

Purpose

This test deals with exactly the same comparisons as Yates' χ^2 Test (p. 285), but applies to cases in which the total number (N) of things observed is 8 to 50. (Yates' χ^2 Test is inaccurate for small numbers.)

Given $N = 8$ to 50, Fisher's Test therefore compares –

(*a*) 2 random *binomial samples* for evidence of difference between their proportions, or

(*b*) *matched observations* made on a single random sample group, for evidence of association between 2 qualities, when both are divided into 2 classes.

There are no restrictions concerning expected numbers.

Principle

This test was first described by the famous English statistician, Professor Sir Ronald A. Fisher (1890–1962), in his book *Statistical Methods for Research Workers* (Oliver & Boyd, 1934, #21.02). We have referred to other aspects of Fisher's work on page 97 *et seq.*

Consider a 2×2 contingency table containing 4 values, *a*, *b*, *c*, and *d*, with marginal totals n_1, n_2, n_3, and n_4, and a grand total N, thus –

a	*b*	n_1
c	*d*	n_2
n_3	n_4	N

Of course, $a + b = n_1$, $c + d = n_2$, $a + c = n_3$, $b + d = n_4$, and $n_1 + n_2 = n_3 + n_4 = N$.

Now, Fisher's Test is based on the discovery that the exact probability of getting any particular values of *a*, *b*, *c*, and *d*, by chance, in such a table is given by the hypergeometric formula –

$$P = \frac{n_1! \times n_2! \times n_3! \times n_4!}{N! \times a! \times b! \times c! \times d!}$$

The '!' signs are not exclamation marks but are mathematical factorial signs. The factorial of a number is the product of that number multiplied by a series of numbers, each being 1 smaller than the preceding number; thus factorial 6, written $6! = 6 \times 5 \times 4 \times 3 \times 2 \times 1 = 720$.

For example, with a table having marginal totals of 5, 1, 4, and 2, let us work out the probability of getting the following interior values –

4	1	5
0	1	1
4	2	6

$$P = \frac{5! \times 1! \times 4! \times 2!}{6! \times 4! \times 1! \times 0! \times 1!}$$

$$= \frac{(5 \times 4 \times 3 \times 2 \times 1)(1)(4 \times 3 \times 2 \times 1)(2 \times 1)}{(6 \times 5 \times 4 \times 3 \times 2 \times 1)(4 \times 3 \times 2 \times 1)(1)(1)(1)}$$

$$= \frac{1}{3} \qquad \text{(Note: } 0! = 1)$$

The probability associated with the above table is thus 1 in 3.

With larger numbers, the calculations become very substantial, even using logarithms of the factorials. We are therefore fortunate that, for up to $N = 50$, the results of such calculations have been published; these results have been reorganized to provide a special set of Tables for interpreting this test. These Tables are extremely simple to use, so do not be put off by their size.

As always, we start with the tentative assumption that there is no significant difference or association in the data presented, and reference to these Tables tells us the probability of this assumption being true. If this probability turns out to be small, we accept that a significant difference or association is present.

Data Required

a, b, c, d = the number of things or instances in each class of the sample or samples.

N = the total number of things or instances observed.

The Tables provided will handle all cases of $N = 8$ to 50. When N is less than 8, a 5% probability level cannot be reached, so a significant difference or association cannot be demonstrated.

When N is more than 50, use Yates' χ^2 Test (p. 285). However, the Tables for Fisher's Test go beyond $N = 50$ for $a = 1$ and $a = 2$, and these should be used whenever possible in preference to Yates' Test (in the interests of accuracy).

Procedure

No calculation is required. Simply arrange the data in a 2 × 2 contingency table in such a way that it satisfies the following 2 rules –

(1) *n_1 must be the smallest, or equal smallest, of the 4 marginal totals*, and
(2) *$a \times d$ must be larger than $b \times c$.*

Start, then, by putting your data in a 2 × 2 table, and add up the marginal totals. If the top right marginal total (n_1) happens to be the smallest or equal smallest of the 4, proceed with the second step, below.

If, however, the smallest or equal smallest marginal total is not in the position of n_1, invert or rotate the data. For instance –

10	16	26
5	1	6
15	17	

Fisher's Test cannot be applied to this table as it stands, because the smallest marginal total (6) is not in the position of n_1. To make it so, the table must be inverted so that it becomes –

5	1	6
10	16	26
15	17	

In the next case the data has to be rotated to make n_1 the smallest marginal total; in doing this manoeuvre one must be careful not to disturb the mutual relationship between the numbers in the table – the marginal totals must remain the same, only their positions are altered.

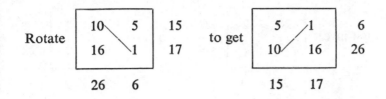

As illustrated in this case (above), a good way to ensure that the marginal totals are not changed is to draw a guide line between 2 diagonally opposite numbers. Then, no matter how you turn the table around, the marginal totals will remain unchanged so long as the guide line keeps this pair of numbers in diagonally opposite corners of the table.

The second and final step is to make sure that $a \times d$ is larger than $b \times c$. The data table may need to be reversed to make this so. For instance –

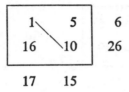

In this arrangement n_1 is the smallest marginal total, but the table is not yet suitable for Fisher's Test because $a \times d$ (i.e. $1 \times 10 = 10$) is not larger than $b \times c$ (i.e. $5 \times 16 = 80$). The table must therefore be reversed, with a guide line for safety, to make it thus –

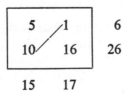

Now n_1 is the smallest marginal total, and $a \times d$ is larger than $b \times c$, so the table is ready for Fisher's Test.

To put the following table in order, it first needs to be inverted to make n_1 the (equal) smallest marginal total, and then reversed to make $a \times d$ larger than $b \times c$ –

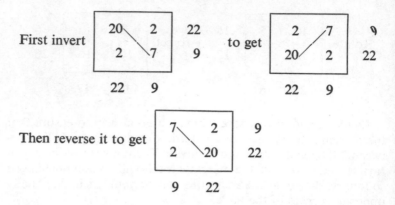

First invert ... to get

Then reverse it to get

If $a \times d = b \times c$, the proportions on the top row are the same as those on the bottom row, so the table shows no difference or association between the 2 sets of observations. Thus –

$$
\begin{array}{|cc|}
\hline
1 & 5 \\
4 & 20 \\
\hline
\end{array}
\qquad 1 \times 20 = 5 \times 4
$$

Or, less obviously –

$$
\begin{array}{|cc|}
\hline
3 & 5 \\
9 & 15 \\
\hline
\end{array}
\qquad 3 \times 15 = 9 \times 5
$$

Interpretation

Having arranged your data table in the way described, turn directly to the Tables provided (p. 297 *et seq.*) for the answer required. Given values of *a*, *b*, and *c*, the higher the value of *d*, the less the likelihood of your tentative negative assumption being correct.

First find the appropriate Table for your value of *a*; then find the right column for your value of *b*; and then find the right row for your value of *c*. In the cell situated where your chosen column and row intersect, you will find the critical values of *d*.

If the observed value of d in your data table is equal to or greater than the numbers given in the appropriate cell of these Tables, the probability of this value occurring by chance (if there is no significant difference or association) is given at the top of the columns. As in the Tables for the Binomial Test, the probability levels indicated tend to be on the conservative side, for an exact probability of say 1·92% has to be shown in the 5% column, merely because it does not reach down to the 1% level. However, it would be quite wrong to imagine that the test is not perfectly accurate; it is simply that the sign 'P = 5%' is to be interpreted as meaning that the probability is 5% or less, but greater than 1%. For practical purposes this is all that is necessary.

A dash (—) in some of the 1% cells indicates that, for N up to 50, no value of d can reach the 1% significance level.

With N up to 50, any data table possessing a value of d which falls *outside the limits* of these Tables (i.e. to the right of or below the occupied cells) has a probability of more than 5%, which indicates that a significant difference or association is not proven.

Further discussion about the interpretation of these critical significance levels was given on page 140.

Note: If you find yourself looking for values of d in a completely unoccupied cell in these Tables, your data table has not been arranged properly.

d TABLES FOR FISHER'S TEST

c ↓	$a = 1$				
	$b = 0$		1		
	P = 5%	1%	5%	1%	
0	19	99			
1	38	198	77	397	
2	57	297	115	595	
3			154	—	

d TABLES FOR FISHER'S TEST – *continued*

c ↓	$b = 0$ P = 5%	1%	1 5%	1%	2 5%	1%
0	5	13				
1	9	22	16	39		
2	12	32	22	—	31	—
3	16	41	29	—	41	—
4	19	50	35	—		
5	23	59	41	—		
6	26	68				
7	30	—				
8	33	—				
9	37	—				

$a = 2$

d TABLES FOR FISHER'S TEST – *continued*

	$a = 3$									
c ↓	$b = 0$		1		2		3		4	
	P = 5%	1%	5%	1%	5%	1%	5%	1%	5%	1%
0	4	7								
1	5	11	9	17						
2	7	15	12	24	16	33				
3	9	18	15	30	21	41	26	—		
4	11	22	18	36	25	—	32	—	39	—
5	12	26	21	—	29	—	37	—		
6	14	29	24	—	34	—				
7	16	33	27	—	38	—				
8	18	37	30	—						
9	19	—	33	—						
10	21	—	36	—						
11	23	—								
12	24	—								
13	26	—								
14	28	—								
15	30	—								
16	31	—								

d TABLES FOR FISHER'S TEST – *continued*

	a = 4											
c ↓	b = 0		1		2		3		4		5	
	P = 5%	1%	5%	1%	5%	1%	5%	1%	5%	1%	5%	1%
0	4	5										
1	4	8	7	11								
2	7	10	9	15	11	20						
3	8	12	10	19	14	25	17	31				
4	9	14	12	22	17	30	21	38	25	—		
5	10	16	14	26	19	35	24	—	29	—	34	—
6	11	19	16	29	22	—	28	—	34	—		
7	12	21	18	33	25	—	31	—				
8	13	23	20	37	28	—	35	—				
9	14	25	22	—	30	—						
10	15	27	24	—	33	—						
11	16	30	26	—								
12	17	32	28	—								
13	18	—	30	—								
14	19	—										
15	20	—										
16	21	—										
17	22	—										
18	23	—										
19	24	—										
20	26	—										

d TABLES FOR FISHER'S TEST – *continued*

			$a = 5$											
c ↓	$b = 0$		1		2		3		4		5		6	
	P = 5%	1%	5%	1%	5%	1%	5%	1%	5%	1%	5%	1%	5%	1%
0		5												
1	5	7	6	9										
2	5	8	7	11	9	15								
3	6	9	8	14	13	18	16	22						
4	6	11	10	16	15	22	18	27	21	32				
5	7	12	14	19	17	25	21	31	24	—	27	—		
6	8	14	15	21	19	28	23	35	27	—	30	—	34	—
7	9	15	17	24	21	31	25	—	29	—	34	—		
8	10	17	18	26	23	35	28	—	32	—				
9	11	18	20	29	25	—	30	—						
10	12	20	21	31	27	—	32	—						
11	13	22	23	33	29	—								
12	14	23	24	—	31	—								
13	15	25	26	—										
14	16	26	27	—										
15	17	28	29	—										
16	19	29												
17	20	—												
18	21	—												
19	23	—												

d TABLES FOR FISHER'S TEST – *continued*

a = 6															
c ↓	*b* = 0		1		2		3		4		5		6		
	P = 5%	1%	5%	1%	5%	1%	5%	1%	5%	1%	5%	1%	5%	1%	
0		6													
1		6	6	8											
2		6	6	10	9	12									
3	6	8	7	12	10	14	12	18							
4	6	11	10	13	11	17	13	21	16	24					
5	6	12	11	15	13	19	15	24	18	28	22	32			
6	7	13	12	17	14	22	17	27	21	32	25	—	30	—	
7	8	14	13	19	15	24	20	30	24	—	29	—			
8	9	15	14	20	17	27	22	33	27	—					
9	11	16	15	22	19	29	24	—	30	—					
10	12	17	16	24	20	32	26	—							
11	12	18	17	26	22	—	29	—							
12	13	19	18	28	24	—									
13	14	20	19	30	26	—									
14	14	21	20	—	27	—									
15	15	22	21	—											
16	16	23	22	—											
17	16	24	23	—											
18	17	25	24	—											
19	18	—													
20	18	—													
21	19	—													
22	20	—													
23	20	—													

d TABLES FOR FISHER'S TEST – *continued*

c ↓	b=0 P=5%	1%	1 5%	1%	2 5%	1%	3 5%	1%	4 5%	1%	5 5%	1%	6 5%	1%	7 5%	1%
0		7														
1		7	7	8												
2		7	7	9	8	10										
3		7	7	10	9	13	10	17								
4	7	9	8	11	10	17	11	19	14	22						
5	7	9	9	13	11	18	13	21	18	24	20	27				
6	7	10	10	15	12	20	16	23	20	26	22	30	24	—		
7	7	11	10	18	14	22	19	25	22	29	24	—	27	—	29	—
8	9	12	11	20	15	24	20	27	23	—	26	—	29	—		
9	9	13	12	21	17	25	22	29	25	—	28	—				
10	10	14	13	22	20	27	24	—	27	—						
11	10	15	14	24	21	29	25	—	29	—						
12	11	16	15	25	23	—	27	—								
13	11	17	16	26	24	—										
14	12	18	17	28	25	—										
15	12	19	18	—	26	—										
16	13	21	20	—												
17	13	22	21	—												
18	14	23	22	—												
19	14	—														
20	15	—														
21	15	—														
22	16	—														
23	16	—														
24	17	—														
25	17	—														

d TABLES FOR FISHER'S TEST – *continued*

c ↓	b=0 P=5%	1%	1 5%	1%	2 5%	1%	3 5%	1%	4 5%	1%	5 5%	1%	6 5%	1%	7 5%	1%
0		8														
1		8	8													
2		8	8		8	9										
3		8		8	8	12	9	14								
4		8	8	10	8	13	10	15	13	18						
5		8	8	13	9	15	13	17	15	20	16	24				
6	8	9	8	14	11	16	14	19	16	23	18	28	20	—		
7	8	9	9	15	14	17	16	22	18	26	20	—	22	—	24	—
8	8	10	10	16	15	19	17	24	19	—	21	—	24	—	27	—
9	8	11	10	17	16	21	18	27	21	—	23	—	27	—		
10	8	12	11	18	17	23	19	29	22	—	25	—				
11	9	15	12	19	18	25	20	—	23	—						
12	9	15	14	20	19	27	22	—	25	—						
13	10	16	16	21	20	—	23	—								
14	10	17	17	22	21	—	24	—								
15	10	18	18	24	22	—										
16	11	18	18	25	23	—										
17	11	19	19	—												
18	12	20	20	—												
19	12	21	21	—												
20	13	21														
21	13	—														
22	13	—														
23	14	—														
24	14	—														
25	15	—														
26	15	—														

d TABLES FOR FISHER'S TEST – *continued*

$a = 9$

c	$b=0$ 5% 1%	1 5% 1%	2 5% 1%	3 5% 1%	4 5% 1%	5 5% 1%	6 5% 1%	7 5% 1%
0	9							
1	9	9						
2	9	9	9					
3	9	9	9 11	9 12				
4	9	9	9 12	9 13	12 16			
5	9	9 11	9 13	12 15	13 19	14 21		
6	9	9 12	10 14	13 19	14 21	16 23	17 25	
7	9	9 12	12 16	14 20	15 23	17 25	20 28	23 —
8	9	9 13	13 18	15 22	16 25	19 27	22 —	26 —
9	9 10	9 14	13 20	16 23	18 26	21 —	26 —	
10	9 12	10 15	14 22	17 25	19 —	24 —		
11	9 13	11 17	15 23	18 26	21 —			
12	9 13	13 18	16 24	19 —	23 —			
13	9 14	14 19	17 25	20 —				
14	9 14	15 21	18 —	21 —				
15	9 15	15 24	18 —	23 —				
16	10 15	16 —	19 —					
17	10 16	16 —	20 —					
18	10 17	17 —	21 —					
19	11 17	18 —						
20	11 18	18 —						
21	11 18	19 —						
22	12 19							
23	12 —							
24	13 —							
25	13 —							
26	13 —							
27	14 —							

d TABLES FOR FISHER'S TEST – *continued*

	a = 10																	
	b = 0		1		2		3		4		5		6		7		8	
c ↓	P = 5%	1%	5%	1%	5%	1%	5%	1%	5%	1%	5%	1%	5%	1%	5%	1%	5%	1%
0		10																
1		10	10															
2		10	10		10													
3		10	10		10		10	11										
4		10	10		10		10	12	11	15								
5		10	10		10	11	10	15	12	17	13	18						
6		10	10		10	13	11	16	13	18	14	20	16	22				
7		10	10	11	11	16	12	17	14	19	15	22	19	25	21	—		
8		10	10	12	11	17	13	19	15	21	18	24	21	—	23	—	25	—
9		10	10	13	12	18	14	20	16	23	20	—	22	—	24	—		
10	10	11	10	14	13	19	15	21	18	25	22	—	24	—				
11	10	11	10	16	13	20	16	23	21	—	23	—						
12	10	12	12	18	14	21	17	24	22	—								
13	10	12	12	18	15	22	18	—	23	—								
14	10	13	13	19	15	23	20	—										
15	10	13	13	20	16	—												
16	10	14	14	21	17	—												
17	10	14	15	22	18	—												
18	10	15	15	—	19	—												
19	10	15	16	—														
20	10	16	16	—														
21	10	16	17	—														
22	11	17	17	—														
23	11	—																
24	11	—																
25	12	—																
26	12	—																
27	13	—																

d TABLES FOR FISHER'S TEST – *continued*

	a = 11																	
	b = 0		1		2		3		4		5		6		7		8	
c ↓	P = 5%	1%	5%	1%	5%	1%	5%	1%	5%	1%	5%	1%	5%	1%	5%	1%	5%	1%
0		11																
1		11		11														
2		11		11		11												
3		11		11		11		11										
4		11		11		11		11	11	14								
5		11		11		11	11	14	11	15	12	16						
6				11	11	12	11	15	11	16	13	18	15	21				
7				11	11	14	11	16	12	17	14	21	17	24	19	—		
8				11	11	15	12	17	13	19	17	24	19	—	20	—	22	—
9			11	12	11	16	12	18	15	22	18	25	20	—	22	—		
10			11	13	12	16	13	19	18	24	19	—	21	—				
11			11	15	12	17	14	21	19	—	20	—	22	—				
12			11	16	13	18	16	23	20	—	22	—						
13			11	16	13	19	18	—	21	—								
14			12	17	14	20	19	—										
15	11	12	12	18	15	22	20	—										
16	11	12	13	18	15	—												
17	11	13	13	19	16	—												
18	11	13	14	20	17	—												
19	11	14	14	—	18	—												
20	11	14	15	—														
21	11	15	15	—														
22	11	16	16	—														
23	11	16																
24	11	—																
25	11	—																
26	11	—																
27	12	—																

d TABLES FOR FISHER'S TEST – *continued*

	a = 12																
	b = 0		1		2		3		4		5		6		7		8
c ↓	P = 5%	1%	5%	1%	5%	1%	5%	1%	5%	1%	5%	1%	5%	1%	5%	1%	5% 1%
0		12															
1		12		12													
2		12		12		12											
3		12		12		12		12									
4		12		12		12		12	12	13							
5		12		12		12		12	12	14	12	15					
6		12		12		12	12	13	12	15	12	17	14	20			
7		12		12	12	13	12	14	12	16	13	20	16	21	17	23	
8		12		12	12	13	12	15	12	19	16	21	17	23	18 —		20 —
9		12		12	12	14	12	16	14	20	17	22	18 —		19 —		21 —
10		12		12	12	15	12	18	16	22	18 —		19 —		21 —		
11		12	12	14	12	16	13	21	17	23	19 —		20 —				
12		12	12	14	12	17	15	22	18 —		20 —						
13		12	12	15	12	18	17 —		19 —								
14		12	12	15	13	19	17 —		19 —								
15		12	12	16	13	21	18 —										
16		12	12	16	14 —		19 —										
17		12	12	17	15 —												
18		12	13	18	16 —												
19	12	13	13	18													
20	12	13	13	18													
21	12	14	14 —														
22	12	15	14 —														
23	12	15															
24	12 —																
25	12 —																
26	12 —																

d TABLES FOR FISHER'S TEST – *continued*

c ↓	b = 0 5% 1%	1 5% 1%	2 5% 1%	3 5% 1%	4 5% 1%	5 5% 1%	6 5% 1%	7 5% 1%	8 5% 1%
					a = 13				
0	13								
1	13	13							
2	13	13	13						
3	13	13	13	13					
4	13	13	13	13	13				
5	13	13	13	13	13	13 14			
6	13	13	13	13	13	13 17	14 18		
7	13	13	13	13	13 15	13 18	15 19	16 21	
8	13	13	13	13 14	13 18	14 19	16 21	17 22	18 —
9	13	13	13	13 15	13 19	15 20	17 22	18 —	19 —
10	13	13	13 14	13 18	15 20	16 21	18 —	19 —	
11	13	13	13 14	13 19	15 21	17 —	19 —		
12	13	13	13 15	15 19	16 —	18 —			
13	13	13 14	13 17	15 20	17 —	19 —			
14	13	13 14	13 19	16 —	18 —				
15	13	13 15	13 20	17 —					
16	13	13 15	13 —	17 —					
17	13	13 16	14 —						
18	13	13 16	15 —						
19	13	13 17							
20	13	13 —							
21	13	13 —							
22	13 14	13 —							
23	13 —								
24	13 —								

d TABLES FOR FISHER'S TEST – *continued*

c ↓	b = 0 P = 5%1%	1 5%1%	2 5%1%	3 5%1%	4 5%1%	5 5%1%	6 5%1%	7 5%1%	8 5%1%
				a = 14					
0	14								
1	14	14							
2	14	14	14						
3	14	14	14	14					
4	14	14	14	14	14				
5	14	14	14	14	14	14			
6	14	14	14	14	14	14 16	14 17		
7	14	14	14	14	14	14 17	14 18	15 19	
8	14	14	14	14	14 16	14 18	14 19	16 21	17 —
9	14	14	14	14	14 17	14 19	15 20	17 —	18 —
10	14	14	14	14 16	14 18	15 20	16 —	18 —	
11	14	14	14	14 17	14 19	16 —	17 —		
12	14	14	14	14 18	15 20	17 —	18 —		
13	14	14	14 15	14 19	16 —	17 —			
14	14	14	14 17	15 19	16 —				
15	14	14	14 18	15 —	17 —				
16	14	14	14 —	16 —					
17	14	14 15	14 —						
18	14	14 15	15 —						
19	14	14 16							
20	14	14 —							
21	14	14 —							
22	14								

d TABLES FOR FISHER'S TEST – *continued*

c	b=0		b=1		b=2		b=3		b=4		b=5		b=6		b=7		b=8	
	a = 15																	
	5%	1%	5%	1%	5%	1%	5%	1%	5%	1%	5%	1%	5%	1%	5%	1%	5%	1%
0	15																	
1	15		15															
2	15		15		15													
3	15		15		15		15											
4	15		15		15		15		15									
5	15		15		15		15		15		15							
6	15		15		15		15		15		15		15	16				
7	15		15		15		15		15		15		15	17	15	18		
8	15		15		15		15		15		15	16	15	18	15	20	16	—
9	15		15		15		15		15	16	15	17	15	19	15	—	17	—
10	15		15		15		15		15	17	15	18	15	—	16	—		
11	15		15		15		15	16	15	17	15	—	16	—				
12	15		15		15		15	17	15	18	15	—	17	—				
13	15		15		15		15	17	15	—	16	—						
14	15		15		15	16	15	18	15	—								
15	15		15		15	17	15	—	16	—								
16	15		15		15	17	15	—										
17	15		15		15	—	15	—										
18	15		15		15	—												
19	15		15															
20	15																	

d TABLES FOR FISHER'S TEST – *continued*

	a = 16																	
c	b = 0		1		2		3		4		5		6		7		8	
↓	P = 5%	1%	5%	1%	5%	1%	5%	1%	5%	1%	5%	1%	5%	1%	5%	1%	5%	1%
0	16																	
1	16		16															
2	16		16		16													
3	16		16		16		16											
4	16		16		16		16		16									
5	16		16		16		16		16		16							
6	16		16		16		16		16		16		16					
7	16		16		16		16		16		16		16		16	17		
8	16		16		16		16		16		16		16		16	19	16	—
9	16		16		16		16		16		16		16	18	16	—	16	—
10	16		16		16		16		16		16	17	16	—	16	—		
11	16		16		16		16		16		16	—	16	—				
12	16		16		16		16		16	17	16	—	16	—				
13	16		16		16		16		16	—	16	—						
14	16		16		16		16	17	16	—								
15	16		16		16		16	—										
16	16		16		16													
17	16		16															
18	16																	

d TABLES FOR FISHER'S TEST – *continued*

c ↓	*b* = 0 P = 5%1%	1 5%1%	2 5%1%	3 5%1%	4 5%1%	5 5%1%	6 5%1%	7 5%1%	8 5%1%
				a = 17					
0	17								
1	17	17							
2	17	17	17						
3	17	17	17	17					
4	17	17	17	17	17				
5	17	17	17	17	17	17			
6	17	17	17	17	17	17	17		
7	17	17	17	17	17	17	17	17	
8	17	17	17	17	17	17	17	17 —	17 —
9	17	17	17	17	17	17	17	17 —	
10	17	17	17	17	17	17	17 —		
11	17	17	17	17	17	17 —			
12	17	17	17	17	17				
13	17	17	17	17					
14	17	17	17						
15	17	17							
16	17								

a = 18 or more
All 2 × 2 data tables with *N* up to 50 have a probability of 1% or less.

These Tables have been adapted from P. Armsen, *Biometrika Journal*, 1955, pp. 506–11.

Examples

Ex. 44. Extensive burns are very apt to get infected. Suppose that, in an endeavour to reduce the incidence of this complication, we decide to try out a new method of treatment. To assess the results of the new treatment, we must have some standard of comparison, and it is obvious that we cannot leave patients untreated to act as controls. Anyway, we're not interested in whether the new treatment is better than nothing, we're interested in whether the new treatment is better than the best existing treatment. So we will treat half of a sample group of patients (a presenting sample, p. 49) with the new treatment, and will compare the results with those of the old treatment applied to the remaining half of the patients. To randomize the treatments, we decide in advance that the first, third, fifth, etc., patient will get the new treatment, while the second, fourth, sixth, etc., will get the old.

Would the following results indicate a significant superiority of one treatment over the other?

	Not infected	Infected	Totals
Old treatment	5	5	10
New treatment	9	1	10
Totals	14	6	20

Calculation: Fisher's Test is applicable because $N = 20$. But as it stands, the data table must be rotated, because n_1 is 10, and this is not the smallest marginal total. We must make $n_1 = 6$, thus –

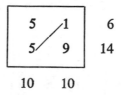

Next, we check that $a \times d$ is larger than $b \times c$; $5 \times 9 = 45$ is in fact larger than $5 \times 1 = 5$, so our data table is now ready for use.

We turn to the Table for $a = 5$ on page 301, and find the column for $b = 1$, and the row for $c = 5$. At the intersection of this column and row we see that d would have to be 14 before the probability of a significant difference between the 2 treatments would be 5%. Our d is only 9, so the conclusion is that a significant difference is not yet proven.

This example illustrates the great practical importance of being able to deal with small numbers, with accurate statistical tests. Remember Darwin's experience (p. 188). Of course, small numbers don't produce high levels of significance very often, but the results can be a very useful guide. And at least it is better than guessing!

Ex. 45. To familiarize yourself with the use of the Tables for Fisher's Test, here are some contingency tables, correctly set out, together with their probability levels. Check the setting out of the tables, and the probability levels shown.

4	0	$P = 5\%$	
1	4		

9	0
3	12

$P < 1\%$

12	0
24	14

$P < 5\%$

2	1
7	40

$P > 5\%$

12	1
3	12

$P = 1\%$

14	9
9	15

$P > 5\%$

Ex. 46. As part of a study to find out whether an artist's personality is expressed in his paintings, an experiment was designed in which monkeys were given a painting test. There were 30 monkeys, 15 of whom were placid and 15 excitable. The expressiveness of their paintings was judged (without knowing their source) by a prominent artist, who was asked to classify the expressiveness of each painting as 'nil', 'low', or 'high'. Here are the results –

Monkeys' nature	Expressiveness of paintings		
	Nil	Low	High
Placid	5	4	6
Excitable	0	4	11

Do these results indicate a significant association between monkeys' excitability and the expressiveness of their paintings?

Data: A glance at the above data Table suggests that the expected numbers in several categories may be less than 5, and if so, the ordinary χ^2 Test would be invalid. By calculating $\dfrac{R \times C}{N}$, we find that all the 'nil' and 'low' classes have expected numbers under 5.

It would not spoil the aim of our experiment if we pooled the results of the 'nil' and 'low' categories, and so produced the following 2 × 2 table –

Monkeys' nature	Expressiveness of paintings		Totals
	Nil or low	High	
Placid	9	6	15
Excitable	4	11	15
Totals	13	17	30

Calculation: The above data table is not suitable for Yates' χ^2 Test, because N is only 30. So we shall apply Fisher's Test.

First we must turn the table around to make $n_1 = 13$ (the smallest of the 4 marginal totals), thus –

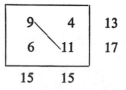

Secondly, we check to see that $a \times d$ is larger than $b \times c$; 9×11 is larger than 4×6, so our table is correctly arranged.

We turn to the Table for Fisher's Test on page 305, which is the one for $a = 9$. Then, for $b = 4$ and $c = 6$, we see that d would have to be 14 to reach the 5% probability level. Our d is only 11, so we conclude that a significant association between monkeys' excitability and their expressiveness in paintings is not proven. (The data was fictitious, of course.)

Questions

Q83. Siegel, in *Non-Parametric Statistics* (McGraw-Hill, 1956, p. 100) quotes some interesting data collected by Lerner, Pool, and Schueller. Of the 15 men who constituted the German cabinet in 1934, 9 were Nazis and 6 were not. All the 6 non-Nazis had begun their careers in non-political occupations, whereas 8 of the 9 Nazis had been employed in political work throughout their careers. Is this difference in proportions statistically significant?

Q84. A sports magazine reported some tests on 2 makes of outboard motors for speed-boats. Three out of 4 of Make *A* had pushed a trial boat along at more than 60 m.p.h., while none of the 4 of Make *B* had managed to reach this speed (on the same boat and under the same trial conditions).

(*a*) Apply Fisher's Test to see if this difference is significant.

(*b*) What do you think of this idea of categorizing measurements into those above and those below a certain figure, and then applying a statistical test for the numbers in each category?

Q85. In an article discussing the χ^2 Test in *Biometrics Journal*, 1954, p. 420, Professor W. G. Cochran stated that in a 2×2 contingency table, Yates' χ^2 Test remains reasonably accurate even when the expected number in one or more cells is less than 5 (but not less than 1), provided that N is more than 40. Test this statement for yourself by applying both Yates' χ^2 Test and Fisher's exact Test to the following data tables in which $N = 50$. What do you conclude?

(*a*)

21	19
9	1

(*b*)

8	5
35	2

(*c*)

22	0
22	6

(*d*)

5	10
24	11

COCHRAN'S TEST (1950)

Purpose

To see if there is a significant difference between 3 or more sets of matched observations, when the observations are divided into 2 categories (such as good and bad).

The matching may be achieved by –

(a) triplicating members of 3 sample groups, carefully, or symmetrically, or by splitting (see p. 118), or

(b) using 1 sample group for different treatments, different methods of testing, different observers, or on different occasions (see p. 120).

Principle

The tentative assumption is made that there is no significant difference between the various sets of observations. On this basis the probability of getting any observed inequality as a result of chance can then be determined by means of this test.

The ordinary χ^2 Test is not applicable in these circumstances, because the sample groups of observations are not independent. The way to modify the χ^2 Test to suit these matched observations was described by Professor William G. Cochran of Harvard University in *Biometrika Journal*, 1950, pp. 256–66.

Data Required

k = number of sets of observations (A, B, C, etc.) being compared. This must be 3 or more.

n = number of individual observations in each set. This must be at least 10.

Plus all the individual results in all the sets of observations.

Procedure

(1) Prepare a calculation table with a fairly wide column for the k sets of observations (A, B, C, etc.), and a row for each of the n test conditions. The table headings required are thus –

Conditions	Observations			y	y^2
	A	B	C		

(2) For this test, all observations must fall into one or other of 2 classes – positive or negative, present or absent, cured or not-cured, etc. Choose one of your classes and call it 'x'. It doesn't matter which class you call 'x', but it is a bit easier if you select the smaller one. Now enter the observed results in your calculation table. Number or otherwise identify each test condition in the first column. This will create n rows in your table. Then enter the tally of x's in the column of observations. It saves time and is less confusing if you simply leave a blank space for each occurrence of 'not-x' (i.e. observations which fell into the class that is not the x class).

(3) Add up the number of x's in each column of observations. Call these totals x_A, x_B, etc.

(4) Square the values of x_A, x_B, etc. (with the help of the Table of Squares at the back of this book). Then add these squares together to make a total which we shall call 'X' –

$$X = x_A{}^2 + x_B{}^2 + x_C{}^2, \text{etc.}$$

(5) Add up the number of x's in each row (across the table), entering the answers in the column headed 'y'.

(6) Add up all the values of y, and call this total 'Y'. Then square the value of Y to get Y^2.

(7) Square each value of y, entering the answers in the last column of the table. Add up all the values of y^2, and call this total 'Z'.

(8) Then calculate χ^2 from Cochran's formula –

$$\chi^2 = \frac{(k - 1)(kX - Y^2)}{kY - Z}$$

Interpretation

The probability of getting, by chance, a value of χ^2 as large as calculated is found by referring to the χ^2 Table on page 276. The degrees of freedom for this test are $k - 1$.

The higher the value of χ^2, the less the likelihood of our no-significant-difference assumption being true. The significance of the various probability levels was discussed on page 140.

Example

Ex. 47. A teacher wanted to find out the best way to demonstrate the proof of a certain mathematical formula to his class. There were 3 possible ways of proving this formula, but perhaps his students would not find them all equally easy to understand. So he explained each of the 3 proofs to his class of 18 maths students, and then asked each student to write down which, if any, of the 3 proofs he had understood. That night he analysed the results – 5 had not understood Method A, 15 had not understood Method B, and 5 had not understood Method C. Does this represent a significant difference between the 3 methods, or might it be reasonably attributed to chance?

Data and Calculation: All 3 methods were tried on each member of the sample group, hence the observations are matched. Now, the above data is not sufficient to deal with matched observations. We must know all the individual results. These are presented in the table below. There are 2 classes of results – understood and not-understood. The latter is the smaller class, so each instance of 'not-understood' has been entered in the table as an 'x'.

| Student | Methods | | | y | y^2 |
	A	B	C		
1	x	x	x	3	9
2		x		1	1
3		x		1	1
4		x		1	1
5		x	x	2	4
6				0	0
7		x		1	1
8	x	x		2	4
9	x	x	x	3	9
10				0	0
11				0	0
12		x		1	1
13		x		1	1
14		x	x	2	4
15		x		1	1
16	x	x		2	4
17	x	x	x	3	9
18		x		1	1
Totals	$x_A = 5$	$x_B = 15$	$x_C = 5$	$Y = 25$	$Z = 51$
Totals squared	25	225	25	625	—

$$X = x_A{}^2 + x_B{}^2 + x_C{}^2$$
$$= 25 + 225 + 25$$
$$= 275$$

We also have $Y = 25$, $Y^2 = 625$, $Z = 51$, and $k = 3$.

$$\chi^2 = \frac{(k-1)(kX - Y^2)}{kY - Z}$$

$$= \frac{(3-1)(3 \times 275, - 625)}{(3 \times 25) - 51}$$

$$= \frac{2(825 - 625)}{75 - 51}$$

$$= \frac{2 \times 200}{24}$$

$$= 16 \cdot 67$$

The degrees of freedom are $k - 1$, which thus equal 2 in the present case. Reference to the χ^2 Table on page 276 shows that this value of χ^2 with 2 degrees of freedom could be expected to occur by chance (if there was no significant difference between the 3 methods) with a probability of less than $0 \cdot 1\%$. We conclude, then, that a very significant difference does exist between the ease of understanding the 3 mathematical proofs.

Question

Q86. A manufacturer of ladies' hats shows 4 of his latest creations (*A*, *B*, *C*, and *D*) to 12 buyers from retail stores, 7 of whom place orders (marked x) as follows –

Buyer	Orders for hats			
	A	*B*	*C*	*D*
1	x	x		
2		x	x	
3			x	x
4		x		x
5			x	
6			x	
7			x	
8–12		nil		

Is there a statistically significant difference between the demand for each of the 4 styles of hats?

RUNS TEST FOR RANDOMNESS

Purpose

To detect a non-random sample.

All tests of significance are carried out on samples, with the aim of inducing conclusions about the parent groups from which the samples have been drawn. For instance, when we compare 2 treatments, our samples must be truly representative of their parent sources, or the comparison would be pointless. What would be gained by comparing my hand-picked results with your hand-picked ones? It might only show who was the better picker! It is to be stressed, then, that significance tests, being based on the Laws of Chance, will only be valid if every individual in a parent group has an equal chance of being chosen to make up the sample; such a sample is called a random sample (p. 41).

The randomness of a sample is thus an extremely important quality. How can one tell if a sample is random or not?

If one or more characteristics of a parent group are known, a sample drawn from that group can be tested for randomness by applying an appropriate test of the kind already described. If the arithmetic mean of the parent group is known, a sample could be tested by 'Student's' t Test (p. 160). Or again, a random sample of people ought to contain approximately the same number of males and females, and any inequality in the sample could be tested with the 50% Probability Test (p. 254). You have already encountered this kind of problem in *Q66* (p. 244). These tests will not, of course, prove that a sample *is* random, but they can certainly indicate when a sample *is not* random.

Well, suppose a sample of 50 people contained 25 males and 25 females. It might well be a random sample on the evidence of its proportions. But what if you discovered that the first 25 people in the sample were all males, and the second 25 people were all females? Or what if the males alternated strictly with the females throughout the whole sample (male, female, male, female, etc.)? Your intuition will tell you that neither of these circumstances would happen by chance except very, very rarely. So you would reject such samples on the grounds that they are almost certainly non-random. Between these 2 extremes will be a large number of possible combinations, and the probability of any particular arrangement can be told by applying the present test.

One of the most ingenious things about this test is that it can be applied to any sample, even when nothing is known about its parent source! It works purely on the *order* of occurrence of observations.

Principle

For this test, the observations in the sample must be in 2 classes. Binomial samples (e.g. males and females) are already in this state. Samples composed of more than 2 classes (e.g. Conservatives, Progressives, and Communists) are easily converted into a binomial state by combining some of the classes. A sample which consists of measurements (inches, gallons, etc.) can also be divided into 2 classes, by splitting the measurements into those above the median (the middle measurement) and those below the median (see *Q84*, p. 317).

Now, knowing the order of the observations in the sample, and having the observations in 2 classes, we can proceed to test the amount of clustering of observations of a similar kind. Some clustering is perfectly normal in a random sequence of events, as can be seen in the Table of Random Numbers (p. 43), but too many or too few clusters point to the sample not being random.

Each cluster of alike observations is called a *run*, so in this test we count the number of runs, and then, taking the proportion of each class in the sample into consideration, we determine the probability of getting the observed number of runs by chance. Our tentative assumption is that the sample is random, so if the observed number of runs is likely to occur by chance only once in 20 times (or less often), we reckon that our tentative assumption is probably not true, and the sample should be rejected as being unsuitable for statistical purposes (i.e. for the purpose of drawing general conclusions).

Many men have been involved in developing this test. The main ones appear to have been W. L. Stevens, who described the way of calculating the exact probabilities for this test (*Ann. Eugenics*, 1939, pp. 10–17), on which the Tables herein are based, and secondly, Professor A. Wald and J. Wolfowitz, who discovered the way to calculate the expected mean and standard deviation of the number of runs (*Ann. Math. Statist.*, 1940, p. 151), which enables a form of z Test to be used for cases beyond the range of the Tables.

Data Required

All the observations constituting the sample, together with the order in which they occurred.

Procedure

(1) With *samples of measurements*, first arrange all the observations in order of size. The value of the middle observation (the one with an equal number of measurements above it and below it) is the median of the sample (p. 74). If there are more than 20 measurements in the sample, it is permissible to guess the median, so long as the numbers in each group turn out to be about equal.

(2) Make a list of the measurements in order of their occurrence in the sample. Below each measurement put a plus sign if its value is equal to or above the true or assumed median, or a minus sign if below it.

(3) With a *binomial sample* (or a multi-class sample converted into a binomial one), list the observations in order of their occurrence, using abbreviations such as plus and minus signs for present and absent, or M for male and F for female, or C for cured and N for not-cured, etc.

(4) Mark the end of each run (i.e. each cluster of alike observations) with a conspicuous line. A run ends whenever the sign changes from plus to minus or vice versa, from male to female or vice versa, etc. A run may consist of 1 observation or many.

(5) Add up the number of runs (r).

(6) Add up the number of observations in each class (plus or minus, etc.). Let the number of the smaller class be called n_1, and the number of the larger class be n_2. If the classes are of equal size, $n_1 = n_2$.

Interpretation

If there are too many or too few runs, our tentative assumption that the sample is random is probably not correct, so we reject the sample as being probably non-random.

(a) *If n_1 and n_2 are up to 20*, refer directly to the r Tables opposite. The numbers of runs shown in these Tables are the deviations from averages which would occur by chance only once in 20 times ($P = 5\%$). Therefore reject your sample as non-random if the observed number of runs is equal to or less than the smaller number in the Tables, or is equal to or more than the larger number in the Tables. If you come across a dash (—) in

r TABLES SHOWING 5% LEVELS FOR RUNS TEST

n_1	n_2	No. of runs		n_1	n_2	No. of runs	
2	2–11	—	—	10	16–18	8	19
2	12–20	2	—	10	19	8	20
3	3–5	—	—	10	20	9	20
3	6–14	2	—	11	11	7	17
3	15–20	3	—	11	12	7	18
4	4	—	—	11	13	7	19
4	5–6	2	9	11	14–15	8	19
4	7	2	—	11	16	8	20
4	8–15	3	—	11	17–18	9	20
4	16–20	4	—	11	19–20	9	21
5	5	2	10	12	12	7	19
5	6	3	10	12	13	8	19
5	7–8	3	11	12	14–15	8	20
5	9–10	3	—	12	16–18	9	21
5	11–17	4	—	12	19–20	10	22
5	18–20	5	—	13	13	8	20
6	6	3	11	13	14	9	20
6	7–8	3	12	13	15–16	9	21
6	9–12	4	13	13	17–18	10	22
6	13–18	5	—	13	19–20	10	23
6	19–20	6	—	14	14	9	21
7	7	3	13	14	15	9	22
7	8	4	13	14	16	10	22
7	9	4	14	14	17–18	10	23
7	10–12	5	14	14	19	11	23
7	13–14	5	15	14	20	11	24
7	15	6	15	15	15	10	22
7	16–20	6	—	15	16	10	23
8	8	4	14	15	17	11	23
8	9	5	14	15	18–19	11	24
8	10–11	5	15	15	20	12	25
8	12–15	6	16	16	16	11	23
8	16	6	17	16	17	11	24
8	17–20	7	17	16	18	11	25
9	9	5	15	16	19–20	12	25
9	10	5	16	17	17	11	25
9	11–12	6	16	17	18	12	25
9	13	6	17	17	19	12	26
9	14	7	17	17	20	13	26
9	15–17	7	18	18	18	12	26
9	18–20	8	18	18	19	13	26
10	10	6	16	18	20	13	27
10	11	6	17	19	19–20	13	27
10	12	7	17	20	20	14	28
10	13–15	7	18				

Adapted from F. Swed and C. Eisenhart, *Ann. Math. Statist.*, 1943, pp. 83–6.

these Tables, it means that a 5% probability level cannot be reached in the particular circumstances, regardless of the number of runs.

(b) *If n_1 and/or n_2 is more than 20,* calculate z from the following formula, which is Wald and Wolfowitz's formula slightly rearranged to keep the numbers in the working-out from getting too big, and with the added refinement of Yates' correction –

$$z = \frac{\left|\dfrac{2n_1n_2}{N} - r\right| \pm c}{\sqrt{\dfrac{2n_1n_2}{N} \times \dfrac{2n_1n_2 - N}{N^2 - N}}}$$

where $N = n_1 + n_2$,

and c = correction factor having a value of

$+0{\cdot}5$ if $r < 2n_1n_2/N$

or $-1{\cdot}5$ if $r > 2n_1n_2/N$.

The critical probability levels of z were given in the z Table on page 154. It is usual to reject any sample as being non-random if $P = 5\%$ or less (i.e. if $z = 1{\cdot}96$ or more).

Examples

Ex. 48. A presenting sample of patients (p. 49) consisted of 28 males and females in the following order –

M F M FF M F MM FFF M F MMM F MM FF MM FFFF

Does this provide evidence that the sample is not random?

Data and Calculation: The above data has been set out in such a way that the number of runs (= clusters) can be counted quite easily. There are 16 runs.

The number of males in the sample is 13, and the number of females 15. We want n_1 to be the smaller of the two, so –

$$n_1 = 13 \quad n_2 = 15 \quad r = 16$$

Armed with these 3 figures, we turn to the r Tables on page 325, and there find that for these values of n_1 and n_2, the critical values of r are 9 and 21. Our value of r is between these limits, so the observed distribution of runs in our sample could be expected by chance with a probability of more than 5%.

We conclude that our sample is not proven to be non-random.

Ex. 49. A taxi driver kept a note of his monthly mileage for a period of a year. Do the following figures indicate a seasonal or any other periodic fluctuation, or might they reasonably be accounted to vary in a random manner?

Jan.	Feb.	Mar.	Apr.	May	June
4,690	4,910	3,520	3,330	3,140	2,850

July	Aug.	Sep.	Oct.	Nov.	Dec.
3,400	3,090	3,480	4,650	3,830	5,270

Data and Calculation: We must first find the median. Here are the mileages in order of size –

2,850	3,090	3,140	3,330	3,400	3,480
3,520	3,830	4,650	4,690	4,910	5,270

The median is seen to lie between 3,480 and 3,520, so we shall assign the median a value of 3,500.

Here again is the data in its original order, with a plus sign beneath each number larger than the median, and a minus sign beneath each one smaller than the median.

4,690	4,910	3,520	3,330	3,140	2,850
+	+	+ /	−	−	−

3,400	3,090	3,480	4,650	3,830	5,270
−	−	− /	+	+	+

As can be seen, the number of runs is 3. There are 6 plus and 6 minus signs, so $n_1 = n_2 = 6$.

Turning now to the *r* Tables for the Runs Test (p. 325), we find that for n_1 and n_2 both equal to 6, the probability of getting as few as 3 runs is 5%. We conclude therefore that the fluctuations are not random, but are probably due to some other factor (such as seasons).

Ex. 50. Would the following series of boys (*B*) and girls (*G*), selected haphazardly for an opinion survey at a coeducational school, be acceptable as a random sample?

G BB GG B G BBB GGGG B G B G BB

G B GGG BB G B GG B GG B G B G BB G

Data: There are 19 boys, 22 girls, and 27 runs. The number of girls is too many to be handled by the r Tables, so we shall calculate z, using –

$$r = 27 \quad n_1 = 19 \quad n_2 = 22 \quad N = n_1 + n_2 = 41$$

Calculation: The value of the correction factor (c) depends on whether r is more or less than $2n_1n_2/N$, so this must be worked out first.

$$\frac{2n_1n_2}{N} = \frac{2 \times 19 \times 22}{41} = \frac{836}{41} = 20 \cdot 39$$

The observed number of runs (r) was greater than this, so $c = -1 \cdot 5$.

$$z = \frac{\left| \dfrac{2n_1n_2}{N} - r \right| - 1 \cdot 5}{\sqrt{\dfrac{2n_1n_2}{N} \times \dfrac{2n_1n_2 - N}{N^2 - N}}}$$

$$= \frac{| 20 \cdot 39 - 27 | - 1 \cdot 5}{\sqrt{20 \cdot 39 \times \dfrac{836 - 41}{1,681 - 41}}}$$

$$= \frac{6 \cdot 61 - 1 \cdot 5}{\sqrt{20 \cdot 39 \times \dfrac{795}{1,640}}}$$

$$= \frac{5 \cdot 11}{\sqrt{9.88}}$$

$$= \frac{5.11}{3 \cdot 14}$$

$$= 1 \cdot 63$$

Reference to the z Table (p. 154) shows this to indicate a probability of just over 10%, so a significant difference from the number of runs that could be expected if the sample was truly random is not proven. We therefore cannot deny that the sample may be random.

Questions

Q87. Here is a list of the winners of the All Star Baseball Game between the American (Am.) and National (N) League from 1933 to 1955. (No game was played in 1945.)

Year	Winner	Year	Winner
1933	Am.	1944	N
1934	Am.	1946	Am.
1935	Am.	1947	Am.
1936	N	1948	Am.
1937	Am.	1949	Am.
1938	N	1950	N
1939	Am.	1951	N
1940	N	1952	N
1941	Am.	1953	N
1942	Am.	1954	Am.
1943	Am.	1955	N

Does this list contain evidence of the action of factors other than chance in determining the outcomes? (From W. A. Wallis and H. V. Roberts, *Statistics – A New Approach*, Free Press, 1960.)

Q88. Instead of finding the true median in *Ex. 49* (p. 327), suppose we had guessed the median to be 4,000. Would this affect the result of the Runs Test? Try it, and see.

Q89. Dr Johnson was a bit worried about the 20 patients he was going to present at the Medical Society. True, he had collected them consecutively, but as he looked over the list, he wondered if they formed a random sample, or rather, a sample equivalent to a random sample. Here was the list of patients, in the order he had encountered them –

Mr Cutler	Mrs O'Toole
Mr Flam	Mrs Price
Mr Harris	Miss Smith
Mrs Donovan	Miss Simonetti
Miss Young	Master Webb
Master Samuels	Mrs Pugh
Mrs Terry	Mrs Romano
Mrs Lobanski	Mrs Gibbs
Mrs Harvey	Miss Brown
Prof. Moriati	Mrs Townsend

(*a*) First he thought, 'Aren't there too many women? After all, it is well known that Coccygeal Grumphitis is equally common in both sexes.' Please help him out. Look up the Guide to Significance Tests at the very end of this book, find the right test to use, and apply it.

(*b*) Then he thought, 'This disease shows no particular racial predeliction. Now, there can't be more than 5% people with Italian names in our city, yet I've collected 3 in my sample. Isn't this too many?' Again, please find the right test to use, and answer his question.

(*c*) Then he mused, 'I've only got 6 males in the sample, and 3 of them are right there at the beginning. Is this a reasonable thing to expect from chance?' You'd better try the Runs Test on this one.

(*d*) By now, poor Dr Johnson was in a highly nervous state. Won't you put him out of his misery by doing just one more test. Do a Runs Test after dividing the patients into 2 groups according to the first letter of their names. Find the median level of the names in the sample, in the same way as you would with numbers. Does this provide evidence of non-randomness?

MISCELLANEOUS QUESTIONS

(See remarks on page 151.)

Q90. In a gross of eggs there is 1 bad one. If you buy a dozen eggs from this batch, what is the probability that you will get the bad egg?

Q91. Over a 10-year period, trains have smashed into cars at an unguarded level-crossing a total of 35 times. A safety device is then installed to warn motorists of approaching trains, and in the next year there are no smashes at all. Can it be said that this proves the safety device is effective?

Just in case you have any difficulty in starting these miscellaneous problems, here is the way to go about it. From the overall pattern of the Guide to Significance Tests at the back of this book, you can see that the first thing is to decide whether the problem in hand deals with measurements, isolated occurrences, or observations divided into classes. You should have no trouble in recognizing the present case as being one of isolated occurrences (accidents). Next, the Guide divides isolated occurrences into 3 groups, and you must therefore decide whether the present problem concerns the comparison of a sample with an average, or whether 2 sample groups are being compared, or whether there are 3 or more samples being compared. Suppose you consider the situation to be the comparison of 2 samples, one of 10 years' duration, the other of 1 year's duration. If, as we tentatively assume, there is no significant difference between the accident *rates* in the 2 samples, the actual *number* of accidents in each sample ought to be in direct proportion to the duration of each sample; this means that the expected number of accidents in the 2 samples will be different (because their durations are different). Referring again to the Guide, and noting that the observed number of accidents in the smaller sample was 0 (which is 'not more than 4'), we are then told to look up the Binomial Test on page 237. Our first job there is to check under 'Purpose' to make sure that this is the right test to use for this data. Unfortunately, it is not, for the duration of the smaller sample is only $\frac{1}{11}$ (= 9·1%) of the combined sample durations, and the Binomial Test only applies (i.e. is only accurate) when this is 10% or more. In the paragraph immediately following, however, we are reminded that 'if the proportion is less than 10%, use Poisson's Test, p. 230'. We would have reached this conclusion at once from the Guide if we had simply looked upon the present problem as being the comparison of a sample and an average. Therefore check that the purpose of Poisson's Test is for data of the kind and dimensions of the problem under discussion; if you are satisfied that it is, check the 'Data Required' section to make sure that a sufficient amount of data is provided, and if so, write down the necessary symbols (x, n, and P_x), equate them to the observed data values ($x = 0$, etc.), and carry on through the 'Procedure' and 'Interpretation' sections. Finally, if you strike a snag, try your hardest to solve it before looking up the answer, for you will then learn twice as much about handling the kinds of problems which occur in real life. You will develop a statistical 'nose'.

Q92. You know, of course, that Hamlet, Lear, and Henry VIII were all keen fishermen. Well, one day they began to argue about the relative merits of their favourite baits for catching cod. To settle the dispute, they took a picnic lunch to a river, and began fishing. Hamlet caught 10 cod and 12 other fish. Lear caught 20 cod and 12 others. Henry VIII caught 30 cod and 12 others. Then they called for their court statistician to assess whether the differences were significant. You be the statistician.

Q93. Five doctors were having dinner together, and the conversation turned to the frequency of night calls. It was suggested that such calls prevented them from getting an average amount of sleep. Each one told his previous night's length of sleep, as follows –

$$5 \quad 6 \quad 4 \quad 5\tfrac{1}{2} \quad 6\tfrac{1}{2} \text{ hours}$$

Does this indicate a significant difference from other professional men, who get an average of 7 hours' sleep a night?

Q94. Miss Neatfingers and Miss Glam are applying for the position of secretary to Mr Bigg, and are invited to take a typing test. This consists of typing out a 4-page letter, and they make the following number of mistakes –

Miss Neatfingers   8
Miss Glam   16

Mr Bigg prefers the appearance of Miss Glam, but the above result seems to be against her. But is the result really significant?

You should be able to advise Mr Bigg of the answer within 60 seconds.

Q95. The Grand Sultan of Mazurka claimed to have fathered 3 times as many boys as girls. He further claimed that this was not just due to chance, but was to be interpreted as a sign of his masculinity. Well, is this a statistically significant difference from the usual 50:50 proportions?

Q96. A comparative trial of 2 underarm deodorants, *A* and *B*, was carried out on 34 people. Seventeen of the group were given *A*, and the other 17 were given *B*. Points were then awarded to each product, according to each person's report; no effect was scored 0, slight to moderate effect was scored 1, and good to excellent scored 2. The total points were then compared. Can you suggest a better design for such an experiment?

Q97. In 1964, 40% shipments of US cotton to Germany were sent to arbitration because of complaints about the quality being substandard, according to *Time Magazine*. Would it signify a real worsening of the situation if 20 out of the first 40 shipments in 1965 were likewise the cause of complaint, or might this simply be a matter of bad luck?

Q98. The 6 honours pupils at Mother Hubbard's Cookery School had challenged the 8 honours pupils at the Good Wives' Training College to a cake-making competition. Would it indicate a significant difference between the 2 groups if the individual marks gained by the competitors were as follows –

Mother Hubbard's	Good Wives'
91	91
92	90
96	91
97	87
97	94
93	95
	88
	89

Q99. A microscopic insect called Demodex was discovered infesting the eyelashes in 34 out of 50 patients (= 68%) with inflamed lid margins, and was thought to be the cause of the inflammation. By contrast, 45 patients with normal eyelids were examined, and Demodex was found in only 22 of them (= 49%).

What conclusion can be drawn from this data? (Data from AMA *Archives of Dermatology*, September 1963.)

Q100. An opinion survey in Madagascar found that 73% of people considered architect's fees to be too high. I asked a random sample of 30 people in Melbourne the same question, and found 15 who thought their fees were too high. Is this a significant difference? (The data is fictitious, of course.)

Q101. Last Christmas the excitement at the Snake Gully Annual Rodeo was even greater than usual because the organizers had managed to get both the Queensland champion and the West Australian champion to compete. The Queenslander won the day, but a lot of people reckoned it was just a matter of luck. Here are their scores for the various events; do you think the differences can reasonably be attributed to luck?

Event No.	I	II	III	IV	V	VI	VII	VIII	IX	X	Totals
Q'land	18	19	18	13	17	20	16	19	12	19	171
W. Aust.	16	19	10	16	18	14	18	15	8	12	146

Q102. A new drink is claimed to hardly ever cause a hangover. To test this wonder beverage, 30 people who were particularly susceptible to hangovers volunteered to first drink 2 ounces of pure alcohol and note the effect, and then a week later to drink 6 glasses of the new drink (containing an equivalent amount of alcohol). If the results were as follows, could the new drink be said to be significantly less prone to cause hangovers than pure alcohol?

		Hangover after alcohol	
		Yes	No
Hangover after { Yes		12	3
new drink { No		13	2

Q103. Which is more important in the making of a criminal, heredity or environment? In a classical study of this problem, Professor J. Lange investigated a group of convicted German criminals who in each case had a twin brother or sister. You know that there are 2 kinds of twins – (a) *identical twins*, who develop from 1 egg cell and who are ger.etically identical and are necessarily of the same sex, and (b) *fraternal twins*, who develop from 2 egg cells and are genetically different and may be of the same or different sexes. Now, it seems reasonable to assume that environmental influences will be as similar in the case of fraternal twins of the same sex as in the case of identical twins, so any difference which develops between fraternal twins of the same sex which is not shown by identical twins must be attributable to genetic differences. The effects of heredity are thus isolated and exposed by the conditions of matching environments and non-matching genetic constitutions. Thus it was that in this particular investigation, 13 convicted criminals were found who each had an identical twin; these twins were sought out, and in 10 cases the twin also proved to be a convicted criminal. By contrast, 17 convicted criminals were found who had a fraternal twin of the same sex; these twins proved to be criminals in only 2 cases. Cast this data into a suitable table, and apply an appropriate test to see if this difference is statistically significant. (From J. Lange, *Crime and Destiny*, Allen & Unwin, 1931, translated by C. Haldane.)

Q104. Over a period of some years, a large store has gained an average of 4 new customer accounts per week. Under a new manager, there are only 30 new accounts opened in the first 3 months. Is this significantly different from the previous average? (Dismiss the manager if the probability of it being due to chance is less than 5%.)

Q105. The owners of a kiln which bakes terracotta roof tiles advertise that an average of 98% of their tiles are perfect. Would you have a right to complain if you bought a load of 800 of their tiles and found that it contained 28 defective tiles?

Q106. A representative of a life assurance company finds that he can sell policies to 20% of the people he interviews. He then takes a course in 'sales psychology', after which he sells 16 policies among the next 50 people interviewed. Does this prove that the course has helped his selling ability, or could this 'post-operative sample' be reasonably ascribed to luck?

Q107. A factory which makes spectacle lenses finds that 10% lenses have to be re-processed (because of scratches and other imperfections) before being dispatched. A special drive is then made to reduce this percentage. On the first day, 200 lenses are produced, and only 10 are imperfect. 'There you are,' said the foreman, '100% improvement!' But is this a real improvement, or might it just be a matter of chance?

Q108. Suppose that a questionnaire is sent to 120 dermatologists, asking whether they prefer X-ray therapy or surgery for treating skin cancers. Of course, they would all reply. Suppose that 71 preferred X-ray therapy, and 49 preferred surgery. Would this be a statistically significant difference?

Q109. Three brands of lubricating oil were tested on 3 cars of the same kind, used for similar work and for the same mileages. At the conclusion of the test period measurements were made of the maximum wear in each cylinder (the cars were a 4-cylinder model). The results were as follows –

Oil	Cylinder wear (in thousandths of an inch)			
A	7·5	7·4	7·8	7·1
B	6·9	7·2	7·4	7·1
C	6·3	7·0	6·5	6·8

Take it that under test conditions like these, previous experience has shown that there is no significant difference in cylinder wear between cars so long as they use the same oil.

Compare these measurements to determine whether there is a significant difference between the amount of wear caused by the 3 oils. If a significant difference is demonstrated, how do oils *A* and *B* compare individually with the control oil, *C*? Are they both significantly inferior to it? (From H. J. Halstead, *An Introduction to Statistical Methods*, Macmillan, 1963.)

Q110. 'The incidence of complications following primary smallpox vaccination in different age groups is summarized in the following table –

Age group (years)	Incidence of complications
Under 1	14 per million
1 to 4	6 per million
5 to 14	24 per million
15 and over	25 per million

'It appears from these figures that the safest time for primary vaccination is between the ages of 1 and 5 years.'

Can you support this statistically? (Data from *Med. Journ. of Australia*, July 1964.)

Q111. Ointment *A* is known to cure 90% patients suffering from tinea within 4 weeks. A new ointment (*B*) is tried out on 115 tinea patients, and cures 110 of them within 4 weeks. Is this significantly different from the results that could be expected with ointment *A*?

Q112. Damon Runyon wrote a lot of short stories about the guys and dolls of Broadway, N.Y. Perhaps you remember his tale about the famous eating contest held in Mindy's Restaurant one night in the summer of 1937. It all starts when Runyon and his friends hear about a Bostonian eater who boasts that he can eat up to 21½ lb of nourishment, even if he is not feeling very hungry. Not being averse to winning some bets on any contest of skill, or otherwise, they challenge the Bostonian to an eating contest with a New Yorker of their acquaintance, who is none other than a character named Nicely-Nicely Jones. Now, although it isn't mentioned in the story, I happen to know that for some time they are keeping a tab on Nicely-Nicely's eating form, and know that his average meal is 17·4 lb, with a standard deviation of 2·5 lb. What actually transpires is told in 'A Piece of Pie' (in *Take It Easy*), but if the Bostonian manages to equal his own previous record (21½ lb), what chance has Nicely-Nicely got of winning the contest?

Q113. There is a township up in Northern Australia where there had been 16 drought years scattered at random through the past 80 years. Then came 3 years of continuous drought. What is the probability that there will be rain in the next year?

Q114. In the Second World War, the Americans determined the best way of training bombing crews by statistically analysing the results of bombing scores after various training schemes. Suppose they wanted to find out if the training time could be reduced without appreciable detriment to the bomber's accuracy. To do this they might have chosen 20 men at random from a large group of men who were about to start bombing training. The chosen 20 could then be divided, again at random, into 2 groups of 10 men each. One group could then have been given standard training, and the other group could have been given a condensed course. Suppose that, after completing their training, all 20 men were given the same bombing accuracy test, with the following results –

	Bombing test Passed Failed	
With standard training	8	2
With condensed training	2	8

Should the idea of condensing the course be dropped, or might this result happen by chance with fair frequency? Accept $P = 5\%$ as critical.

Q115. The trumpeter said, 'People who are really musical love all kinds of music. They just don't like jazz or classics, they like both.' We won't argue with such a man, of course, because his definition of a person who is really musical is one who likes all kinds of music. However, we could ask a random sample of people about their tastes, in an endeavour to find out if music lovers tend to be broad or narrow in their musical interests. I tried this once on a sample of 60 people, with the following results –

		Classical music		
		Love	Like	Hate
Jazz {	Love	3	7	5
	Like	0	4	11
	Hate	4	4	22

Extract from this table whatever information you need to test whether the numbers of those who like or love both kinds of music are significantly different from those who like or love only one kind or the other.

Q116. If 12 sharks have been sighted at Bondi Beach in the past 10 years, would the sighting of 5 sharks in 1 year represent a positive increase in the number of sharks in the area, or might it reasonably be due to chance?

Q117. Ten tests are performed with an insect spray to find out if its efficiency is increased significantly when the amount of spray is increased beyond a 'recommended' concentration in the air of 0.0005%. Do the following results support the manufacturer's claim that 'a little of this spray will do the same job as a lot'?

Concentration of spray in the air (percentage)	No. of flies killed in 2 minutes (out of 100 per test)
0·0005	80
0·0007	78
0·0009	85
0·0014	89
0·0020	83
0·0040	93
0·0060	90
0·0080	89
0·0100	89
0·0150	94

Q118. Three hundred patients suffering from chronic backache were divided at random into 3 equal-sized groups. The first group was given Treatment *A*, which relieved 48% of them. The second group was given Treatment *B*, which relieved 61%, and the third group was given Treatment *C*, which was successful in 71%. Do these figures indicate a statistically significant difference between the efficacy of the 3 treatments?

Q119. During an influenza outbreak, a firm found that 6 out of their office staff of 25, and 6 out of their factory staff of 45, contracted influenza. Does this prove beyond reasonable doubt that the office staff are really more susceptible to this infection than the factory staff? Is your answer trustworthy?

Q120. The English scientist, Sir William Crookes, discovered the element thallium in 1861. He did 10 tests to determine its atomic weight, and got the following results –

203·628   203·632   203·636   203·638   203·639
203·642   203·644   203·649   203·650   203·666

(Data from H. J. Halstead, *An Introduction to Statistical Methods*, Macmillan, 1963.)

Were these results significantly different from the modern figure of 204·39?

Q121. The research worker said, 'Most people can't distinguish one brand of beer from another.' He then proceeded to demonstrate his point. He lined up 12 volunteers (who each claimed could recognize the flavour of his favourite brand), and gave each one a taste of 4 different beers. Only 5 lived up to their claim and got the right answer; the other 7 failed to identify their favourite brand. The research worker beamed, 'There you are. It's just like I said.' But you had better test this data to assess the part played by chance. What is the probability of a person guessing his favourite brand? What number of volunteers would you expect to have given the right answer if they had no power of discrimination at all? Was the observed number significantly different from that expected number?

Q122. In the course of a year, 360 carburettors became faulty in a batch of 4,000 cars, so the manufacturer changed to a different kind of carburettor. In the next year a further 4,000 cars were produced, and in these, 260 carburettors became faulty. Is this a statistically significant improvement?

Q123. A doctor writing in the *Medical Journal of Australia* (February 1964) reported on 83 patients suffering from hydatid cysts. There were 47 men and 36 women in the group. Do these figures indicate that men get these cysts significantly more often than women, or might the observed disparity be fairly attributed to chance?

Q124. I wanted to buy an electric shaver recently. I tried out 2 models for 4 shaves each, and noted the time taken for each shave. Do the following results indicate a significant difference in the shaving times of the 2 shavers?

Shaver	Shaving time (minutes)			
A	4	$4\frac{1}{4}$	5	$4\frac{1}{2}$
B	$4\frac{3}{4}$	$5\frac{1}{2}$	$4\frac{1}{2}$	$5\frac{1}{4}$

Q125. A chemist invents a new detergent which can be made in 4 forms – liquid, powder, tablet, and pressure-pack spray. Before proceeding to manufacture all of these versions, he decides to find out how they appeal to housewives. He conducts a survey in which he gives a sample of all 4 versions to a number of housewives, who are chosen in a random manner; he asks each selected housewife to try each version a number of times, and then to indicate her order of preference. Here are the first 5 results he got; do they indicate a significant difference between the 4 versions of the detergent, or should the investigation be continued further?

Tester	Liquid	Powder	Tablet	Spray
Mrs A	1	3	2	4
Mrs B	1	2	4	3
Mrs C	2	1	4	3
Mrs D	$1\frac{1}{2}$	$1\frac{1}{2}$	3	4
Mrs E	2	1	4	3

Q126. The Speedie Taxi Company owns 12 cars of X make and 6 of Y make (please excuse the algebraic symbols, but it's to avoid lawsuits). All cars are used equally, but over a period of 6 months, the X cars required mechanical attention (apart from routine servicing) on 28 occasions, whereas the Y cars did so only 8 times. Could this difference be reasonably attributed to chance?

Q127. Horse racing fans often maintain that inside starting positions are advantageous on circular race-tracks. Remembering how the great discoveries of Galileo and de Moivre had started, Professor Siegel felt quite justified in looking into the matter. One hundred and forty-four races later, he had collected the following frequency table –

Starting post positions	1	2	3	4	5	6	7	8
No. of wins observed	29	19	18	25	17	10	15	11

Does this data support the popular claim, or not? (Data from Sidney Siegel, *Non-Parametric Statistics*, McGraw-Hill, 1956.)

Q128. A comparative trial of sulphas and penicillin was carried out on 16 patients suffering from Z fever. Eight of the patients were given sulpha tablets, and the other 8 were given penicillin injections. The penicillin-treated group recovered more quickly than those on sulphas, to a statistically significant degree. However, the data in the report also revealed that the patients' ages were as follows –

Treatment	Age (years)							
Penicillin	11	42	21	23	58	17	29	19
Sulphas	32	45	53	26	42	64	38	51

Notice that the minimum age of the sulpha-treated group was 32. Perhaps the penicillin-treated group was essentially younger than the sulpha-treated group; if so, it would be wrong to compare the 2 groups, because the quicker recovery of the penicillin

group might then be due, not to the treatment being more effective, but to the younger persons in the group who may well tend
to recover from a fever more quickly than older people. Apply a
suitable test to see if the 2 groups are fit to be compared.

Q129. Peggy (to husband): 'I've got news for you, Harry. Last
week I timed myself when I vacuumed the house. It took me 25
minutes. Then I borrowed Jean's new "floating" vacuum cleaner,
and it was so much more efficient that the same work took me
only 17 minutes. I want a new vacuum cleaner like Jean's.'

Helpful Harry: 'But the difference might only be a matter of
chance. Pass me Langley's *Practical Statistics*, and I'll work out
the probability. You can have a new cleaner if the difference is
only likely to occur by chance once in 20 times, or less.'

Did Peggy get a new vacuum cleaner?

Q130. It was only a small school, but the headmistress was
'modern' in the sense that she was very interested in trying out
new teaching methods. One system she tried was the use of
French television programmes as an aid to the teaching of that
language. She randomly divided the French class into 2 groups,
one of which was given the TV lessons, while the other received
the conventional course. There were only 10 students taking
French that first year, so no definite conclusion was apparent at
the annual exams (both groups sat for the same exam). But in
the second year, a further 8 students took on the subject. If the
results were then as follows, would it indicate a statistically
significant difference between the 2 teaching methods, or might
the differences be readily attributable to chance?

| Students | Marks at 2nd annual exam with – | |
	Conventional teaching	Television supplement
First year	81	72
	59	70
	48	63
	46	51
Second year	92	97
	87	95
	65	88
	61	80
	53	72

Q131. These days, when singers of pop tunes often earn more than prime ministers, plenty of people are seeking their success formula. Let us look at the 'top 40' songs from the hit parade of a year ago, and see if there is any relationship between the number of sales of these songs and the attractiveness of their record album covers. We shall classify each record cover as being either 'good' or 'bad' according to our considered judgement, and will divide the sales of the top 40 records into those that have sold more than half a million and those that have sold less.

(*a*) If the results were as follows, would it indicate that a significant association existed between these 2 variables?

	No. of songs with –	
	More than $\frac{1}{2}$ million sales	Less than $\frac{1}{2}$ million sales
Cover $\begin{cases} \text{Good} \\ \text{Bad} \end{cases}$	3 2	7 28

(*b*) Can you suggest a more sensitive design for this investigation, without increasing the sample size?

Q132. Uncle John was very proud of his 12 apple trees, stretched out in a row along his back garden wall. One day he 'phoned me to come over straight away; the apples on 4 of his trees had developed some strange spots on them. He felt sure these spots were a sign of some contagious disease, and to support his theory, pointed to the way the affected trees were clustered, as follows –

Sp. Sp. Sp. N Sp. N N N N N N N

– where *Sp.* = spotted apples, and *N* = normal. I explained to Uncle John that although I knew a bit about diseases of human skin, I was quite at sea concerning diseases of apple skin. However, I was able to test the clustering to see if it was likely to be a chance effect or not. See if you agree with my answer.

Q133. A newspaper reported that 12% less people were smoking now, compared with a year ago. A well-conducted survey of 200 adults had been carried out; 118 of these people had been smokers a year ago, whereas only 94 of them were currently smokers (a decrease of 24 per 200 = 12%). Further information was supplied; 78 of the sample group said they were non-smokers then and now, 28 had stopped smoking, 4 had started smoking, and 90 had been and still were smoking. Tabulate this data properly, and then determine the probability that the observed decreased incidence of smoking could be due to chance.

Q134. The standard test for cholesterol in the blood is a fairly complicated one, so the laboratory was most interested to try out the new simple test that was reported to be equally accurate. They compared the 2 tests as follows. Each sample of blood sent in for cholesterol estimation was divided into 2 portions; one portion was used for carrying out the old standard test, and the other portion was tested with the new test. Here are the results (expressed in mg%); do they indicate a significant difference, or might the differences be fairly ascribed to chance? So far as the laboratory is concerned, the decision is one of considerable importance (they don't want to change to an inferior test), so they will take $P = 5\%$ or less as indicating a significant difference.

Blood sample	Old test	New test
A	256	249
B	235	230
C	196	195
D	307	314
E	287	282
F	239	236
G	213	215
H	349	336
I	264	249
J	263	256
K	360	360
L	520	508
M	481	492

Blood sample	Old test	New test
N	198	199
O	206	201
P	271	266
Q	200	204
R	195	195
S	346	336
T	405	401
U	247	242
V	250	241
W	281	275
X	303	299
Y	298	300

Q135. The following data is provided by a manufacturer of a new kind of antiseptic dressing which we shall call XYZ –

Dressing	No. of injuries	Av. healing time (days)
XYZ	400	7
Conventional	400	12

Is the reduced healing time with the new dressing statistically significant?

Q136. The normal level of magnesium in the blood is 2·00 milli-equivalents per litre, with a standard deviation of 0·29. Would the finding of an average of 2·42 in a group of 15 schizophrenic patients represent a significant difference? (From *Med. Journ. of Australia*, February 1964.)

Q137. At the annual shareholders' meeting of the Syntho-Diamond Company, the managing director said he was pleased to report that they had captured a further 6% of the gem market. 'How do you know?' asked a shareholder. The managing director replied, 'Last year we conducted a survey of 1,000 people who owned jewellery. They were selected quite haphazardly, so that we believe them to be a random sample. Of this group, 400, that is 40%, owned one or more of our artificial diamonds. Last week we conducted another such survey, and found 460 out of 1,000 owning our gemstones; this represents a rise to 46%.' This didn't impress the shareholder, who retaliated, 'It looks to me as though that difference might easily be due to chance.' The managing director, who shared H. G. Wells' belief in the need for a knowledge of Statistics as part of his general education (p. 20), was able to answer this doubt within a few minutes. Can you?

Q138. The quality of house-paints can be tested by accelerated ageing tests. For example, to compare 4 paints (*A*, *B*, *C*, and *D*) we could apply each paint to a sheet of wood and to a sheet of galvanized iron, and cut each sample into 3 parts. One part of each sample can then be treated by heat, another part can be kept immersed in water, and the third part can be kept as controls. At the end of the test period, the percentage deterioration of the tested portions can be assessed by comparison with the control portions, each paint against its own control. Suppose we got the following results, would this be sufficient evidence to say that one paint is superior to the others?

Test conditions	Brand of paint			
	A	B	C	D
Wood + Heat	75%	65%	90%	80%
Wood + Water	50%	60%	75%	75%
Iron + Heat	80%	70%	85%	80%
Iron + Water	30%	40%	50%	45%

Q139. It was quite obvious from the first that the Martians were of 3 sexes – male, female, and neuter. After examining the first arrivals (that brave little band of 12), Professor Malkovitch formulated a theory that the population proportions of the Martian males to females to neuters must be 1:1:4.

I was there when their next space ship landed. It was a much bigger one than the first, and contained 24 males, 12 females, and 60 neuters. Having been selected for the trip by ballot, this group could be accepted as a random sample of Martians. Are these proportions consistent with the professor's theory within fair limits of chance variation, or do they refute it?

Q140. You will recall that Hammond and Horn's sample group of smokers and non-smokers was criticized as being selected in a way which did not ensure that it was a random sample (p. 38). Part of the complaint was that the proportion of smokers in the sample seemed to be lower than in the general population. For example, a Gallup poll in 1954 revealed that 57% of men in the 50 to 60 years age group were currently cigarette smokers, compared with 48·5% of the men in this age group in the sample. However, the proportion of smokers varies quite a lot in different age groups, so it is a pity that no figures are available for precisely the 50 to 70 years age group of Hammond and Horn's sample. Let us therefore invent some figures for the purpose of this question.

Suppose that the US Bureau of the Census chose a random sample of 12,000 white males aged 50 to 70, and found that 58·3% of this sample group were regular cigarette smokers. Some degree of sampling error is, of course, unavoidable, so it would be rash to presume that this represented the absolute proportion in the entire US population, but at least we could be sure that it was a true random sample. Now, Hammond and Horn's sample of 188,000 white American males aged 50 to 70 contained 57·4% of regular cigarette smokers, which is a mere 0·9% less than in the Bureau's sample.

We can now pose this question: if a truly random sample shows 58·3% smokers, what is the probability of getting 57·4% smokers in another random sample from the same population? If the probability turns out to be remote, we will conclude that the second sample is not a random one.

Here is a tabulation of this data, converted into round numbers to make your calculation a bit easier –

	Non-smokers and occas. smokers	Regular smokers	Totals
True random sample	5,000	7,000	12,000
Suspect sample	80,000	108,000	188,000

Admittedly a difference of 0·9% doesn't seem much, but we have seen how small differences tend to be significant when large numbers are involved. So apply a suitable test, and see what you find.

Q141. While we're on the hazards of smoking, this question may have some significance. Suppose that in a certain city there were an average of 3 fires a week started by people falling asleep while smoking in bed. The Fire Authorities decide to have an advertising campaign in an attempt to reduce this shameful figure. In the first 6 weeks after the campaign there are only 9 such fires. Is this significantly different from the previous average, or might it merely be a matter of chance?

Q142. Seventy-five people with fair, sun-sensitive skins were given 3 anti-sunburn remedies (*A*, *B*, and *C*) to try out. Afterwards, they were asked which remedy they had found to be the most effective.

(*a*) Do the following results indicate a significant difference between the 3?

Prefer *A*   33
Prefer *B*   22
Prefer *C*   20
—————
Total   75

(*b*) Can you suggest another design for this experiment which would utilize more of the available information?

Q143. A study of 'leading US admirals' revealed that 14 out of 31 had come from the upper quarter of their classes at Annapolis. Can one conclude from this that scholastic standing at Annapolis has something to do with becoming a 'leading admiral'? (From W. A. Wallis and H. V. Roberts, *Statistics – A New Approach*, Free Press, 1960.)

Q144. Would you accept the following series of males (*M*) and females (*F*) obtained in the order shown, in a house-to-house survey, to be a random sample of city people?

MMMM FFFFF M FFF MM FF MMM
    F MMM FFFFFF MM FFFFF MMMM

Q145. Mr Jacko, the secretary of the Kangaroo Ping-Pong Club, is offered a new brand of ping-pong (table tennis) ball, at a very attractive special 'club' price. The salesman assures him that these new balls will withstand, on the average, 11·0 lb steady weight without breaking. So, like any good club secretary would do, Mr Jacko plunges his hand into a large carton of the balls, and withdraws a random sample of 6 of them. He silently resolves to buy the balls if the average of his sample implies that the salesman's statement is true, but not if the probability of this being the case is only 5% or less.

He then proceeds to test the breaking point of the 6 balls, with the following results –

9·5 8·0 11·0 11·5 8·5 9·0 lb

Did he buy the balls?

Q146. A wealthy business man decided to take up Culture. He advertised that he wanted to buy 5,000 books to make a library, and specified that the books must have an average of at least 450 pages, with a standard deviation of not more than 100 pages. One applicant turned up with a van full of books. The business man took a random sample of 25 of these, and found their average to be 410 pages. Is it probable that the van-load is up to the required standard?

Q147. The heading in the business magazine was: 'New Test Detects No-Hopers'. The article described a new aptitude test which, given to people applying for clerical jobs, was claimed to determine their suitability or otherwise in a considerable proportion of cases. They quoted an instance of a large firm which had tested 50 consecutive new clerical employees. Of these, 15 had been classified as unsuitable by the aptitude test, and within 3 months 10 of these employees had been dismissed. By contrast, of the 35 who had been judged as suitable by the test, only 5 had been dismissed during the first 3 months' period. Do these figures support the claim that this aptitude test is useful (i.e. is there a significant association between the test forecasts and the dismissal rates), or might the quoted figures be reasonably ascribed to chance? Start by putting the data into an appropriate table.

Q148. A manufacturer of one of the sulpha drugs (*A*) advertised that his drug gave higher concentrations of the drug in body tissues than 2 alternative sulphas (*B* and *C*). This claim was based on results obtained by 3 independent laboratories (each had tested 1 drug), which were reported as follows –

| | Sulpha drug | | |
	A	B	C
Arith. mean tissue levels of sulpha (mg %)	2·32	1·8	1·7

Is the observed difference between *A* and its competitors statistically significant, or might these differences be simply a matter of chance? (Accept P = 5% as significant in this case.)

Q149. A teacher wanted to assess the effect of watching television on exam results. He asked 80 pupils how much TV they watched, and then noted whether they passed or failed their exams. Suppose the results were as follows, would the differences be significant?

| | Av. hours per week spent watching TV | | |
	Less than 1½	From 1¾ to 7	More than 7
Passed	13	34	13
Failed	2	6	12

Q150. It is sometimes maintained that women sleep less soundly after having children than they did beforehand. Suppose we asked 60 women with children, and found –

| No. of children | Present sleep compared with before having children – | | |
	Worse	Same	Better
1	25	5	0
2	10	4	1
3 or more	5	7	3

What inference can you draw from this data?

Q151. In a piece of research reported in a French medical journal in 1962, the effectiveness of 2 tranquillizers was compared, using 30 patients suffering from anxiety symptoms. Each patient was first given one drug, then the other, each for a period of 2 weeks; the order was randomized. The results were –

Drug	None	Degree of improvement Slight	Good	Very Good	Totals
A	7	5	14	4	30
B	12	7	7	4	30

(*a*) Comment on the statement, 'The greater potency of *A* over *B* is shown most clearly by the lesser number showing no response (7 as against 12), and the greater number showing a good response (14 as against 7).'

(*b*) Does this investigation show one tranquillizer to be significantly better than the other?

Q152. In the *Journal of the American Dental Association* (1948, pp. 28–36), the Director of the US National Institute of Dental Research, Dr Francis Arnold Jr, reported the incidence of tooth decay in samples of children aged 12 to 14 years, in different areas of the USA. Against each group was noted the fluoride content of the drinking water in the area. Does the following excerpt warrant any firm conclusions about the possible association between tooth decay and the fluoride content of the water supply?

Town	No. of children examined	No. of decayed teeth found	∴ Av. No. of decayed teeth per child	Fluoride content of water supply (parts per million)
Hereford, Texas	60	88	1·47	3·1
Colorado Springs, Colo.	404	994	2·46	2·6
Elmhurst, Ill.	170	428	2·52	1·8
Joliet, Ill.	447	1,444	3·23	1·3
Kawanee, Ill.	123	422	3·43	0·9
Pueblo, Colo.	614	2,530	4·12	0·6
Marion, Ohio	263	1,462	5·56	0·4
Vicksburg, Miss.	172	1,010	5·87	0·2
Oak Park, Ill.	329	2,375	7·22	0·0
Elkhart, Ind.	278	2,288	8·23	0·1
Escanaba, Mich.	270	2,368	8·77	0·2
Michigan City, Mich.	236	2,446	10·36	0·1

Q153. A student was asked a set of 100 chemistry questions to which the answers were either 'yes' or 'no'.

(*a*) If we are prepared to accept a risk of being wrong at most once in 500 times, what is the least and the most he would be likely to answer correctly by pure guesswork (as if he tossed a coin to get the answers)?

(*b*) What is the most he might score if he knew half of the questions correctly, and guessed the other half?

Q154. In the AMA *Archives of Dermatology*, October 1958, there is a report concerning 2 new lotions which had been formulated for treating a certain kind of skin rash. The lotions were tested on 177 patients who had this rash on both sides of their bodies, so that each patient was able to compare the effectiveness of the 2 lotions quite accurately by applying Lotion *A* to one side of the body, and Lotion *B* to the other side. The results were –

No difference between *A* and *B*	96
Much better on *A* than *B*	19
Moderately better on *A* than *B*	36
Much better on *B* than *A*	5
Moderately better on *B* than *A*	21

Total = 177 patients

The statistician's report on these figures was that the difference between the results of *A* and *B* was not significant at the 5% level (i.e. the probability of getting these results by chance if there was no significant difference between *A* and *B* is more than 5%). Do you agree with this conclusion?

Q155. Refer back to Fig. 10 (p. 132). You can see that 16 patients preferred Treatment *A*, as against 4 who preferred Treatment *B*. At that stage, the sequential trial was stopped, and Treatment *A* declared to be significantly better than Treatment *B*. What probability level was this chart designed to show?

Q156. The farmer was sceptical about the claims made for a certain new fertilizer which was 'sure to increase your apple yield; some reports have been up to 50%'. He had tried these new-fangled ideas before, and had always ended up disappointed. Yet he could not afford to ignore this claim in case, for once, it was true.

To test the fertilizer, he chose 12 apple trees of the same age and variety, which were conveniently growing in 2 adjacent groups, with 6 trees in each group. He treated one group with the new fertilizer, and treated the other group in his normal manner. In the course of a year he collected the following yields (in lb) of apples from each tree –

| Normal treatment | 135 | 151 | 109 | 115 | 118 | 146 |
| New treatment | 137 | 143 | 132 | 153 | 167 | 120 |

The figures reveal that the average yield for the group which received normal treatment is 129 lb, whereas the average of the group which was treated with the new fertilizer is 142 lb. So far, so good. 'But,' thought the farmer, 'there's quite a bit of variation between the yield of the trees in each of the 2 groups, so perhaps this result is only due to luck.' Decide whether you would consider the trees to be matched pairs or independent samples, and then apply a suitable test to advise the farmer whether his suspicion is justified or not. First test the means, and then test the spread of the measurements (in case one group is significantly more uniform than the other).

Q157. It is very inconvenient, to say the least, when sailors get seasick, so Captain McIntosh thought he should try out the latest remedy, 'Sicko' pills. He bought 200 pills (1 for each member of his crew), and 200 others that looked exactly the same but which were made of pure lactose (milk sugar). He would use the latter as a control, for he knew that they had no specific effect on seasickness. He put all the pills in his medicine cupboard, and waited for a storm.

During the first storm, Captain McIntosh did nothing except note the names of the 30 men (out of 200, remember) who became seasick.

At the first warning of the next storm, he got each and every man aboard to swallow a lactose pill (he didn't tell anyone what they were or what they were for), and again he noted the names of those who became sick.

Finally, when the next storm approached, he gave each man a 'Sicko' pill, and again noted the results.

All 3 storms being similar in severity and duration, he felt it was quite fair to compare the results of each storm with each other. So he sorted out the results and found –

15 men were sick only when they took no pill.
3 other men were sick only when they took a 'Sicko' pill.
1 other man was sick only when he took a lactose pill.
8 other men were sick both when they took no pill, and when they took a lactose pill.
4 other men were sick both when they took no pill, and when they took a 'Sicko' pill.
3 other men were sick both when they took a lactose pill, and when they took a 'Sicko' pill.
3 other men were sick on all 3 occasions.

Can you tell –

(a) Was there a significant difference between the numbers of men seasick during the 3 storms?

(b) Were the 'Sicko' pills significantly better than the lactose control pills?

(c) How could this investigation have been planned so that it would have remained valid if the storms had proved unequal in severity?

Q158. With a certain kind of packing, 99·5% of eggs used to arrive at their destination perfect and unbroken. A new way of packing them is then tried, and of the first 3,000 eggs, 99·8% arrived perfect. Is this a significant improvement?

Q159. The première of Longhair's 2nd Symphony got a mixed reception. Some of the audience thought it was a great work, others hated it intensely. Does the following analysis reveal a significant difference between various groups who heard the work?

	Opinion of Longhair's 2nd Symphony		
	Good	Bad	Undecided
Members of the orchestra	75	20	5
Musicians in the audience	120	70	10
University students and graduates	25	20	5
Others in the audience	310	290	50

Q160. A manufacturer of chocolate peppermint creams consulted a market research firm for advice as to how he could improve his sales. One of the first things they did was to get a customer appraisal of the package in which the chocolates were sold. They designed 3 alternative packages, and devised a scheme to see if any one package appealed to customers as being 'better value' than the others. This scheme entailed having an investigator visiting 20 confectionary shops in different areas, and stopping the first person over the age of 14 emerging from each shop, showing that person one or other of the 4 packages, and asking them how much they would expect to pay for such a product. This design avoided the risk of annoying anyone, because each person was only asked a single brief question (in contrast to showing them all 4 packages and asking them to price or grade them all). In this way, 20 potential customers were interviewed, and each of the 4 packages was evaluated independently by 5 people. The results are tabulated below; is there a significant difference between the package values, or might the differences be reasonably likely to occur by chance?

Package	Evaluations (cents)				
A	60	40	40	60	50
B	70	70	60	70	75
C	55	70	60	55	70
D	55	45	35	40	50

Q161. 'The shortage of slaves is getting quite beyond a joke. Most Roman patricians are extremely unhappy about it. It all began when Julius imposed those darned import restrictions 5 years ago. We haven't got a quarterly price index yet, but look what's happened to the cost of living. Take the price of slaves. Look, here's a typical bill that my agent in Delos sent me 5 years ago – 15 slaves consisting of 5 labourers (priced at 975, 950, 1,000, 1,100, and 1,020 denarii), 5 craftsmen (who cost me 1,850, 1,925, 1,910, 1,890, and 1,900 d.), and 5 housemaids (2 at 450 d., and 3 at 600 d.). Now, here's the bill I got from him to-day, also for 5 labourers (at 1,275, 1,250, 1,200, 1,050, and 1,200 d.), 5 craftsmen (at 1,875, 1,950, 1,940, 1,800, and 1,940 d.), and 5 housemaids (2 at 500, 2 at 650, and 1 at 550 d.). What do you make of that, eh?'

Q162. The Betta Hamma Co. invented a new type of pneumatic drill. To compare its performance with a conventional drill, they decided to use 3 operators (Al, Ben, and Charlie), each of whom would be timed doing the same job with both the old and new drill on 3 test materials – asphalt, bricks, and concrete. Would the following times prove the new drill to be significantly better than the old type?

Material	Operator	Old drill	New drill
Asphalt	Al	2′ 30″	2′ 20″
	Ben	2′ 15″	2′ 10″
	Charlie	2′ 30″	2′ 20″
Bricks	Al	4′ 15″	4′ 10″
	Ben	4′ 10″	3′ 55″
	Charlie	4′ 15″	4′ 05″
Concrete	Al	7′ 20″	6′ 45″
	Ben	7′ 00″	6′ 55″
	Charlie	7′ 25″	7′ 25″

Q163. Over a period of 30 years there were 33 separate floods in Weepy Valley. In an attempt to cure this recurrent problem, a river which was thought to be the main cause of the floods was then partly diverted. In the next 10 years the valley was only flooded 5 times. Does this prove the diversion scheme was successful, or might such figures be merely due to chance?

Q164. In the AMA *Archives of Dermatology*, June 1958, there is a report of an investigation into the treatment of chronic dermatitis with 2 kinds of tar – the conventional tar (*T*), and a colourless synthetic substitute (*S*). The trial dealt with each of these tars in 2 ointment bases – a vanishing cream (*C*), and a thick paste (*P*). All 4 combinations (*TC, TP, SC,* and *SP*) were tested. There were 220 patients in the trial; all had chronic dermatitis on both sides of their bodies, and each patient was given at least 2 of the 4 formulae to try out on comparable areas of their skin rash (actually, 38 tried 3 formulae, 32 tried all 4 formulae, and the rest tried 2 formulae each). Allocations were, of course, randomized. From the following results, can you tell –

(*a*) Is one tar better than the other?

(*b*) Is one ointment base better than the other?

Group I patients	Prefers TC	Prefers SC	No difference
	15	37	68

Group II patients	Prefers TP	Prefers SP	No difference
	13	34	73

Group III patients	Prefers TC	Prefers TP	No difference
	10	13	28

Group IV patients	Prefers SC	Prefers SP	No difference
	7	12	45

Q165. The following is based on an actual case. A manufacturer of ladies' nightdresses decided to conduct a survey of his customers to find out their tastes regarding the next season's colours, so that he could order his raw materials in plenty of time and as accurately as possible. He therefore interviewed the buyers of the 20 firms who together represented about 90% of his accounts. Each was shown 8 colours, and asked to state the proportions of each colour that they anticipated they would be buying. When the results of the survey were added up, the expected demand for each of the colours was in the following proportions –

Rose	Mauve	Jade	Blue	Fawn	White	Grey	Cream
20	17	12	12	8	3	3	1

Are these differences statistically significant?

Q166. Would the differences between the cloudiness of London and New York be statistically significant if the amount of cloud in the sky at noon each day in July for the years 1964–8 inclusive were as follows?

	Amount of Cloud in Sky (to nearest 5%)											Totals
	0%	10%	20%	30%	40%	50%	60%	70%	80%	90%	100%	
No. of days in *London* with stated %	30	11	6	5	3	4	5	6	9	14	62	155
No. of days in *New York* with stated %	26	17	10	7	5	4	7	10	15	20	34	155

Q167. Suppose that, at the start of a weight-lifting course, 50 consecutive men and 50 consecutive women applicants, all between 10 and 11 stone body weight, were compared to see if men were physically stronger than women. For convenience, the results were tabulated as shown below. What do you notice about the distribution of the two sets of results? Is the difference statistically significant?

	Maximum weight (lbs.) that was lifted						Totals
	20–29	30–39	40–49	50–59	60–69	70–79	
No. of men	—	2	8	17	18	5	50
No. of women	3	5	12	17	11	2	50

Q168. The number of petals on 3 varieties of Ranunculus was counted, on 100 flowers of each variety, with the following results. Are the differences between the varieties likely to be due to normal fluctuations of sampling, or do they represent significant differences?

	No. of Petals per Flower							Total No. of Flowers
	5	6	7	8	9	10	11	
Variety A	82	11	4	2	—	1	—	100
Variety B	78	15	2	2	1	1	1	100
Variety C	74	19	5	—	2	—	—	100

Q169. How to become an art sleuth. Like all creative artists, composers of music develop certain personal characteristics in their works. One such characteristic is the number of melody notes in each bar of music. Now suppose you buy an old unsigned manuscript of a waltz which you suspect is an unknown work by Johann Strauss, and if so, very valuable. You count the number of melody notes per bar of several genuine Strauss waltzes and compare the frequency distribution with a similar count of the unknown work. Would the following results support your high hopes?

	No. of Melody Notes per Bar							Total No. of Bars
	0	1	2	3	4	5	6 or more	
Strauss waltzes	5	32	133	114	67	22	15	388
Unknown waltz	6	60	62	96	33	7	18	282

Q170. (a) A bag of marbles contains 1 black and 19 white marbles, and another bag contains 1 black and 9 red marbles. In a single, blindfolded draw from each bag, what is the probability of drawing both black marbles? Express your answer as a percentage.

(b) Two independent reports have come to hand concerning the value of Ammoniated Tincture of Virgin Moss for treating elephantiasis in camels. The first report was made on a small 'pilot' series, and showed the new Tincture to be probably superior to the old treatment, with a Yates' χ^2 of 3·84, D.O.F. = 1, P = 5%. The second report was of a larger trial; it gave a 'not significant' result with a Yates' $\chi^2 = 2·71$, D.O.F. = 1, P = 10%. Does this second report, based on more cases, nullify the conclusion of the first report? Can you combine the 2 reports to arrive at a new probability level? If this differs from your answer to (a) above, state why.

(c) Dr Smith found that infants fed on baked beans gained more weight than average; a zM Test gave $z = 1·96$, P = 5%. Dr Jones tried to confirm this, but his results gave $z = 1·64$, P = 10%. What's your conclusion?

(d) The man with the big earphones at the ham radio felt his marrow freeze as he intercepted a strange voice: 'Calling Stellar XII. Calling Stellar XII. This is Explorer I, reporting from Southern Hemisphere of Earth. Happy to report that new delta-ray gun is working beautifully. Have just killed 3 Earthmen with it. Lethal dose in first case was 10δ, in second case 11δ, in third case 12δ. Is this significantly different from fatal dose required for Moonmen?'

'This is Stellar XII. Received your message clearly, Explorer I. The mean lethal dose of your sample is 11·00, standard deviation 1·00. Comparison with mean lethal dose for Moonmen (which is 13·30δ), using Student's t Test gives $t = 3·98$, $P > 5\%$, so the difference is not statistically significant.'

'Hullo, hullo. Calling Stellar XII. This is Explorer II, from Northern Hemisphere of Earth. Delighted with our new δ-guns. Our party has just killed 6 Earthmen. Great fun. Red syrup squirted everywhere. Lethal doses as follows – 9, 10, 11, 12, 13, and 14δ. How does this compare with results on Moon-people?'

'Stellar XII replying to Explorer II in Northern Hemisphere of Earth. The mean of your sample is 11·50, standard deviation 1·87. Comparison with mean lethal dose for Moonmen using Student's t Test gives $t = 2·36$, $P > 5\%$, so a significant difference is not proven.'

'Thank you, Stellar XII. Now, how about pooling Explorer I's results with mine. Perhaps the larger sample will show significance.'

'Good idea, Explorer II. Will do. It now gives Student's t . . .' and static interrupted the message.

What was the answer?

ANSWERS

Q1. $0·091 = \frac{1}{11} = 1$ chance in 11.

Q2. $\frac{1}{16} =$ once in 16 such trials.

Q3. $\frac{1}{1296} =$ once in 1,296 times. Notice that rolling a pair of dice twice is exactly equivalent to rolling a single die 4 times.

Q4. 3 chances in 8.

Q5. One would go broke. Remember that as the number of tosses increases, the difference between the number of heads and tails will increase, even though their proportions grow closer and closer to the 50% mark (see p. 23).

Q6. $\frac{4}{52} \times \frac{3}{51} \times \frac{2}{50} \times \frac{1}{49} = \frac{1}{270,725}$, that is, 1 chance in over 270,000.

Q7. $\frac{1}{2} = 50\%$.

Q8. (a) *ABC, ACB, BCA, BAC, CAB, CBA.* Therefore the probability of getting *ABC* is 1 in 6.

(b) $\frac{1}{3} \times \frac{1}{2} \times \frac{1}{1} = \frac{1}{6}$.

Q9. $0·046 = 4·6\%$, that is about 5 cancers would be missed in each batch of 100 tests.

Q10. (a) $\frac{1}{25}$.   (b) $\frac{16}{25}$.   (c) $1 - (\frac{1}{25} + \frac{16}{25}) = \frac{8}{25}$.

Q11. Patients numbered 5, 4, 9, 8, 11, and 3 go into the first group; the remainder go into the second group.

Q12. 24, 37, 4, 18, and 50.

Q13. 1, 7, 5, 8, 2, 3, 6, 9, 4, in that order.

Q14. Arithmetic mean = 2·70. Standard deviation = 1·64. Snedecor's check = 1·67. Method 1 is not much of a shortcut in a simple case like this, but you will certainly find it so when the data is more complicated.

Q15. Using Method 2, arithmetic mean = 25·45, standard deviation = 2·97. Snedecor's check = 3·5.

Q16. Using Method 3 (I used class intervals of 3–4·9, 5–6·9, etc.), the approximate arithmetic mean = 11·6, and the standard deviation = 4·0. Snedecor's check = 4·4.

Q17. (a) 4.   (b) 4.

Q18. (a) 3. If you thought it was 1, you forgot to arrange the numbers in order from smallest to largest.

(b) 11·5. Notice that the median doesn't have to be the *middle observation*; it can be a value *between* 2 observations.

(c) 51.

(d) 14. Notice that the median value can be shared by more than 1 observation in the series.

Q19. (a) 46.   (b) 44·5.   (c) 42 and 46.

Q20. Lower quartile = $2,500. Upper quartile = $3,875 Davies' test could be inaccurate here because there are less than 50 observations. With grouped data, notice that the quartiles may not divide the observations into proper proportions; thus only 1 employee ($= \frac{1}{23}$ instead of $\frac{1}{4}$) earns less than the lower quartile, and 12 employees ($= \frac{12}{23}$ instead of $\frac{3}{4}$) earn more than it (compare median, p. 75).

Q21. Davies' coefficient = $+0.265$, therefore the arithmetic mean is indicated.

Q22. Davies' coefficient = -0.21, therefore distribution is approximately logarithmic; geometric mean and standard deviation would be suitable for descriptive purposes; logarithmic mean and standard deviation would be suitable for comparative tests. Logarithmic mean = 0·956 (antilog of this = geometric mean = 9·0). Logarithmic standard deviation = 0·198 (antilog of this = geometric standard deviation = 1·6). Notice incidentally that it would be impossible to work out the arithmetic mean of the data as presented, because the upper limit of the last class is unspecified – it might have been 100 hours! The median would have been a suitable substitute if the arithmetic mean had been called for, in the absence of a specified upper limit.

Q23. Arithmetic mean would be appropriate here. Davies' test should not be applied in this case because the data is distributed too symmetrically. (Actually, Davies' coefficient works out to be -0.02, when in fact the data is not a normal logarithmic distribution at all. However, with symmetrical distributions the geometric mean is generally close to the arithmetic mean if Davies' coefficient equals or is less than the critical $+0.20$; for instance, in the present case the arithmetic mean is $3s.$ $10\frac{1}{2}d.$, while the geometric mean is $3s.$ $9d.$) The class ranges are poorly chosen, for most of the boys would have received a round number of shillings ($3s.$, $4s.$, $5s.$, etc.), so these quantities should have been in the centres of each class, thus $6d.–1s.$ $5d.$, $1s.$ $6d.–2s.$ $5d.$, etc., for the reason described on page 63.

Q24. Arithmetic mean, because although the pattern of the frequency distribution looks somewhat logarithmic, Davies' coefficient turns out to be $+1.0$.

Q25. By Method 2, arithmetic mean = $54·151''$. Mode and median $= 54·15''$. Similarity between all 3 is due to the symmetry of the distribution of the measurements, as illustrated in Fig. 2 (p. 34).

Q26. 0·6 laughs per minute is the combined arithmetic mean.

Q27. The harmonic mean is needed in this instance. So work out how many shares you bought, and you will see that the average price per share is $13·33.

Q28. 80·0%.

Q29. Quibbler is right in the sense that the only meaningful average of a set of numbers which increase by geometric progression is that number which occurs at the midpoint of the observation period (18 months in the present case). This number is the geometric mean, and 190 is the geometric mean of 6, and 6,020, as calculated by the method described on page 78. Zarathustra was using the arithmetic mean to aid his plea for his nefarious scheme.

Q30. The range between the average maximum and minimum temperatures. Lindyville might range from 20° to 106°, so that you would freeze by night and roast by day, whereas Thornton might range only from 52° to 78°, a much more comfortable climate for camping.

Q31. (*a*) Quoting the percentage to 3 decimal places is not justifiable when you consider that the total number of car owners interviewed must have been less than 1,000. This is a misleading trick, suggesting a degree of precision in excess of the truth.

(*b*) The sample seems to have merely been collected haphazardly. It may or may not be equivalent to a true random sample. It certainly would not be random if the interviewer had consciously or unconsciously chosen well-to-do looking persons predominantly.

(*c*) It is almost certain that some people would not tell the truth about their financial affairs to a stranger.

Q32. (*a*) The comparison of the 2 groups is only valid if the presenting sample is divided up in a random manner, i.e. so that each patient has an equal chance of getting either treatment. Strict alternation of successive patients is generally acceptable as a means of randomization, because the order in which patients arrive for treatment is usually only a matter of chance. But notice that the effectiveness of alternation as a method of randomization is entirely dependent on the assumption that the order of arrival is random. A safer way, which guarantees the randomization, is to allot patients to one or other group using a Table of Random Numbers, as was done in *Q11* (p. 44).

(*b*) The difference in response to the 2 treatments is so marked that the doctors would soon be able to note the regular alternation in effectiveness of the treatments. They would then know whether the next patient would be due for the effective or the ineffective treatment.

(*c*) By lumping all types of headache together, it runs the risk of overlooking the fact that one particular kind of headache (of which there might have been only a few cases in the whole sample group) may have responded better to the new remedy than to aspirin. This is an instance of asking too many questions at once.

Q33. The name of the organization who conducted the survey, when and where it was done, the size of the sample, how the sample was chosen, and the number of non-responses encountered.

Q34. Yes, provided that his 40 'typical' cases are not just a biased selection but are actually equivalent to a random sample of children who habitually watch horror films, and provided that other observations made on children who only *occasionally* watch horror films show that their crime rate is no less than the horror film fans. The presentation of 'typical' cases is not uncommon in medical fields, and it is by no means rare to find that careful study of a few such cases is more illuminating than a cursory study of a lot of cases.

Q35. I found that 880 pages measured 1 inch, so the average thickness of each page was $\frac{1}{880}$ inch = 0·00114 inch. To determine the thickness of gold leaf, which is gold beaten so incredibly thin that a thousand leaves together make up only about $\frac{1}{300}$ inch, the famous English scientist, Michael Faraday, had to modify the above experimental design. What he did was to weigh a group of 2,000 gold leaves, each $3\frac{3}{8}$ inches square, and from the known density of gold, he was then able to calculate quite simply that the average thickness of the leaves was 3·55 millionths of an inch!

Q36. There are no controls, so perhaps the improved performances were due to other factors.

Q37. By questioning several hundred people who bought JJJ tyres 2 years ago, and making up a 'sample group' of the 20 luckiest customers – a dishonest trick.

Q38. It is not valid. The precision of a randomly chosen sample, taken as a whole, is a function of its size, not the proportion it bears to the parent group (p. 46).

Q39. (i) $I_{ABC} = 0$.

(ii) $I_{AB} = +2$ minutes. As 2 groups of patients had $A + B$, this figure must be halved to get average interaction per patient, which is therefore $+1$ minute. Dr Ay's average induction time with the Butobarbitone was thus 1 minute more than Dr a'Beckett's average.

(iii) $I_{AC} = 0$.

(iv) $I_{BC} = 0$.

(v) $4(A - a) = +20$ minutes (allowing for AB interaction), so $(A - a) = +5$ minutes. This means that Dr Ay's average induction time was 5 minutes more than Dr a'Beckett's.

(vi) $4(B - b) = -8$ minutes (allowing for AB interaction), so $(B - b) = -2$ minutes. We cannot tell whether the bicarbonate (b) had any effect (by suggestion), because there was no control group tested without it, but at least we can say that in the absence of AB interaction (i.e. in the hands of Dr a'Beckett) the Butobarbitone decreased the induction time by an average of 2 minutes, compared with the bicarbonate time. With Dr Ay the Butobarbitone had a positive interaction of 1 minute, so it would have decreased his induction time by only 1 minute ($-2 + 1 = -1$).

(vii) $4(C - c) = +4$ minutes, so $(C - c) = 1$ minute. The AB interaction does not enter the formula for calculating $C - c$, as it cancels itself out. Since $c = 0$, this answer means that the effect of C was to increase the induction time by an average of 1 minute per patient.

Q40.

	A	a	α	
B	1	2	3	1st period
b	2	3	1	2nd period
β	3	1	2	3rd period

In case you didn't get this one right, the experiment is carried out in 3 parts, each of say 2 weeks' duration, according to the plan shown above. In the first fortnight, cow 1 gets A and B, cow 2 gets a and B, and cow 3 gets α and B. In the second fortnight, cow 1 gets α and b, cow 2 gets A and b, and cow 3 gets a and b. Another combination is used in the third fortnight. In this way each cow gets a turn at each combination. Addition of the milk yields in the columns gives the effects of A and a, averaged over all 3 cows, which can be compared with the control, α. Likewise, the rows give the effect of B, b, and β. The totals of each cow can be extracted from the table and added separately to compare their individual performances during the period of the experiment. This particular design does not enable interactions to be studied.

In practice, it would be necessary to make a random choice out of the 12 possible plans which would, like the above one, disperse the various factors in a balanced design. For instance, other possible plans include –

	A	a	α
B	1	2	3
b	3	1	2
β	2	3	1

	A	a	α
B	1	3	2
b	2	1	3
β	3	2	1

	A	a	α
B	2	1	3
b	1	3	2
β	3	2	1

$Q41$. $z = 1\cdot60$, so $P > 10\%$. Therefore the metal could be copper, but this test cannot prove it. After all, it could be one of hundreds of alloys with a similar melting point. Contrast this with the way in which this test was able to reject the possibility of the unknown metal being gold in *Ex. 1* (p. 154).

N.B. The sign '$>$' means 'is larger than'. Its opposite, '$<$', means 'is smaller than'.

$Q42$. Mean $= 1,072°$C. $z = 3\cdot20$, so $P < 0\cdot2\%$. In other words, there is a probability of less than 1 in 500 that the observed difference between the means is due to chance. The additional information therefore virtually cancels out the possibility of the unknown metal being copper. Let this drive home the point that the probability of more than 10% found in $Q41$ does not imply that a sample *does* belong to a certain parent group, but only that it *could* belong to that group; on the other hand, a probability of 1% or less points to the essential dissimilarity of the sample and the parent group.

$Q43$. $z = 2\cdot00$, so $P < 5\%$. The observed difference between the means is probably significant.

$Q44$. $s = 0\cdot707$, $t = 6\cdot34$, $P < 1\%$ (for $n = 5$). Difference is now statistically significant. This shows the general desirability of continuing any investigation which shows a probability of between 5% and 10%, until it becomes either less than 5% or greater than 10%. The practical significance of this small increase in output per minute may be worth while if the press is working all day. Thus this increase of 2 copies per minute will amount to an extra 960 copies in each 8-hour day.

Q45. $s = 5 \cdot 20$, $t = 3 \cdot 65$, $P < 5\%$ (for $n = 4$). The observed difference is therefore probably significant, for such a set of 4 telephone bills could be expected to arise by chance less than one in 20 times. However, it is possible that the increase was not due to my secretary, but to other causes (e.g. perhaps the charge for telephone calls was increased about the same time as this secretary began working for me).

Q46. Can't tell. Lack of information about the dispersion of the parent group (i.e. its standard deviation, S) prevents the zM Test being applied, and lack of details about the sample group prevents the calculation of the sample standard deviation (s), so 'Student's' Test cannot be applied. We noted the same sort of thing in the case of the 2 local anaesthetics (p. 56).

Q47. $R = 28$, $P = 10\%$. Significant difference not proven.

Q48. $R = 13$, $P < 5\%$. The difference is probably significant.

Q49. $z = 2 \cdot 67$, $P < 1\%$. Difference is statistically significant at the 1% probability level. (This is a good way to sum up the answer, because for some experiments a 5% level is accepted as being sufficient.)

Q50. $R = 4\frac{1}{2}$, $P > 10\%$. Significant difference is not proven.

Q51. $R = 24$, P just under 5%. This result is virtually identical with that obtained by the t Test. The exact probabilities are $P = 4 \cdot 97\%$ according to the t Test, and $P = 4 \cdot 2\%$ according to Wilcoxon's Test.

Incidentally, notice the way in which Darwin recorded his results in uniform units of $\frac{1}{8}$ inch. Thus he wrote '$23\frac{4}{8}$ inches' so that we would know his measurements were correct to the nearest $\frac{1}{8}$ inch. This is decidedly better than writing '$23\frac{1}{2}$ inches', which implies 'correct only to the nearest $\frac{1}{2}$ inch'.

Q52. Wilcoxon's Sum of Ranks Test gives $R = 97\frac{1}{2}$, $n_A = n_B = 10$, so $P > 10\%$, which means that a significant difference is not proven. This is in contrast with the paired tests, which led to an answer of $P = 5\%$. With independent samples, it would have taken much larger sample groups to reach such a probability level.

Q53. The difference is not statistically significant, $R = 16$, $P > 5\%$.

Q54. $R = 38$, $z = 1 \cdot 81$, so $P > 5\%$. A significant difference between the 2 detergents is not proven. (This value of z actually indicates a probability of $7 \cdot 0\%$; testing this data with the more sensitive technique of Analysis of Variance yields a probability of $5 \cdot 0\%$, which implies that the differences are probably significant. However, Analysis of Variance is a great deal more complicated than Wilcoxon's Test, so it is better left in the hands of the statisticians. Let us be content with the simpler technique and the approximation which it provides.)

Q55. $D^2 + T = 21$, $P > 10\%$. Significant correlation is not proven.

Q56. $D^2 + T = 66$, $P < 5\%$. Inverse correlation is probably significant. $z = 1 \cdot 98$, $P < 5\%$. The usefulness of this answer for forecasting was borne out by the uniform continuation of this inverse correlation from 1957 to 1963.

Q57. $\chi^2 = 1.3$, P $> 10\%$. A significant difference is not proven. Under these circumstances, we would not expect any particular pair of samples to show a significant difference. But we shall proceed because it illustrates how to do the selected comparison test. The rank totals of Fifi and Maria were $24\frac{1}{2}$ and $32\frac{1}{2}$, so the difference, $d = 8$. The number of observations in each sample, $n = 4$. Therefore –

$$K = \frac{d - 0.8}{n \cdot \sqrt{n}} = \frac{8 - 0.8}{4 \cdot \sqrt{4}} = \frac{7.2}{8} = 0.90$$

Referring to the left section of the K Table on page 220, we see that for $k = 3$, the calculated value of K must reach 2.89 to indicate a probability of 5%. Our calculated value is much smaller than this, so P $> 5\%$, indicating that the observed difference between Fifi and Maria is not statistically significant.

Before leaving the subject of comparing unmatched samples of measurements, it should be mentioned that there are a number of significance tests which use the *range* (p. 53) in place of the standard deviation as a measure of dispersion. The range is such a simple thing to calculate that these tests are gaining in popularity, especially in factories where they are a boon for checking the quality of mass-produced articles.

One of these is **Lord's Range Test** (*Biometrika Journal*, 1947, pp. 41–67, with extensions by Patnaik, *Biometrika Journal*, 1950, pp. 78–87, and Moore, *Biometrika Journal*, 1957, pp. 482–9). It deserves description here, not only because it is a good shortcut test, but more particularly because it will compare very small samples. It therefore serves instead of Wilcoxon's Sum of Ranks Test (for comparing 2 unmatched samples of measurements) when the number of measurements in each sample group is only 2 to 4, or instead of Kruskal and Wallis' Test when there are 3 or 4 sample groups, each containing 2 to 4 measurements, again extending down into regions too small to be handled satisfactorily by a ranking test.

Lord's Range Test thus compares the means of $k = 2$ to 4 random samples of unmatched measurements, each consisting of 2 to 4 measurements (n). You should make a note of this on pages 166 and 212. However, unlike the ranking tests, Lord's Test does assume that the samples are taken from parent groups whose individual measurements can be expected to vary in the pattern of the Normal Curve (like my desk measurements, p. 34) and whose standard deviations are similar.

The test is performed as follows –

(1) Calculate the arithmetic mean of each sample group (p. 52); call the larger or largest mean M, and the smaller or smallest mean m. With more than 2 sample groups, the middle means are not used, although they are not ignored by the test.

(2) Determine the range (i.e. the difference between the largest and smallest measurement) of each sample group; call these r_A, r_B, etc.

(3) Then calculate the ratio (L) of the 'range of the sample means' to the 'sum of the sample ranges', thus –

$$L = \frac{M - m}{r_A + r_B + r_C, \text{ etc.}}$$

It can be seen that the value of L will increase as the difference between the sample means increases, and as the range of the samples decreases. The significance of L is given in the accompanying Table.

L TABLE FOR LORD'S RANGE TEST

No. of Samples k	No. of Measurements in each Sample n		Values of L Indicating $P = 5\%$	$P = 1\%$
2	2	2	1·71	3·96
	2	3	0·92	1·56
	2	4	0·73	1·24
	2	5	0·62	1·01
	2	6	0·55	0·87
	2	7	0·50	0·78
	3	3	0·64	1·05
	3	4	0·51	0·81
	4	4	0·41	0·62
3	2		1·17	2·20
	3		0·48	0·70
	4		0·31	0·44
4	2		0·88	1·46
	3		0·38	0·53
	4		0·25	0·33

Values for 2 samples adapted from P. G. Moore, *Biometrika Journal*, 1957, p. 487; for 3 and 4 samples adapted from H. A. David, *Biometrika Journal*, 1951, p. 408, and J. Pachares, *Biometrika Journal*, 1959, pp. 465–6, using graphical and linear interpolations and compounding the data in the manner shown to me by Mr B. D. Craven of the Mathematics Department, Melbourne University.

Example. Suppose you suspect that there is more extraneous matter in supplies of raw cotton from source *A* than from source *B*, and you take 3 random samples from batches of cotton from each source and determine the percentage of rubbish in each instance. Would the following results be sufficient evidence to confirm your suspicion, or might they reasonably have arisen by chance from identical supplies?

Source	Percentage of impurities		
A	1·05	1·18	1·13
B	0·75	1·01	0·64

We are comparing unmatched samples from 2 sources (so $k = 2$), and each sample group consists of 3 measurements (so $n = 3$ and 3). Wilcoxon's Sum of Ranks Test will not deal with so few observations, so we apply Lord's Range Test. We determine the probability that there is no real difference between the 2 supplies.

$$\text{Mean of } A = \frac{1\cdot05 + 1\cdot18 + 1\cdot13}{3} = 1\cdot12 \quad (=M)$$

$$\text{Mean of } B = \frac{0\cdot75 + 1\cdot01 + 0\cdot64}{3} = 0\cdot80 \quad (=m)$$

$$\text{Range of } A = 1\cdot18 - 1\cdot05 = 0\cdot13 \quad (=r_A)$$

$$\text{Range of } B = 1\cdot01 - 0\cdot64 = 0\cdot37 \quad (=r_B)$$

$$\therefore L = \frac{1\cdot12 - 0\cdot80}{0\cdot13 + 0\cdot37} = \frac{0\cdot32}{0\cdot50} = \underline{\underline{0\cdot64}}$$

Reference to the *L* Table shows that for $k = 2$ and $n = 3$ and 3, this value of *L* indicates P = 5% exactly; the difference between the supplies is therefore probably significant. (Naturally, this is a two-sided probability, despite our suspicion of the direction of the difference, if any, for the reason explained on page 145.)

When the same data is analysed by the standard form of *t* Test which compares 2 sample means, the answer proves to be P = 5·1%, showing that there is no need to sneer at this very simple range test.

Now try Lord's Test yourself, using the above data for *A*, but with only 2 measurements for *B*, namely 0·81 and 0·71. This yields $L = 1\cdot57$. Reference to the *L* Table shows that for $k = 2$, with $n = 2$ and 3, this value of *L* indicates a probability of just under 1% (the critical value for 1% is 1·56). Accordingly, our inference would be that the observed differences between the two sources are very unlikely to be due to chance. Once again it is reassuring to find that the standard *t* Test gives an almost identical answer ($t = 5\cdot86$, with 5·84 as the 1% value).

For a bit more practice with this delightful test, try it on the data of

Q57 (p. 221). You should get $L = 0.05$, which for $k = 3$ and $n = 4$ indicates $P > 5\%$, in agreement with Kruskal and Wallis' Test.

A further note concerning the accuracy of this test will be found in the answer to *Q109* (p. 378).

Q58. $\chi^2 = 11.96$, $P < 5\%$. The seasonal influence is thus statistically significant at the 5% level. Corrected for 4 pairs of ties, $\chi^2 = 12.31$, P nearly 1%.

Regarding the comparison between Games I and III, $R_1 = 6\frac{1}{2}$ and $R_{III} = 18\frac{1}{2}$, so $d = 12$. Then as $n = 4$, we calculate –

$$F = \frac{d}{\sqrt{n}} = \frac{12}{\sqrt{4}} = 6.00$$

The left side of the *F* Table on page 228 is appropriate for this comparison; it shows the 5% critical value of F (for $k = 6$) to be 7.54. Our calculated value of F does not reach this level, so $P > 5\%$ and the difference between these 2 games is not statistically significant.

Q59. Friedman's Test showed that *if* there was no significant difference between all 4 players, the observed results could arise purely by chance with fair frequency (more than once in 10 times). Therefore, although there *may* be a real difference between them, it simply is *not proven* with the data so far available. Further data may, of course, produce evidence that a significant difference exists, but no handicapping is warranted as yet. It is not right to compare merely the best with the worst scores (p. 213) by the same standards as you would judge the group as a whole. After all, Andy might have chosen any one of $\frac{1}{2}(4 \times 3) = 6$ different pairs (*AB, AC, AD, BC, BD,* or *CD*) for such a comparison, so if Wilcoxon's Sum of Ranks Test was applied, the difference between Bill and Derek would have to reach a probability level of $\frac{1}{6} \times 5\% = 0.8\%$ to suggest that the difference between them was probably significant.

However, we can easily apply the appropriate selected comparison test. The calculation Table on page 225 shows Bill's rank total $= 10\frac{1}{2}$, and Derek's rank total $= 18$. The difference, d, is therefore $7\frac{1}{2}$.

$$F = \frac{d}{\sqrt{n}} = \frac{7.5}{\sqrt{6}} = 3.06$$

The left hand section of the *F* Table (p. 228) shows that, for $k = 4$, the calculated value of F must reach 4.70 to reach the 5% significance level. The conclusion is that the selected comparison test applied to Bill and Derek shows $P > 5\%$, which confirms that a significant difference between their scores is not proven by the data available.

Q60. $\chi^2 = 3.94$, $k = 3$, $n = 8$, $P > 5\%$. A significant difference is not proven.

Q61. $P < 1\%$. Difference is statistically significant. This strongly suggests that the preliminary trial was inadequate.

Q62. No, $P > 10\%$. A significant difference is not proven.

Q63. Yes, $P < 1\%$.

Q64. Not yet, P > 5%. However, the result is obviously close to the 5% level, so it would be desirable to continue the trial. In fact, the true binomial probability for this data is 4·86%, compared with the Poisson probability of 5·69%. Although the difference between these results is less than 1%, we have caught it straddling the 5% mark. Notice that for binomial situations, Poisson's Test tends to be slightly weak (which at least is safer than the reverse).

Q65. Yes, difference is probably significant, P < 5%.

Q66. Yes, 5% $n = 10·34$, and 1% $n = 8·34$, so probability of $n = 12$ is more than 5%. A significant difference from average is not proven.

Q67. Significant difference is not proven, P > 5%. Expected numbers under null hypothesis are 12 and 4.

Q68. $z = 4·14$, P < 0·2%. Difference is highly significant. Advise Pablo to start looking for a new fishing spot.

Q69. $z = 1·68$, P > 5%. Difference from average is not proven to be significant. We kept the die.

Q70. $z = 3·92$, P < 0·2%. The difference is highly significant. Assign more police to the northern half of the city immediately!

Q71. $R = 7$, $n = 14$, so P < 1%. This is a significant result, in contrast with the 50% Probability Test, which showed P > 5%.

Q72. With $x = 37$, the probability of $n = 104$ is < 1%. The difference between the observed result and that expected by chance is therefore significant; such a result is most unlikely to be due to guessing. However, it does not *prove* the young man to be clairvoyant, for he might have been guessing and we might have struck that one occasion in about 250 when luck was on his side.

However, even in an apparently simple experiment like this, careful attention must always be paid to the design, in order to be sure that the proposed statistical test will be valid. Thus the above answer presumes that the probability of correctly guessing the colour of each card is 50%. Is this really the case? Here we strike trouble, for the young man wrote down his answers (so that they could be checked properly) but this offered him the opportunity to equalize the number of times he said 'red' and 'black', in which case the analysis would become a good deal more complicated. It would have been a much better design to have presented him with one card at a time, drawn randomly from a shuffled pack, with replacement of the card and re-shuffling before drawing again. There would then be absolutely no doubt about the 50% Probability Test being valid.

Q73. A significant difference is not proven, P > 10%.

Q74. Not necessarily dismal. A significant difference between the pro-Potkins voters on the one hand, and the combined pro-Smith and undecided voters on the other hand, is not proven by this sample since $\chi^2 = 3·64$, P > 5%.

Q75. A significant difference is not proven, P > 10%.

Q76. $\chi^2 = 32·67$. The difference is highly significant, P < 0·2%. In other words, the number of people who came away with a better

opinion of her acting ability was significantly more than those who came
away with a poorer opinion than they had held previously.

Q77. (a) $\chi^2 = 4\cdot54$, D.O.F. $= 2$, P $> 10\%$. Significant difference is
not proven.

(b) No. With 3 samples one might equally have chosen $\frac{1}{3}(3 \times 2) = 3$
different pairs (p. 213). Therefore if the probability of no significant
difference between A and C is less than one third of $5\% = 1\cdot67\%$,
there may well be a difference, so further specific observations to test
A and C alone might be worth while.

Q78. $\chi^2 = 8\cdot00$, D.O.F. $= 2$, P $< 5\%$. Reprimand him!

Q79. $\chi^2 = 4\cdot84$, D.O.F. $= 2$, P $> 5\%$. A significant difference is
not proven.

Q80. No. Yates' $\chi^2 = 3\cdot65$, P $> 5\%$, significant difference is not
proven.

Q81. (a) Yes. Yates' $\chi^2 = 28\cdot19$, P $< 0\cdot2\%$, difference is very significant.

(b) No. Combined $\chi^2 = 42\cdot74$ for D.O.F. $= 6$, P $< 0\cdot1\%$, difference
is still very significant.

Note regarding Yates' χ^2 Test. When determining χ^2, Yates' correction is only helpful when there is 1 degree of freedom; this correction
must not be used with 2 or more degrees of freedom (as in the ordinary
χ^2 Test). Furthermore, Yates' correction should be omitted in all cases
in which you are going to add χ^2 values from independent trials, because the combined χ^2 value will be interpreted with 2 or more degrees
of freedom.

Consequently, the value of χ^2 calculated with Yates' correction in
Q81 (a) was correct for assessing the significance of the data in the
table on page 271, but for the purpose of combining the outcome of
this trial with the other 5 trials, the calculation should be done again,
adding the values of $\frac{D^2}{E}$ rather than $\frac{Y^2}{E}$; this gives $\chi^2 = 29\cdot61$ (with 1
degree of freedom). This new value is the right one to use for getting the
combined probability of the 6 trials. Addition now yields a total
$\chi^2 = 44\cdot16$ with 6 degrees of freedom, which is stronger and more
accurate than the $42\cdot74$ quoted above (although it doesn't alter the
significance in the present instance).

You may care to draw attention to this refinement by noting that
the χ^2 values in the Table on page 283 were calculated without Yates'
correction.

Q82. No. Yates' $\chi^2 = 3\cdot4$, P $> 5\%$. Significant association is not
proven.

Q83. Yes, the difference is significant. With $a = 6$, etc., the probability of getting, purely by chance, a value of d as large as 8 is less than
1%.

Q84. (a) A significant difference is not proven. For $a = 3$, etc., the
probability of getting $d = 4$ is $>5\%$.

(b) It is quite legitimate, but should be reserved for cases in which measurement tests are inapplicable (as in the present case, in which incomplete data is provided). It does not utilize all the possible information, hence gives weaker answers (i.e. larger probabilities) than would be given by measurement tests.

However, this principle is put to good use in a **Median Test,** which *compares the medians of 2 independent samples* (say, the results of an aptitude test conducted on 2 tribes). Medians were described on page 74. First pool the samples, as in Wilcoxon's Sum of Ranks Test (p. 168), and find the median of the combined data. Then draw up a 2 × 2 table, and show in its cells the number of observations above and below the common median, for each sample, thus –

No. of observations	Tribe A	Tribe B
At or above common median	20	16
Below common median	30	6

The table so formed is then analysed by Fisher's Test or Yates' χ^2 Test. Wilcoxon's Sum of Ranks Test compares the arithmetic means of the 2 samples, so make a note of this additional test on page 166.

The same idea can be extended to comparisons of the medians of more than 2 samples, using an ordinary χ^2 Test on the table so formed (you may care to note this on p. 212).

While on the subject of medians, there is a very simple approximate test for *comparing the median of a sample with that of a large parent group*. This is appropriate when data is presented in an 'open-ended' frequency table (as in *Q22*, pp. 89 and 361), from which an arithmetic mean cannot be calculated, or in any instance in which the data units belong to a scale about which there is no certainty concerning the equality of its intervals (p. 52), so that the arithmetic or geometric mean might be quite inaccurate. The test simply involves seeing how many observations there are between the median of the sample and the position which the median of the large parent group would occupy in the sample series. Given a sample of n observations, we know (p. 80) that the position of its median will be $\frac{1}{2}(n + 1)$. For instance, suppose we have a sample of 7 observations, as follows –

$$14 \quad 14 \quad 15 \quad 17 \quad 18 \quad 20 \quad 21 \text{ units}$$

The position of the sample median will be $\frac{1}{2}(7 + 1) = 4$. This means that the fourth observation (17 units) is the sample median. Now if the median of a large parent group was 15 units, its position in relation to the above sample series would be 3. Had the parent median been 13, its position would have been 0; had it been 16, its position would have been $3\frac{1}{2}$; had it been 14, its position would have been $1\frac{1}{2}$ (like an average rank value). When the parent median happens to be larger than the sample median, it is simpler to count down from the large end of the sample series, so that if the parent median was 20, it would earn a positional rating of 2. This does not affect the answer, as we are only concerned with the difference between the two median positions.

We therefore find the difference (d) between the sample and parent median positions thus –

$$d = \frac{n+1}{2} - M_i$$

where n = number of observations in the sample, and
   M_i = position of the parent median.

As usual, we start by assuming that there is no significant difference between the two medians, and calculate the probability of getting d as large as observed, by chance, from the following formula –

$$z = \frac{2d}{\sqrt{n}}$$

The significance of this value of z is then given by the z Table on p. 154. If this indicates a small probability, we reject our original assumption and accept instead that the observed difference is significant. Make a note of this test on p. 160.

For practice, try this Median Test on the data of $Q44$ (p. 165). In this case it is fair to assume that the parent median equals the parent mean (because the measurements would almost certainly be distributed symmetrically around the mean). You should get $z = 6/\sqrt{5} = 2\cdot68$, so $P < 1\%$ (which agrees with the result of the t Test).

Another example is provided by the data of $Ex.$ 3 (p. 156). The parent median was earlier calculated to be $5\cdot57$ weeks, which earns a positional rating of $6\cdot215$ on the sample group (or $8\cdot785$ if you work from the other end). Either way this gives $z = 2\cdot57/\sqrt{14} = 0\cdot69$, $P > 10\%$. This conclusion is the same as that given by the zM Test, although it is a weaker answer due to the fact that the Median Test utilizes less of the available data.

$Q85.$ (a) 1 $cell$ has $E < 5$. Yates' $\chi^2 = 3\cdot25$, $P > 5\%$ (actually $7\cdot2\%$). Fisher's Test with $a = 9$, etc., gives $P = 5\%$ (actually $3\cdot7\%$, see remarks on page 297).

(b) 1 $cell$ has $E < 5$. Yates' $\chi^2 = 6\cdot21$, $P < 5\%$ (actually $1\cdot3\%$). Fisher's Test with $a = 5$ etc. gives $P = 1\%$ (actually $0\cdot9\%$).

(c) 2 $cells$ $have$ $E < 5$. Yates' $\chi^2 = 3\cdot52$, $P > 5\%$ (actually $6\cdot0\%$). Fisher's Test with $a = 6$, etc., gives $P < 5\%$ (actually $2\cdot8\%$).

(d) All $cells$ $have$ $E > 5$. Yates' $\chi^2 = 4\cdot01$, $P < 5\%$ (actually $4\cdot6\%$). Fisher's Test with $a = 10$, etc., gives $P < 5\%$ (actually $3\cdot0\%$).

$Conclusion:$ With $N = 50$, Yates' χ^2 Test is inaccurate when 2 cells have $E < 5$, but it could be used with discretion if only 1 cell has

$E = 1$ to 5 (e.g. it could be trusted to imply 'probably significant' if $P = 1\%$, or 'significant' if $P = 0.2\%$). The accuracy of Yates' Test increases with larger values of N.

Q86. $\chi^2 = 4.20$, D.O.F. $= 3$, $P > 10\%$. The difference between the orders for the various hats is not proven to be significant.

Q87. 50% Probability Test shows $P > 10\%$. Runs Test shows $n_1 = 9$, $n_2 = 13$, $r = 12$, $P > 5\%$. Therefore proportions and order are not proven to be anything other than random.

Q88. With assumed median of 4,000, $n_1 = 4$ and $n_2 = 8$, and the probability of the fluctuations being random becomes greater than 5%. Generally speaking, the Runs Test has its maximum power when $n_1 = n_2$.

Q89. (a) 50% Probability Test for $x = 6$ and $n = 20$ shows $P > 10\%$. Hence a significant difference is not proven.

(b) Poisson's Test shows that for $x = 3$, a value of $E = 1$ has a probability of $>5\%$. Hence a significant difference between what was observed and what was expected is not proven.

(c) Runs Test for $n_1 = 6$ and $n_2 = 14$ shows the probability of getting $r = 8$ is $>5\%$. The sample could still be random.

(d) Hey, wait a minute! It's not hard to do a Runs Test (it now shows $P = 5\%$) but if you examine any random sample from 20 points of view, you're due to find, by chance, 1 test significant at the 5% level. After all, when a sample is rejected on the grounds that such a sample could be expected by chance only once in 20 times, this means that on an average of once in 20 times it will, in fact, be truly due to chance, and you will be, on this particular occasion, rejecting it unfairly. So if you want to use the Runs Test more than once on any 1 sample, you had better reduce the probability level required, proportionately. For instance, if you are going to accept the better of 2 such tests, you must demand that the probability of one test or the other be 2.5% (instead of 5%). For n_1 and n_2 up to 20, the exact probabilities for various values of r are given in Swed and Eisenhart's Tables (*Ann. Math. Statist.*, 1943, pp. 70–82). These Tables show that in the present case, where $n_1 = n_2 = 10$, getting a value of $r = 6$ has an exact probability of 3.7% (it shows as $P = 5\%$ in our Tables because the next value, $r = 7$, moves up to $P = 10.2\%$). In the present case, we have applied 4 tests for randomness on this one sample, so one of the tests would have to indicate a probability of $\frac{1}{4} \times 5\% = 1.25\%$, to be equivalent to a $P = 5\%$ with a single test. The present answer (3.7%) does not reach this mark, so we can finally assure Dr Johnson that there is still no evidence to prove that his sample is not a random one.

Anyway, even if this had proved to be significant, it would only have been valid if it had been decided to test this particular feature *before* studying the sample details. As it is, Dr Johnson has committed the fallacy of testing chance oddities. He has examined the sample, and chosen certain unusual features for testing. If he had thrown a handful of coins into the air, the probability of them landing in the positions that they do may be a billion to one, but this doesn't imply that some factor other than chance is acting to cause this particular pattern. Such a probability would only be significant if the pattern had been predicted before the experiment; only then could we say that it was extremely unlikely to be due to chance. (See p. 142.)

One final point. It is only necessary for a presenting sample to be equivalent to a true random sample if the whole sample is going to be compared with another sample or samples (say, from another city). However, most presenting samples are divided into 2 or more groups for the purpose of comparing different treatments; in this case even if the presenting sample as a whole is not random, the validity of the comparison between the treatments can be ensured by allotting the treatments in a random manner. See answer to *Q32* (*a*) on page 362.

Q90. $P = \frac{1}{12} = 8\cdot3\%$.

Q91. Poisson's Test for $x = 0$, $E = 3\cdot5$, shows P slightly greater than 5%, so a significant difference is not proven. However, the answer is so close to the 5% mark that it would be desirable to continue the investigation for a further 6 or 12 months, or until the result becomes more definite, one way or the other, before arriving at a conclusion for practical action (see *Ex. 4* and *Q44*, pp. 163 and 165).

Q92. 'Your Royal Highnesses, $\chi^2 = 10\cdot0$, which for 2 degrees of freedom, indicates a probability of less than 1%. The observed difference between the number of cod caught by the 3 baits is therefore most unlikely to be due to chance, so the difference must be attributed to the baits themselves.'

Notice that this is a good example of the need to define clearly what is being sought before the experiment is started (p. 98). The above answer is only concerned with the number of cod caught – it ignores the number of other fish, as this was not the point of concern. Had the question been: 'Is there a significant difference between the total number of fish caught (regardless of variety)?', one would apply the χ^2 Test to the 3 totals (22, 32, and 42), which gives an answer of $\chi^2 = 6\cdot25$, D.O.F. $= 2$, $P < 5\%$, so the difference is probably significant.

Alternatively, had the question been: 'Is there a significant difference between the proportions of cod and other fish caught by the 3 baits?', the comparison would have been between the 3 binomial samples (Hamlet 10 and 12, Lear 20 and 12, Henry VIII 30 and 12), and the χ^2 Test then gives $\chi^2 = 4\cdot15$, D.O.F. $= 2$, $P > 10\%$, so a significant difference is not proven.

Q93. Sample mean $= 5\cdot40$ hours, sample standard deviation $= 0\cdot962$ hours, so 'Student's' $t = 3\cdot73$, $P < 5\%$. The difference is thus probably significant.

Q94. The 50% Probability Test applies here because we are comparing 2 independent sets of isolated occurrences in which, since both applicants typed the same 4-page letter, the expected number of typing mistakes ought to be about the same in both cases, if there is no significant difference between them. The 50% Probability Table shows P > 10%, so Miss Glam got the job.

Q95. Can't tell. We would need to know the actual number of children he had in order to apply the 50% Probability Test. The 'sample size' might have been 4, in which case it would not be significant, or it might have been 400, in which case it would be highly significant.

Q96. Yes, paired comparison by symmetrical matching (p. 119). Treat one armpit of each of the 34 testers with *A*, and the other armpit with *B*, determining in each individual person which armpit shall get *A* and which *B* either by strict alternation of successive testers, or by coin tossing or other random procedure. The data could then be analysed by Wilcoxon's Signed Ranks Test, as in *Q71* (p. 265).

Q97. Assuming that the 1964 percentage has been based on a large enough series to be considered an 'average', the Table for Binomial Test (in column for $P_x = 40\%$) shows P > 5%, so a significant difference in the quality is not proven.

Q98. Wilcoxon's Sum of Ranks Test gives $R = 28$, so P < 5%. The difference between the 2 sets of marks is probably significant.

Notice that, being a ranking procedure, this Test remains valid in spite of the fact that we have no way of telling whether the exam marks were on an equi-intervalled scale or not. A *t* Test (of the sort mentioned on p. 167), to compare the arithmetic means of the 2 schools, could give an erroneous answer in this case, for the reasons mentioned on page 52.

Q99. Yates' χ^2 Test gives $\chi^2 = 2\cdot83$, so P > 5%. A significant difference in the incidence of Demodex in the 2 groups of patients is thus not proven. The idea that Demodex may be associated with some cases of lid inflammation is therefore not borne out by statistical analysis of the data presented. Of course, a larger series of cases might alter this verdict.

Q100. Binomial Test shows that for $x = 15$ and $P_x = 27\%$, a value of $n = 30$ has a probability of 1%. The difference between the 2 surveys is therefore significant. (The exact probability for this data is $1\cdot18\%$, so our interpolation has given a good approximation.)

Q101. If you applied Wilcoxon's Signed Ranks Test and found $R = 7\frac{1}{2}$, P > 5%, meaning that a significant difference between the 2 champions was not proven, your arithmetic was correct, but your choice of test was rash. Admittedly the question involves the comparison of matched measurements; both champions competed in the same set of events under conditions which would have been as alike as possible. However, the measurements being compared were scores of ability, and this means that we cannot be sure that they were based on a scale which increased by uniform intervals. For instance, it may have been much easier to have gained the first 3 points (0 to 3) than the topmost 3 (17 to 20) in any one event. But, you may say, *Q98* involved similar measurements, and we saw above that Wilcoxon's

Sum of Ranks Test was valid there, because it was a ranking test. Wilcoxon's Signed Ranks Test is also a ranking test, but it differs in that it involves ranking *numerical differences* between the matched observations, and accordingly the observations need to be measurements belonging to an equi-intervalled scale.

To make this point quite clear, look at the results of the first event. The difference in scores was 2. Can we be sure that this difference is really smaller than the difference of 3 which occurred in the fourth event? Certainly it is smaller numerically, but does this also represent a smaller difference in ability? As different regions of the scale are involved, the answer is no, we can't be sure. It might even happen that to rise from a score of 13 to 16 could be easier than to rise from 16 to 18.

In the present question, therefore, one should use the 50% Probability Test (as in *Ex. 35*, p. 263), and ask what is the probability of the West Australian winning but 3 of the 9 events in which a difference in scores is present, if there is really no significant difference between the 2 champions. The 50% Probability Table shows (for $x = 3$) that $P > 10\%$, so it must be admitted that the results could easily be due to luck. This probability level is weaker than that obtained with Wilcoxon's Signed Ranks Test, but it has the advantage of being more trustworthy under the circumstances.

Q102. From the data table provided, we see that 13 testers had a hangover after alcohol but not after the new drink, whereas only 3 had a hangover after the new drink but not after alcohol. The remaining testers showed no difference between their response to the 2 drinks. Hence $x = 3$, $y = 13$, $n = 16$, and the 50% Probability Table shows $P < 5\%$. The difference is probably significant.

Q103. Fisher's Test for $a = 10$, $b = 2$, etc., shows $P < 1\%$. The difference is highly significant.

Q104. Poisson's Test shows $P < 1\%$. Dismiss the manager!

Q105. Poisson's Test shows $P = 1\%$. There is only 1 chance in 100 that the batch is up to the advertised standard, so go ahead and complain.

Q106. Binomial Test shows $P = 5\%$. The difference is therefore probably significant.

Q107. With $x < E$, and $x > 4$, the zI Test (with $c = 0.2$) applies. This gives $z = 2.31$, $P < 5\%$. The reduction in imperfect lenses is probably significant. However, it would be worth while examining a further batch of 200 lenses before jumping to conclusions (compare *Ex. 20*, p. 234).

Q108. 50% Probability Test (McNemar's Formula) gives $\chi^2 = 3.68$, $P > 5\%$. The difference is not significant at the 5% probability level.

Q109. Kruskal and Wallis' Test is applicable. The observations are not matched, despite the fact that the way they are presented in the data table may make them appear so. For it is not reasonable to expect the 1st cylinders of the 3 cars to be more alike than the 4 cylinders of each individual car. $\chi^2 = 7.65$, P slightly $< 1\%$. The observed differences are therefore significant. (More precisely, this value of χ^2 indicates a probability of 0.8%, which is virtually identical with the result obtained by the more complicated technique of Analysis of Variance.)

If you have not already done so, apply Lord's Range Test (p. 366) to this data, just for practice. It yields a value of $L = 0.42$, which indicates that the probability does not quite reach down to the 1% level. On this data, the range test proves slightly weaker than the ranking test and the full Analysis of Variance test.

Regarding the individual comparisons, let us deal with A and C first. The difference between their rank totals, d, is 28, whence $K = 3.40$. The right-hand part of the K Table (p. 220) is relevant in this instance (because C is a control); our calculated value of K indicates a probability of almost 1%, which is a statistically significant difference.

However, with B and C, $K = 2.02$, P $> 5\%$, so a significant difference between these 2 oils is not proven.

Q110. It is not possible to apply a statistical test on these proportions. One must have the actual numbers in each sample for χ^2 Test. *Q95* was similar. (The doctor did, in fact, provide full data in the article from which this question was excerpted. However, insufficient data is encountered only too commonly, so you should be prepared for it.)

Q111. $x < E$, $x > 4$, so zI Test (with $c = 0.2$) applies. $z = 1.96$, P $= 5\%$. The difference is probably significant.

Q112. zM Test gives $z = 1.64$, P $= 10\%$. However, this is a 2-sided probability (p. 143), so it means that on 10% occasions Nicely-Nicely will either exceed his average by the necessary 4.1 lb or will eat 4.1 lb less than his average. Therefore, his chance of winning the contest will be half this probability, i.e. 5%. His chances will be greater, of course, if the Bostonian eats less than his previous record, so this answer of 5% is really a minimum figure.

Q113. The best estimate of the probability of drought is now $\frac{19}{83}$, which is 23%, so that the probability of rain is the complementary percentage, 77%. Contrast the case of the 'gambler's fallacy' (p. 27) in which the probability is a fixed one.

Q114. Fisher's Test shows P $= 5\%$. The difference is therefore such that the standard training should be continued.

Q115. 50% Probability Test (Table) for $x = 14$ and $n = 38$ shows P $> 10\%$. The difference is not proven to be significant.

Q116. Poisson's Test shows P $< 1\%$. The difference is significant. Beware of sharks!

Q117. Spearman's Correlation Test gives $D^2 + T = 36$, P $< 5\%$. Direct correlation is probably significant. In other words, increasing the spray concentration probably does kill more flies. Notice, by the way, that this correlation test would still have been valid if there were a different number of flies in each experiment, for the percentages of flies killed can be ranked in the usual manner.

Some readers may have thought that a χ^2 Test was appropriate for this question. Taking each of the 10 test conditions as independent binomial samples consisting of the stated number of killed flies and a number of flies that were not killed (the latter being found by subtracting the number killed from 100), it is then possible to see whether there is a significant difference between the proportions in all 10 samples. This yields $\chi^2 = 22.63$, with D.O.F. $= 9$, P $< 1\%$. However, although this indicates a significant difference between the proportion of flies killed by the various concentrations of insect spray, it does not tell us whether or not this difference is related to the concentration of insect spray in the various tests. In other words, it reveals a difference but not correlation.

It is to be hoped that no one did a χ^2 test as follows. Set up a tentative negative assumption that there is no significant difference between the number of flies killed by the various concentrations of insect spray. Add the number of flies killed in the 10 tests, and divide by 10 to get the arithmetic mean, which is 87. Now apply a χ^2 test to this 10 $\times$ 1 table to see if the observed numbers differ significantly from the overall mean of 87. This yields $\chi^2 = 2.94$, D.O.F. $= 9$, P $> 10\%$, indicating that a significant difference between the numbers killed in each category is not significant. This is quite wrong, because it completely ignores the number of flies that were not killed. It treats a set of binomial samples as if they were a set of isolated occurrences. I have recently seen exactly the same kind of error in a reputable medical journal. If this error does not strike you as being obvious, work out the above χ^2 values for yourself and the fallacy will become crystal clear.

Q118. $\chi^2 = 11.09$, D.O.F. $= 2$, P $< 1\%$. The difference is significant.

Q119. Yates' χ^2 Test gives $\chi^2 = 0.64$, P $> 10\%$. This result is so definitely 'significant difference not proven' that it can be trusted even though $E < 5$ in one cell (see *Q85*, p. 317, and its answer, p. 373). Further observations to make all E's > 5 would be necessary if the probability had been anywhere near 5% (say, 3% or 7%).

Q120. Sample mean $= 203.6424$, sample S.D. $= 0.0108$. 'Student's' $t = 218.7$, P $< 0.2\%$. The difference is extremely unlikely to be due to chance. Techniques have improved since 1861.

Q121. Binomial Test with $P_x = 25\%$, $E = 3$, shows P $> 5\%$. A significant difference from expectation is not demonstrated. The research worker's statement is not disproved.

Q122. Yates' χ^2 Test gives $\chi^2 = 17.13$, P $< 0.2\%$. Difference is highly significant.

Q123. 50% Probability Test (Table) shows P $> 10\%$. The difference between the number of men and women is not proven to be significant. This suggests that men and women are equally susceptible to hydatid cysts, at least unless further evidence comes to hand disproving it.

Q124. Wilcoxon's Sum of Ranks Test gives $R = 12\frac{1}{2}$, P $> 10\%$. A significant difference between the 2 shavers is not proven. Likewise, Lord's Range Test gives $L = 0.28$, P $> 5\%$.

Q125. Friedman's Test gives $\chi^2 = 9.78$, P $< 5\%$. Corrected for ties, $\chi^2 = 9.98$, P $< 1\%$, so further investigations are unwarranted.

Note that with this data, the selected comparison test is much weaker than the overall Friedman's Test. The difference in rank totals between the best and worst samples is $9\frac{1}{2}$, which gives $F = 4.24$, and indicates P $= 10\%$. Therefore always do Friedman's Test first, and follow up with the selected comparison test if necessary.

Q126. $x < E$, $x > 4$, so zI Test (with $c = 0.2$) applies. $z = 1.34$, P $> 10\%$. The difference is not proven to be significant, so may well be due to chance.

Q127. E for each cell $= 18$. $\chi^2 = 16.33$, D.O.F. $= 7$, P $< 5\%$. Therefore the difference between the starting post positions is probably significant, but more data would be needed to make the matter more definite. As it stands, the popular claim is to be considered probably correct.

Q128. Wilcoxon's Sum of Ranks Test gives $R = 47\frac{1}{2}$, so P $< 5\%$. Thus the difference in the 2 groups' ages is probably significant. This implies that the patients were probably not divided up at random, so it would be unwise to draw any general conclusions from this trial. Further testing of the treatments with properly randomized sample groups is desirable.

Q129. First of all, Harry tried the 50% Probability Test, until he realized that this was only applicable to occurrences of things, and he was comparing measurements (minutes). He then concluded (correctly) that he did not have enough data for a significance test. Peggy said, 'Well, if it helps you to decide, you can take it that 25 minutes has been the *average* time for me to clean the house with my old vacuum.' Is this enough information now? If so, proceed to test; if not, state what would be enough. For answer, see p. 391.

Q130. Wilcoxon's Stratified Test gives $R = 35$, $z = 1.78$, P $> 5\%$. A significant difference between the 2 teaching methods is not proven, but as the result is between P $= 5\%$ and 10%, and the number of students is small, it would obviously be worth while to continue the investigation for another year or so, as the result might become significant when based on more students. See also answer to *Q98*, page 376.

Q131. (a) Although only $\frac{1}{8}$ of the records sold more than $\frac{1}{2}$ million copies, $\frac{1}{4}$ of the group had good covers. Even so, Fisher's Test (for $a = 3, b = 2$, etc.) shows P $> 5\%$, so a significant association is not proven.

(b) Yes, the design could have been improved by using the median number of sales (instead of the arbitrary $\frac{1}{2}$ million sales), and then applying Fisher's Test as before. This was pointed out in connection with the hexachlorophene ointment trial (p. 114); calculate Yates' χ^2 for those numbers, and then repeat the calculation with 77 patients in each group and the same cure rates (80% and 89%). You will find that the value of χ^2 is 0.57 higher with the equal sized groups (in terms of probabilities, P $= 13\%$ becomes P $= 9\%$, an appreciable strengthening).

Another way of improving the design would be to see if there was any significant correlation between the sales of individual songs and the beauty of the record covers, ranked in order of preference, using Spearman's Correlation Test. However, it might be considered that ranking 40 album covers in order of beauty was making the matter entirely too personal and subjective, so that even a highly significant statistical result may have very little practical significance.

Q132. Runs Test gives $r = 4$, so P > 5% (actually 22%). Thus if 4 spotted trees were scattered at random among 8 normal ones, the difference between the observed number of runs and average could be expected by chance about once in 5 times. The degree of clustering therefore gives no support to Uncle John's theory of contagion. It is true that we are here interested only in clustering, i.e. in the possibility that there are too few runs, but we must still use a 2-sided probability (p. 143).

Q133. 50% Probability Test (Table) shows that with $x = 4$, the probability of $n = 32$ is <0·2%. The decreased number of smokers is highly significant.

Q134. Wilcoxon's Signed Ranks Test gives $R = 51\frac{1}{2}$, $z = 2·63$, P < 1%. The difference is significant, so the laboratory should not change to the new test.

Q135. Data provided is insufficient to apply Wilcoxon's Sum of Ranks Test. The difference may be highly significant, but we would need to know the dispersion around the averages for any kind of significance test. The situation is the same as that described on page 56.

Q136. zM Test gives $z = 5·60$, P < 0·2%. Actually, $z = 4·9$ indicates a probability of 1 in a million, so the result in the present case is extremely remote. The difference is thus very significant statistically, although the practical significance of the finding is not yet known.

Q137. Yates' χ^2 Test gives $\chi^2 = 7·10$, P < 1%. The increase is significant. This question emphasizes once again the cardinal importance of the sampling technique used, for this result is only true if the 2 samples were genuinely random. It would not be good enough if the interviewer had collected all his information 'at the railway station, because all sorts of people travel by train'.

Q138. Friedman's $\chi^2 = 9·375$, P = 1%. The observed differences are significant, due chiefly to paint C which contributed the greatest share towards this value of χ^2.

The selected comparison test applied to paints B and C gives $F = 4·75$, P < 5%. Likewise, between A and C, $F = 4·50$, P > 5%. Therefore although there is a significant difference between the 4 paints taken as a whole, the only additional conclusion available from the data provided is that paints B and C are probably significantly different, considered as individuals, while a significant difference is not proven between A and C, or between C and D.

Q139. $\chi^2 = 5·25$, D.O.F. = 2, P > 5%. The professor's theory is not denied.

Q140. Yates' $\chi^2 = 3\cdot59$. This does not quite reach the 5% probability level, so a significant difference between the 2 samples is not proven: the suspect sample could be random. Note that with very large samples like these, the accuracy of Yates' χ^2 Test can be relied upon to be virtually perfect.

Notice that this test is valid even though it was suggested *after* examining the results, because the issue is directly concerned with the purpose of the investigation. Had it been applied to the incomes, number of children, religions of the men in the sample group, or such like (comparing these with the national averages), it would have been irrelevant to the question of whether there was a significant difference between the health of smokers and non-smokers, so such tests for randomness of the sample would only have been valid if it had been decided to apply them *before* looking at the results (see p. 142). With these remarks in mind, have another look at *Q128* (p. 341).

Q141. Poisson's Test shows P < 5%. The reduction in fires is probably significant.

Q142. (*a*) $\chi^2 = 3\cdot92$, D.O.F. = 2, P > 10%. A significant difference between the 3 remedies is not demonstrated.

(*b*) Yes, get the testers to rank the 3 remedies in order of preference, and use Friedman's Test, as in *Q60* (p. 229).

Q143. Binomial Test for $P_x = 25\%$ shows P < 5%. The observed difference from expectation is therefore probably significant.

Q144. Sample is to be considered not random, because Runs Test gives $z = 2\cdot51$, P < 5%. Compare the data with that in *Ex. 50* (p. 327) where the proportions and numbers were the same, but the number of runs was *greater* than expected (there were too few runs in the present question).

Q145. Sample mean = 9·58, sample S.D. = 1·39, 'Student's' t = 2·50, which for $n = 6$ indicates P > 5%. He bought the balls.

Q146. zM Test gives $z = 2\cdot00$, P < 5%. The supply is probably not up to standard. He didn't buy this batch of books.

Q147. Fisher's Test for $a = 10$, $b = 5$, etc., gives P < 1%. The association is significant. The aptitude test works. (Note: the data was fictitious.)

Q148. Lack of information about the degree of dispersion about the means makes it impossible to assess the part played by chance. Significance tests (such as Kruskal and Wallis' Test) need more data.

Q149. As the data stands, a χ^2 Test will not be quite accurate because one of the cells has an expected number of 3·75. However, if you proceeded in spite of this, $\chi^2 = 10\cdot28$ with D.O.F. = 2, so P < 1% (actually 0·59%), so the observed differences are significant.

It would be wiser to re-group the data, combining those who watched up to 7 hours per week, making a 2 × 2 contingency table in which all E's are >5. Then Yates' $\chi^2 = 8\cdot55$, P < 1% (actually 0·35%).

Q150. As in the previous question, re-grouping of the data is necessary to make E in all cells equal to 5 or more. This is attained by pooling the mothers who slept the same or better after having children; this does not interfere with the aim of the investigation. $\chi^2 = 11\cdot25$, D.O.F. $= 2$, P $< 1\%$ (actually $0\cdot36\%$), so the difference is significant, and the popular claim is upheld. However, if you over-do such pooling of groups, and also combine the mothers of 2 children with those having 3 or more children, you end up with a 2×2 table thus –

No. of children	Sleeping	
	Worse	Same or better
1	25	5
2 or more	15	15

Yates' χ^2 for this table is $6\cdot07$, P $< 5\%$ (actually $1\cdot4\%$), so strength has been lost by unnecessary over-pooling.

Q151. (*a*) Naughty! Mustn't hand-pick figures for comparison (p. 213).

(*b*) Can't tell. A χ^2 Test should not be applied to this data because the results have been presented in the wrong kind of table. As it stands, the contingency table has a structure suitable for unmatched samples, but the results were not those of independent samples; they have been made on 1 sample group of patients, hence are matched observations. We would therefore need the results re-cast into a 2×2 contingency table for matched observations (p. 272), and Fisher's Test would then give the answer we are seeking. However, the data provided in this question cannot be converted into this kind of table; one would need access to all the original results to compile a table for matched observations.

Q152. Spearman's Correlation Test applied to the last 2 columns of the data table provided gives $D^2 + T = 553$, P $< 1\%$. Therefore the observed inverse correlation between the average number of decayed teeth per child and the fluoride content of the water supply is statistically significant. It does not prove that fluorides prevent tooth decay, but it has certainly stimulated a great deal of research into the matter.

Q153. (*a*) The answer to this question is to be found in the 50% Probability Table. We find there that if $x = 34$, $n = 100$ has a probability of $0\cdot2\%$ (which is 1 in 500). Bearing in mind that this Table gives 2-sided probabilities, this means that a candidate might get as few as 34 or as many as $100 - 34 = 66$ answers correct by sheer guesswork, once in 500 times.

(*b*) Similarly, if $x = 13$, $n = 50$ has a P $< 0\cdot2\%$. Therefore he might get as many as $50 - 13 \doteq 37$ right by guesswork, plus the 50 he knew, making a total of 87 altogether. These results are a mute comment on this type of examination.

Q154. Comparing those who prefer *A* with those who prefer *B*, by means of the 50% Probability Test (Table), indicates the probability of no significant difference is only 0·2%. The difference between the effectiveness of *A* and *B* is therefore highly significant. We beg to differ with the statistician's conclusion (which was based on an unconventional test).

Q155. P < 5%, as told by the 50% Probability Test (Table).

Q156. The individual trees are not matched, only the 2 *groups* are comparable, so Wilcoxon's Sum of Ranks Test is appropriate. This gives *R* = 31, P > 10%, so a significant difference between the means (which, of course, *may* be present) is not proven by this particular experiment. Siegel and Tukey's modification, without adjustment to equalize the sample means, gives *R* = 35; with adjustment, increasing each measurement in the 'normal treatment group' by the difference between the sample means (which was 13), gives *R* = 36. Either way, P > 10%, so a significant difference between the dispersion of the 2 sets of observations is not proven.

Q157. (*a*) Cochran's χ^2 = 15·24, D.O.F. = 2, P < 0·1%. The difference in the numbers sick during the 3 storms is highly significant.

If you had trouble starting, realize that we are comparing 3 sets of observations (*A* = no pills, *B* = lactose, and *C* = 'Sicko' pills) and that the results of each are in 2 groups (sick and not-sick). So prepare a calculation table as described for Cochran's Test (p. 318), and you should find the numbers fall into place quite easily.

(*b*) The table you have prepared for calculating Cochran's Test enables the required information to be extracted without trouble. You will find that 9 were not sick with 'Sicko' pills but were with lactose, as against 7 who were not sick with lactose but were with 'Sicko'. The 50% Probability Test (Table) shows this has P > 10%, so a significant difference between the 'Sicko' pills and the lactose is not proven. Taken in conjunction with the result of Cochran's Test, this implies that the effect of both pills was probably largely psychological.

(*c*) The design could have divided the crew into 3 groups by ballot; at the approach of the first storm, 1 group would get no pills, the second group would get a lactose pill, and the third group would each get a 'Sicko' pill. At the approach of each subsequent storm, the groups would be rotated, so that the group which got no pills the first time would get a lactose pill the second time, and a 'Sicko' pill the third time, and so with the other 2 groups. At the end of the 3 storms, every man would have tried each of the 3 'treatments', and the results could be analysed by Cochran's Test. Then if 1 storm was appreciably milder or stronger than the others, all 3 groups would show a proportionate decrease or increase in seasickness (as the case may be), and the experiment would not be ruined.

Q158. Poisson's Test shows P < 5%, so the improvement (or rather, the *difference* from expectation) is probably significant.

Q159. χ^2 = 32·02, D.O.F. = 6, so P < 0·1%. The observed differences are highly significant. The main contribution to this high value of χ^2 is from members of the orchestra, who in general gave this work a better reception than the audience (perhaps because they knew the music better as a result of rehearsing it). Notice that we have chosen to overlook the fact that 1 cell has $E = 3.5$, because this cell contributes so little to the total value of χ^2 that re-grouping the classes would make no appreciable difference to the answer.

Q160. Kruskal and Wallis' Test gives $\chi^2 = 12.08$. As 4 samples are being compared, we refer this answer to the ordinary χ^2 Table (p. 276), where for D.O.F. = 3, we get P < 1%. The differences are therefore significant. The lion's share of this value of χ^2 is obviously due to package *B*, a point which can be nicely confirmed by the appropriate selected comparison test.

A point about the ties. In their article describing this Test, Kruskal and Wallis give a method of correcting the slight weakness that ensues when there are many *tied observations*. In the present example, 17 of the 20 measurements are tied, and applying this correction only raises the value of χ^2 from 12.08 to 12.44. The effect on the probability level is thus very small. However, you may wish to apply this correction if you get an answer just greater than P = 5% or 1%; it consists of dividing your calculated value of χ^2 (computed as described on p. 215) by the factor *T*, which is –

$$T = 1 - \frac{6t_2 + 24t_3 + 60t_4 + 120t_5 + (x^3 - x)t_x}{N(N^2 - 1)}$$

Where t_x = number of ties involving x observations, and N = total number of observations in all the samples or sets of measurements. Thus in the present case we get –

$$T = 1 - \frac{(6 \times 1) + (24 \times 2) + (60 \times 1) + (120 \times 1)}{20(20^2 - 1)}$$

$$= 1 - 0.0293$$

$$= 0.9707$$

The corrected value of χ^2 is therefore –

$$\frac{12.08}{0.9707} = 12.44$$

Q161. Assuming that the samples are random, Wilcoxon's Stratified Test gives $R = 64$, P < 5%. The observed rise in the price of slaves is therefore probably significant, and hardly likely to be due to chance.

Q162. If you applied Wilcoxon's Stratified Test, you should have got $R = 23$, which for $k = 3$ and $n = 3$ indicates P < 5%, so the difference between the 2 drills is probably significant.

However, a stronger answer is obtainable by utilizing the fact that the comparisons were all strictly paired and matched, which permits the application of Wilcoxon's Signed Ranks Test. This gives $R = 0$ for $n = 8$, $P = 1\%$, so the difference is statistically significant.

However, both of these results would be valid only if one could be sure that there was no significant interaction between operators and drills. For instance, suppose A1 handled the new drill much better than the other operators on all 3 materials; this bias would then be repeated 3 times, and would distort the results unduly in favour of the new drill. Analysis of Variance would be necessary to assess such interactions, failing which it would have been safer to use 9 operators for the experiment (3 different ones for each material) to obviate the risk of being misled in this way. The experiment would then be less sensitive, but the results of the Wilcoxon Tests would be more trustworthy (in the sense of indicating what will happen when other operators use the drills).

Q163. $x < E$, $x > 4$, so zI Test applies. $z = 1\cdot61$, $P > 10\%$. The observed reduction in number of floods is not proven to be significant.

Q164. (a) To compare the efficacy of the ordinary tar (T) with that of the synthetic tar (S), it is tempting to pool the results of Group I and Group II, as in a factorial experiment. However, the design of this particular trial was such that some patients who were in Group I were also in Group II (this applies to the 32 patients who tried all 4 formulae), hence to combine these 2 groups would involve mixing some matched with some unmatched observations. This is not permissible. It is therefore necessary to analyse the results as they stand, separately; the 50% Probability Table shows that in both Group I and Group II, the observed results could be due to chance if there was no significant difference between T and S with a probability of less than 1%. The differences are therefore significant, both with the cream base and with the paste base.

(b) Similarly, $P > 10\%$ for the data in both Group III and Group IV, tested separately, so there is no significant difference proven between the cream base and the paste base.

Q165. A χ^2 Test would not be applicable to this data, because the results have been quoted only as proportions (not as the actual numbers of things).

Furthermore, a significance test would be pointless here, for the phrase 'statistically significant' only has meaning in respect of a random sample, and then implies 'beyond the bounds of chance'. In the survey under discussion, we are told that the sample represented about 90% of the manufacturer's outlets, so the precision of the results would approach that which would apply if a total count had been performed. At the most, it could only be about 10% wrong, even if all the remaining (non-interviewed) firms had tastes which were diametrically opposed to those interviewed. The manufacturer would therefore be correct in ordering his raw materials on the basis of the results obtained, bearing in mind, however, that it will be the tastes of the general purchasing public (rather than that of the retailers) which will ultimately determine the demand.

Q166. Unlike the above question, the percentages of cloudiness do not constitute the data to be tested, but are here merely class-makers. The comparison is between counts of days, so the χ^2 Test is applicable. It is first necessary to fuse the 40% and 50% categories to raise the expected numbers in all cells to 5 or more. Then $\chi^2 = 21 \cdot 05$, D.O.F. = 9, so there is considerably less than a 5% probability that the observed differences are simply a matter of chance.

Note particularly the frequency distribution of the data in this question. A frequency chart of this data has quite the opposite shape of the Normal Distribution shown in Fig. 2 (p. 34). The degree of cloudiness in the sky is the classical example of this rare type of distribution, which because of its maximum frequencies at the extremes and minimum near the centre, is called a *U-shaped distribution*. The arithmetic mean and standard deviation are poor descriptive measures of such a pattern of frequencies (compare p. 73), so in these cases you must only use those significance tests which do not call for the mean and standard deviation. The χ^2 Test is 'distribution-free' (p. 167), so can safely be applied to any kind of distribution.

Q167. Once again, the χ^2 Test is the correct one for this case. After clubbing the end cells together to make a 4 × 2 data table with $E = 5$ or more in all cells, you get $\chi^2 = 7 \cdot 18$, D.O.F. = 3, P > 5%, so the difference between the men and women is not proven to be significant. Here, too, the distributions are not symmetrical like the Normal Distribution, but are mirror images of the Normal Logarithmic Distribution (Fig. 8, p. 85), with their modes on the right of their mid-ranges.

This question is concerned with the comparison of two samples of measurements (maximum weights lifted). Had all the individual measurements been quoted, we could have tested for a difference in the means of the two samples with Wilcoxon's Sum of Ranks Test (as in *Q128*), and/or tested for a difference in their dispersions with Siegel and Tukey's Test (p. 167), and because we would then be utilizing all the data instead of a mere summary of the data, the test would be more powerful and may well show P < 5%. Something is always lost whenever measurements are grouped or abbreviated, but at least the χ^2 Test is sensitive to differences both in the main location of frequencies and in their degree of spread.

Q168. The distribution of the frequencies in these samples dies away in a logarithmic pattern, called a *J-shaped distribution* because the frequency curve is like the mirror image of a *J*. Grouping of the categories for 7–11 petals to get the necessary minimum expected numbers leads to $\chi^2 = 2 \cdot 56$, D.O.F. = 4, and a probability of more than 1 in 10 that the differences between the varieties are simply due to the ordinary fluctuations of sampling from a common parent source. Note that in contrast with *Q167*, all the data is provided in this case. Why not use Kruskal and Wallis' Test then? We could list out Sample *A* as having a value of 5 eighty-two times, a value of 6 eleven times, etc. The trouble is that this leads to a colossal number of ties, and even though a correction can be made for any reasonable number of ties (p. 385), the test was really designed for dealing with measurements made on a continuous scale (such as length, weight, etc.), and we would therefore be stretching the test beyond fair limits, with consequent inaccuracy of the answer, if we applied it to the present mass of data made on a purely whole-number scale.

Q169. It is to be hoped that you didn't pay more than a shilling for that old manuscript. Working correct to 2 decimal places throughout gives $\chi^2 = 39.86$, which for D.O.F. = 6 indicates P < 0.1% (the actual probability is about 1 in a million). So there is not much chance that the unknown work was written by Strauss. Judging from the size of the contributions to the total χ^2 value, the main discrepancy is in the bars with single notes. One cell has $E = 4.63$, but this was allowed to pass in the above calculation because it was close to 5 and because this cell contributed very little to the total χ^2 value.

Of course, one of the main things here is to realize that significance tests could never prove that it *was* a work by Strauss, only that it probably *wasn't*. We met a similar situation in *Q41-2* (p. 158).

Wilcoxon's Sum of Ranks Test and Siegel and Tukey's Test must not be applied to this kind of data, for they were designed to deal with measurements made on a continuous scale.

The numerical data for this question was not fictitious; the 'unknown' composer was not Strauss. But it does show one of the ways in which Statistics can be applied to the Humanities.

Q170. (*a*) By the Multiplication Law (p. 24), P = 0.5%.

(*b*) As mentioned on page 283, χ^2 is an additive value, so to get the probability of the combined reports, we add the values of χ^2 and the degrees of freedom. When dealing with 2 × 2 data tables, this should be done after re-calculating each value of χ^2 without Yates' correction (p. 371), for this makes the combined probability answer both stronger and more accurate. Unfortunately, the necessary data to do this is not provided in the present question, so we must be satisfied with the weaker answer obtained by adding the Yates' χ^2 values. This gives $\chi^2 = 6.55$ for D.O.F. = 2. Reference to the χ^2 Table on page 276 shows this to indicate a probability of less than 5% (actually P = 3.8%). This means that the difference between the old and the new treatment, judged by the results of the 2 trials, is probably significant. The fact that the second report was based on a larger number of cases has already been taken into account in calculating the value of χ^2 for that trial; it does not have the effect of weighting the probability of the combined results in favour of the second trial rather than the first, but merely helps to create a larger combined sample capable of attaining a more remote probability level in the same way as getting 200 heads in

1,000 tosses of an unbiased coin is a much rarer occurrence than getting 2 heads in 10 tosses.

At first, it may strike you as puzzling as to why this combined answer of P = 3·8% should be different from the answer to (a) above, which gave P = 0·5%. But the 2 situations are not really analagous. When we pick a marble from a bag, the probability of the outcome is quite straightforward. When we compare an old with a new treatment, the situation is mathematically much more complex, for in comparing the 2 sample groups, we are really estimating the probability of them being derived from one and the same parent group. The situation is like drawing a sample of marbles from 2 different bags and then, from the composition of the 2 samples, estimating the probability that the parent bags have identical compositions. The situation becomes even more complicated when we add the results of 2 such trials. For although the trials are independent in the sense that they are carried out on different occasions and perhaps by different investigators, and are furthermore independent in the sense that the outcome of the first trial exerts no influence on the outcome of the second trial, the fact remains that the second trial is of the nature of a repetition of the first, so it is natural to expect the outcomes to be related in their similarities. The Multiplication Law only applies to situations in which the outcomes are completely unrelated. It would therefore be a case of false analogy to compare the present problem to that of the first part of this question (above). A simple model is an excellent way of demonstrating the principles of a statistical situation (e.g. we used it on Jenner's experiment, p. 95), but the model must, of course, fit the situation faithfully.

(c) The problem is a similar one to that of (b) above. However, z values can't be added together like χ^2 values, so we must learn a new trick for *combining the probabilities of any 2 or more independent experiments*. This can be done, rather ingeniously, by converting each individual probability (regardless of how calculated) into a χ^2 value, and then adding all the χ^2 values and degrees of freedom in the usual way. These conversions are accomplished by means of the following formula –

$$\chi^2 \text{ (with 2 D.O.F.)} = -2 \log_e \frac{P\%}{100}$$

$$= 4·605(2 - \log_{10}P\%)$$

We shall use the second line of this formula because it only involves looking up common logarithms. When a probability is less than 1%, its logarithm will be a bar number, in which case remember that $2 - \bar{1}·6990 = 2·3010$ (which means, in ordinary numbers, $100 \div \frac{1}{2} = 200$).

The only catch is that you will have to refer to source Tables to get the exact probabilities of each individual experiment (e.g. P = 5·36%; plain P > 5% won't do), but the rest of the calculation is quite easy.

Thus, in the present case, to combine $P = 5 \cdot 00\%$ and $P = 10 \cdot 00\%$ into a single overall probability, we proceed as follows –

Individual probabilities $(P\%)$	$\log P\%$	$2 - \log P\%$	$4 \cdot 605(2 - \log P\%) = \chi^2$	D.O.F.
5·00	0·6990	1·3010	$4 \cdot 605 \times 1 \cdot 3010 = 5 \cdot 991$	2
10·00	1·0000	1·0000	$4 \cdot 605 \times 1 \cdot 0000 = 4 \cdot 605$	2

$$\therefore \text{Combined } \chi^2 = 10 \cdot 596 \text{ with 4 D.O.F.}$$

You can see that the χ^2 values and the degrees of freedom have been added together to get a combined $\chi^2 = 10 \cdot 60$ with D.O.F. = 4. This is now referred to the χ^2 Table (p. 276), which shows $P < 5\%$ (actually $P = 3 \cdot 1\%$). Therefore, the overall conclusion, based on both trials, is that a diet of baked beans probably does cause a greater weight gain than average.

(d) Providing that you can feel sure that both samples are derived from the same parent group (and this is a reasonable assumption in the present instance), it is quite in order to combine the samples. There is a short-cut for calculating the arithmetic mean of the combined samples from the separate means of each sample, but there is no short-cut for combining the standard deviations in the same way (for each sample standard deviation is calculated around the mean of the sample, and this tends to vary for each sample, even when derived from the same parent source). It is therefore necessary to calculate the standard deviation from the original data afresh. We then find that the mean of the 2 samples combined is $11 \cdot 33\delta$, and the S.D. of the combined samples is $1 \cdot 58\delta$. 'Student's' $t = 3 \cdot 74$, which for $n = 9$ indicates a probability of no significant difference of less than 1% (more precisely, $P = 0 \cdot 6\%$).

By combining the 2 samples, which separately gave $P > 5\%$ in each case, we find that the difference between the lethal dose for Earthmen and that required for Moonmen is now significant at the 1% level.

Had the 2 samples been collected or tested in different ways, it would not be reasonable to pool the results in the above manner. In that case, we would be obliged to combine the individual probabilities, as described on page 389. To do this, we must know the exact probabilities of $t = 3 \cdot 98$ for $n = 3$ (the Southern sample), and of $t = 2 \cdot 36$ for $n = 6$ (the Northern sample). These can be found by referring to a Graph of t values in F. Wilcoxon and R. Wilcox, *Some Rapid Approximate Statistical Procedures* (Lederle Laboratories, 1964, p. 55), or to special t Tables such as Table 9 in *Biometrika Tables for Statisticians*, Vol. 1 (CUP, 1966). These tell us that the probability associated with the first sample was $5 \cdot 8\%$, and that of the second sample was $6 \cdot 5\%$. Now have a go at calculating the combined probability yourself. The answer is $\chi^2 = 11 \cdot 16$, D.O.F. = 4, $P < 5\%$ (actually $P = 2 \cdot 5\%$). This is a weaker result than that obtained by pooling the data of the 2 samples which gave $P = 0.6\%$), because it utilizes less of the available facts.

Unfortunately, the static persisted, so the ham never did find out what happened at the end of this radio play!

Q129 (concluded). Not enough information yet. Harry would need to know either (*a*) the S.D. of the average for a zM Test, or (*b*) 3 measurements with the new vacuum for a 'Student's' t Test.

P.S. Peggy did get a new vacuum cleaner, but not by the application of Statistics; *vide* Aristophanes' *Lysistrata*.

SQUARES

	0	1	2	3	4	5	6	7	8	9
1·0	1·000	1·020	1·040	1·061	1·082	1·102$_5$	1·124	1·145	1·166	1·188
1·1	1·210	1·232	1·254	1·277	1·300	1·322$_5$	1·346	1·369	1·392	1·416
1·2	1·440	1·464	1·488	1·513	1·538	1·562$_5$	1·588	1·613	1·638	1·664
1·3	1·690	1·716	1·742	1·769	1·796	1·822$_5$	1·850	1·877	1·904	1·932
1·4	1·960	1·988	2·016	2·045	2·074	2·102$_5$	2·132	2·161	2·190	2·220
1·5	2·250	2·280	2·310	2·341	2·372	2·402$_5$	2·434	2·465	2·496	2·528
1·6	2·560	2·592	2·624	2·657	2·690	2·722$_5$	2·756	2·789	2·822	2·856
1·7	2·890	2·924	2·958	2·993	3·028	3·062$_5$	3·098	3·133	3·168	3·204
1·8	3·240	3·276	3·312	3·349	3·386	3·422$_5$	3·460	3·497	3·534	3·572
1·9	3·610	3·648	3·686	3·725	3·764	3·802$_5$	3·842	3·881	3·920	3·960
2·0	4·000	4·040	4·080	4·121	4·162	4·202$_5$	4·244	4·285	4·326	4·368
2·1	4·410	4·452	4·494	4·537	4·580	4·622$_5$	4·666	4·709	4·752	4·796
2·2	4·840	4·884	4·928	4·973	5·018	5·062$_5$	5·108	5·153	5·198	5·244
2·3	5·290	5·336	5·382	5·429	5·476	5·522$_5$	5·570	5·617	5·664	5·712
2·4	5·760	5·808	5·856	5·905	5·954	6·002$_5$	6·052	6·101	6·150	6·200
2·5	6·250	6·300	6·350	6·401	6·452	6·502$_5$	6·554	6·605	6·656	6·708
2·6	6·760	6·812	6·864	6·917	6·970	7·022$_5$	7·076	7·129	7·182	7·236
2·7	7·290	7·344	7·398	7·453	7·508	7·562$_5$	7·618	7·673	7·728	7·784
2·8	7·840	7·896	7·952	8·009	8·066	8·122$_5$	8·180	8·237	8·294	8·352
2·9	8·410	8·468	8·526	8·585	8·644	8·702$_5$	8·762	8·821	8·880	8·940
3·0	9·000	9·060	9·120	9·181	9·242	9·302$_5$	9·364	9·425	9·486	9·548
3·1	9·610	9·672	9·734	9·797	9·860	9·922$_5$	9·986	10·05	10·11	10·18
3·2	10·24	10·30	10·37	10·43	10·50	10·56	10·63	10·69	10·76	10·82
3·3	10·89	10·96	11·02	11·09	11·16	11·22	11·29	11·36	11·42	11·49
3·4	11·56	11·63	11·70	11·76	11·83	11·90	11·97	12·04	12·11	12·18
3·5	12·25	12·32	12·39	12·46	12·53	12·60	12·67	12·74	12·82	12·89
3·6	12·96	13·03	13·10	13·18	13·25	13·32	13·40	13·47	13·54	13·62
3·7	13·69	13·76	13·84	13·91	13·99	14·06	14·14	14·21	14·29	14·36
3·8	14·44	14·52	14·59	14·67	14·75	14·82	14·90	14·98	15·05	15·13
3·9	15·21	15·29	15·37	15·44	15·52	15·60	15·68	15·76	15·84	15·92
4·0	16·00	16·08	16·16	16·24	16·32	16·40	16·48	16·56	16·65	16·73
4·1	16·81	16·89	16·97	17·06	17·14	17·22	17·31	17·39	17·47	17·56
4·2	17·64	17·72	17·81	17·89	17·98	18·06	18·15	18·23	18·32	18·40
4·3	18·49	18·58	18·66	18·75	18·84	18·92	19·01	19·10	19·18	19·27
4·4	19·36	19·45	19·54	19·62	19·71	19·80	19·89	19·98	20·07	20·16
4·5	20·25	20·34	20·43	20·52	20·61	20·70	20·79	20·88	20·98	21·07
4·6	21·16	21·25	21·34	21·44	21·53	21·62	21·72	21·81	21·90	22·00
4·7	22·09	22·18	22·28	22·37	22·47	22·56	22·66	22·75	22·85	22·94
4·8	23·04	23·14	23·23	23·33	23·43	23·52	23·62	23·72	23·81	23·91
4·9	24·01	24·11	24·21	24·30	24·40	24·50	24·60	24·70	24·80	24·90
5·0	25·00	25·10	25·20	25·30	25·40	25·50	25·60	25·70	25·81	25·91
5·1	26·01	26·11	26·21	26·32	26·42	26·52	26·63	26·73	26·83	26·94
5·2	27·04	27·14	27·25	27·35	27·46	27·56	27·67	27·77	27·88	27·98
5·3	28·09	28·20	28·30	28·41	28·52	28·62	28·73	28·84	28·94	29·05
5·4	29·16	29·27	29·38	29·48	29·59	29·70	29·81	29·92	30·03	30·14

Numbers at top of columns are second decimal place numbers. Thus $1{\cdot}40^2 = 1{\cdot}960$; $1{\cdot}42^2 = 2{\cdot}016$

($\tfrac{1}{10}$ these numbers)^2 = $\tfrac{1}{100}$ answers shown. E.g. $0{\cdot}14^2 = 0{\cdot}0196$

($10 \times$ these numbers)^2 = $100 \times$ these answers. E.g. $14^2 = 196{\cdot}0$

SQUARES – continued

	0	1	2	3	4	5	6	7	8	9
5·5	30·25	30·36	30·47	30·58	30·69	30·80	30·91	31·02	31·14	31·25
5·6	31·36	31·47	31·58	31·70	31·81	31·92	32·04	32·15	32·26	32·38
5·7	32·49	32·60	32·72	32·83	32·95	33·06	33·18	33·29	33·41	33·52
5·8	33·64	33·76	33·87	33·99	34·11	34·22	34·34	34·46	34·57	34·69
5·9	34·81	34·93	35·05	35·16	35·28	35·40	35·52	35·64	35·76	35·88
6·0	36·00	36·12	36·24	36·36	36·48	36·60	36·72	36·84	36·97	37·09
6·1	37·21	37·33	37·45	37·58	37·70	37·82	37·95	38·07	38·19	38·32
6·2	38·44	38·56	38·69	38·81	38·94	39·06	39·19	39·31	39·44	39·56
6·3	39·69	39·82	39·94	40·07	40·20	40·32	40·45	40·58	40·70	40·83
6·4	40·96	41·09	41·22	41·34	41·47	41·60	41·73	41·86	41·99	42·12
6·5	42·25	42·38	42·51	42·64	42·77	42·90	43·03	43·16	43·30	43·43
6·6	43·56	43·69	43·82	43·96	44·09	44·22	44·36	44·49	44·62	44·76
6·7	44·89	45·02	45·16	45·29	45·43	45·56	45·70	45·83	45·97	46·10
6·8	46·24	46·38	46·51	46·65	46·79	46·92	47·06	47·20	47·33	47·47
6·9	47·61	47·75	47·89	48·02	48·16	48·30	48·44	48·58	48·72	48·86
7·0	49·00	49·14	49·28	49·42	49·56	49·70	49·84	49·98	50·13	50·27
7·1	50·41	50·55	50·69	50·84	50·98	51·12	51·27	51·41	51·55	51·70
7·2	51·84	51·98	52·13	52·27	52·42	52·56	52·71	52·85	53·00	53·14
7·3	53·29	53·44	53·58	53·73	53·88	54·02	54·17	54·32	54·46	54·61
7·4	54·76	54·91	55·06	55·20	55·35	55·50	55·65	55·80	55·95	56·10
7·5	56·25	56·40	56·55	56·70	56·85	57·00	57·15	57·30	57·46	57·61
7·6	57·76	57·91	58·06	58·22	58·37	58·52	58·68	58·83	58·98	59·14
7·7	59·29	59·44	59·60	59·75	59·91	60·06	60·22	60·37	60·53	60·68
7·8	60·84	61·00	61·15	61·31	61·47	61·62	61·78	61·94	62·09	62·25
7·9	62·41	62·57	62·73	62·88	63·04	63·20	63·36	63·52	63·68	63·84
8·0	64·00	64·16	64·32	64·48	64·64	64·80	64·96	65·12	65·29	65·45
8·1	65·61	65·77	65·93	66·10	66·26	66·42	66·59	66·75	66·91	67·08
8·2	67·24	67·40	67·57	67·73	67·90	68·06	68·23	68·39	68·56	68·72
8·3	68·89	69·06	69·22	69·39	69·56	69·72	69·89	70·06	70·22	70·39
8·4	70·56	70·73	70·90	71·06	71·23	71·40	71·57	71·74	71·91	72·08
8·5	72·25	72·42	72·59	72·76	72·93	73·10	73·27	73·44	73·62	73·79
8·6	73·96	74·13	74·30	74·48	74·65	74·82	75·00	75·17	75·34	75·52
8·7	75·69	75·86	76·04	76·21	76·39	76·56	76·74	76·91	77·09	77·26
8·8	77·44	77·62	77·79	77·97	78·15	78·32	78·50	78·68	78·85	79·03
8·9	79·21	79·39	79·57	79·74	79·92	80·10	80·28	80·46	80·64	80·82
9·0	81·00	81·18	81·36	81·54	81·72	81·90	82·08	82·26	82·45	82·63
9·1	82·81	82·99	83·17	83·36	83·54	83·72	83·91	84·09	84·27	84·46
9·2	84·64	84·82	85·01	85·19	85·38	85·56	85·75	85·93	86·12	86·30
9·3	86·49	86·68	86·86	87·05	87·24	87·42	87·61	87·80	87·98	88·17
9·4	88·36	88·55	88·74	88·92	89·11	89·30	89·49	89·68	89·87	90·06
9·5	90·25	90·44	90·63	90·82	91·01	91·20	91·39	91·58	91·78	91·97
9·6	92·16	92·35	92·54	92·74	92·93	93·12	93·32	93·51	93·70	93·90
9·7	94·09	94·28	94·48	94·67	94·87	95·06	95·26	95·45	95·65	95·84
9·8	96·04	96·24	96·43	96·63	96·83	97·02	97·22	97·42	97·61	97·81
9·9	98·01	98·21	98·41	98·60	98·80	99·00	99·20	99·40	99·60	99·80

These Tables have been adapted from R. A. Fisher and F. Yates, *Statistical Tables for Biological, Agricultural and Medical Research*, Table 27 (Oliver & Boyd, 1963), by permission of the authors and publishers.

SQUARE ROOTS

	0	1	2	3	4	5	6	7	8	9
1·0	1·00	1·00	1·01	1·01	1·02	1·02	1·03	1·03	1·04	1·04
10·	3·16	3·18	3·19	3·21	3·22	3·24	3·26	3·27	3·29	3·30
1·1	1·05	1·05	1·06	1·06	1·07	1·07	1·08	1·08	1·09	1·09
11·	3·32	3·33	3·35	3·36	3·38	3·39	3·41	3·42	3·44	3·45
1·2	1·10	1·10	1·10	1·11	1·11	1·12	1·12	1·13	1·13	1·14
12·	3·46	3·48	3·49	3·51	3·52	3·54	3·55	3·56	3·58	3·59
1·3	1·14	1·14	1·15	1·15	1·16	1·16	1·17	1·17	1·17	1·18
13·	3·61	3·62	3·63	3·65	3·66	3·67	3·69	3·70	3·71	3·73
1·4	1·18	1·19	1·19	1·20	1·20	1·20	1·21	1·21	1·22	1·22
14·	3·74	3·76	3·77	3·78	3·79	3·81	3·82	3·83	3·85	3·86
1·5	1·22	1·23	1·23	1·24	1·24	1·24	1·25	1·25	1·26	1·26
15·	3·87	3·89	3·90	3·91	3·92	3·94	3·95	3·96	3·97	3·99
1·6	1·26	1·27	1·27	1·28	1·28	1·28	1·29	1·29	1·30	1·30
16·	4·00	4·01	4·02	4·04	4·05	4·06	4·07	4·09	4·10	4·11
1·7	1·30	1·31	1·31	1·32	1·32	1·32	1·33	1·33	1·33	1·34
17·	4·12	4·14	4·15	4·16	4·17	4·18	4·20	4·21	4·22	4·23
1·8	1·34	1·35	1·35	1·35	1·36	1·36	1·36	1·37	1·37	1·37
18·	4·24	4·25	4·27	4·28	4·29	4·30	4·31	4·32	4·34	4·35
1·9	1·38	1·38	1·39	1·39	1·39	1·40	1·40	1·40	1·41	1·41
19·	4·36	4·37	4·38	4·39	4·40	4·42	4·43	4·44	4·45	4·46
2·0	1·41	1·42	1·42	1·42	1·43	1·43	1·44	1·44	1·44	1·45
20·	4·47	4·48	4·49	4·51	4·52	4·53	4·54	4·55	4·56	4·57
2·1	1·45	1·45	1·46	1·46	1·46	1·47	1·47	1·47	1·48	1·48
21·	4·58	4·59	4·60	4·62	4·63	4·64	4·65	4·66	4·67	4·68
2·2	1·48	1·49	1·49	1·49	1·50	1·50	1·50	1·51	1·51	1·51
22·	4·69	4·70	4·71	4·72	4·73	4·74	4·75	4·76	4·77	4·79
2·3	1·52	1·52	1·52	1·53	1·53	1·53	1·54	1·54	1·54	1·55
23·	4·80	4·81	4·82	4·83	4·84	4·85	4·86	4·87	4·88	4·89
2·4	1·55	1·55	1·56	1·56	1·56	1·57	1·57	1·57	1·57	1·58
24·	4·90	4·91	4·92	4·93	4·94	4·95	4·96	4·97	4·98	4·99
2·5	1·58	1·58	1·59	1·59	1·59	1·60	1·60	1·60	1·61	1·61
25·	5·00	5·01	5·02	5·03	5·04	5·05	5·06	5·07	5·08	5·09
2·6	1·61	1·62	1·62	1·62	1·62	1·63	1·63	1·63	1·64	1·64
26·	5·10	5·11	5·12	5·13	5·14	5·15	5·16	5·17	5·18	5·19
2·7	1·64	1·65	1·65	1·65	1·66	1·66	1·66	1·66	1·67	1·67
27·	5·20	5·21	5·22	5·22	5·23	5·24	5·25	5·26	5·27	5·28
2·8	1·67	1·68	1·68	1·68	1·69	1·69	1·69	1·69	1·70	1·70
28·	5·29	5·30	5·31	5·32	5·33	5·34	5·35	5·36	5·37	5·38
2·9	1·70	1·71	1·71	1·71	1·71	1·72	1·72	1·72	1·73	1·73
29·	5·39	5·39	5·40	5·41	5·42	5·43	5·44	5·45	5·46	5·47
3·0	1·73	1·73	1·74	1·74	1·74	1·75	1·75	1·75	1·75	1·76
30·	5·48	5·49	5·50	5·50	5·51	5·52	5·53	5·54	5·55	5·56

Numbers at top of columns are additional decimal place numbers.
Thus $\sqrt{2\cdot90} = 1\cdot70$; $\sqrt{2\cdot92} = 1\cdot71$; $\sqrt{29\cdot0} = 5\cdot39$; $\sqrt{29\cdot2} = 5\cdot40$
$\sqrt{\frac{1}{10}\text{ smaller numbers}} = \frac{1}{10}$ larger answer. E.g. $\sqrt{0\cdot29} = 0\cdot539$
$\sqrt{10 \times \text{ larger numbers}} = 10 \times$ smaller answer. E.g. $\sqrt{290} = 17\cdot0$
$\sqrt{100 \times \text{ larger numbers}} = 10 \times$ larger answer. E.g. $\sqrt{2920} = 54\cdot0$

SQUARE ROOTS – *continued*

	0	1	2	3	4	5	6	7	8	9
3·1	1·76	1·76	1·77	1·77	1·77	1·77	1·78	1·78	1·78	1·79
31·	5·57	5·58	5·59	5·59	5·60	5·61	5·62	5·63	5·64	5·65
3·2	1·79	1·79	1·79	1·80	1·80	1·80	1·81	1·81	1·81	1·81
32·	5·66	5·67	5·67	5·68	5·69	5·70	5·71	5·72	5·73	5·74
3·3	1·82	1·82	1·82	1·82	1·83	1·83	1·83	1·84	1·84	1·84
33·	5·74	5·75	5·76	5·77	5·78	5·79	5·80	5·81	5·81	5·82
3·4	1·84	1·85	1·85	1·85	1·85	1·86	1·86	1·86	1·87	1·87
34·	5·83	5·84	5·85	5·86	5·87	5·87	5·88	5·89	5·90	5·91
3·5	1·87	1·87	1·88	1·88	1·88	1·88	1·89	1·89	1·89	1·89
35·	5·92	5·92	5·93	5·94	5·95	5·96	5·97	5·97	5·98	5·99
3·6	1·90	1·90	1·90	1·91	1·91	1·91	1·91	1·92	1·92	1·92
36·	6·00	6·01	6·02	6·02	6·03	6·04	6·05	6·06	6·07	6·07
3·7	1·92	1·93	1·93	1·93	1·93	1·94	1·94	1·94	1·94	1·95
37·	6·08	6·09	6·10	6·11	6·12	6·12	6·13	6·14	6·15	6·16
3·8	1·95	1·95	1·95	1·96	1·96	1·96	1·96	1·97	1·97	1·97
38·	6·16	6·17	6·18	6·19	6·20	6·20	6·21	6·22	6·23	6·24
3·9	1·97	1·98	1·98	1·98	1·98	1·99	1·99	1·99	1·99	2·00
39·	6·24	6·25	6·26	6·27	6·28	6·28	6·29	6·30	6·31	6·32
4·0	2·00	2·00	2·00	2·01	2·01	2·01	2·01	2·02	2·02	2·02
40·	6·32	6·33	6·34	6·35	6·36	6·36	6·37	6·38	6·39	6·40
4·1	2·02	2·03	2·03	2·03	2·03	2·04	2·04	2·04	2·04	2·05
41·	6·40	6·41	6·42	6·43	6·43	6·44	6·45	6·46	6·47	6·47
4·2	2·05	2·05	2·05	2·06	2·06	2·06	2·06	2·07	2·07	2·07
42·	6·48	6·49	6·50	6·50	6·51	6·52	6·53	6·53	6·54	6·55
4·3	2·07	2·08	2·08	2·08	2·08	2·09	2·09	2·09	2·09	2·10
43·	6·56	6·57	6·57	6·58	6·59	6·60	6·60	6·61	6·62	6·63
4·4	2·10	2·10	2·10	2·10	2·11	2·11	2·11	2·11	2·12	2·12
44·	6·63	6·64	6·65	6·66	6·66	6·67	6·68	6·69	6·69	6·70
4·5	2·12	2·12	2·13	2·13	2·13	2·13	2·14	2·14	2·14	2·14
45·	6·71	6·72	6·72	6·73	6·74	6·75	6·75	6·76	6·77	6·77
4·6	2·14	2·15	2·15	2·15	2·15	2·16	2·16	2·16	2·16	2·17
46·	6·78	6·79	6·80	6·80	6·81	6·82	6·83	6·83	6·84	6·85
4·7	2·17	2·17	2·17	2·17	2·18	2·18	2·18	2·18	2·19	2·19
47·	6·86	6·86	6·87	6·88	6·88	6·89	6·90	6·91	6·91	6·92
4·8	2·19	2·19	2·20	2·20	2·20	2·20	2·20	2·21	2·21	2·21
48·	6·93	6·94	6·94	6·95	6·96	6·96	6·97	6·98	6·99	6·99
4·9	2·21	2·22	2·22	2·22	2·22	2·22	2·23	2·23	2·23	2·23
49·	7·00	7·01	7·01	7·02	7·03	7·04	7·04	7·05	7·06	7·06
5·0	2·24	2·24	2·24	2·24	2·25	2·25	2·25	2·25	2·25	2·26
50·	7·07	7·08	7·09	7·09	7·10	7·11	7·11	7·12	7·13	7·13
5·1	2·26	2·26	2·26	2·26	2·27	2·27	2·27	2·27	2·28	2·28
51·	7·14	7·15	7·16	7·16	7·17	7·18	7·18	7·19	7·20	7·20
5·2	2·28	2·28	2·28	2·29	2·29	2·29	2·29	2·30	2·30	2·30
52·	7·21	7·22	7·22	7·23	7·24	7·25	7·25	7·26	7·27	7·27
5·3	2·30	2·30	2·31	2·31	2·31	2·31	2·32	2·32	2·32	2·32
53·	7·28	7·29	7·29	7·30	7·31	7·31	7·32	7·33	7·33	7·34
5·4	2·32	2·33	2·33	2·33	2·33	2·33	2·34	2·34	2·34	2·34
54·	7·35	7·36	7·36	7·37	7·38	7·38	7·39	7·40	7·40	7·41

SQUARE ROOTS – *continued*

	0	1	2	3	4	5	6	7	8	9
5·5 55·	2·35 7·42	2·35 7·42	2·35 7·43	2·35 7·44	2·35 7·44	2·36 7·45	2·36 7·46	2·36 7·46	2·36 7·47	2·36 7·48
5·6 56·	2·37 7·48	2·37 7·49	2·37 7·50	2·37 7·50	2·37 7·51	2·38 7·52	2·38 7·52	2·38 7·53	2·38 7·54	2·39 7·54
5·7 57·	2·39 7·55	2·39 7·56	2·39 7·56	2·39 7·57	2·40 7·58	2·40 7·58	2·40 7·59	2·40 7·60	2·40 7·60	2·41 7·61
5·8 58·	2·41 7·62	2·41 7·62	2·41 7·63	2·41 7·64	2·42 7·64	2·42 7·65	2·42 7·66	2·42 7·66	2·42 7·67	2·43 7·67
5·9 59·	2·43 7·68	2·43 7·69	2·43 7·69	2·44 7·70	2·44 7·71	2·44 7·71	2·44 7·72	2·44 7·73	2·45 7·73	2·45 7·74
6·0 60·	2·45 7·75	2·45 7·75	2·45 7·76	2·46 7·77	2·46 7·77	2·46 7·78	2·46 7·78	2·46 7·79	2·47 7·80	2·47 7·80
6·1 61·	2·47 7·81	2·47 7·82	2·47 7·82	2·48 7·83	2·48 7·84	2·48 7·84	2·48 7·85	2·48 7·85	2·49 7·86	2·49 7·87
6·2 62·	2·49 7·87	2·49 7·88	2·49 7·89	2·50 7·89	2·50 7·90	2·50 7·91	2·50 7·91	2·50 7·92	2·51 7·92	2·51 7·93
6·3 63	2·51 7·94	2·51 7·94	2·51 7·95	2·52 7·96	2·52 7·96	2·52 7·97	2·52 7·97	2·52 7·98	2·53 7·99	2·53 7·99
6·4 64·	2·53 8·00	2·53 8·01	2·53 8·01	2·54 8·02	2·54 8·02	2·54 8·03	2·54 8·04	2·54 8·04	2·55 8·05	2·55 8·06
6·5 65·	2·55 8·06	2·55 8·07	2·55 8·07	2·56 8·08	2·56 8·09	2·56 8·09	2·56 8·10	2·56 8·11	2·57 8·11	2·57 8·12
6·6 66·	2·57 8·12	2·57 8·13	2·57 8·14	2·57 8·14	2·58 8·15	2·58 8·15	2·58 8·16	2·58 8·17	2·58 8·17	2·59 8·18
6·7 67·	2·59 8·19	2·59 8·19	2·59 8·20	2·59 8·20	2·60 8·21	2·60 8·22	2·60 8·22	2·60 8·23	2·60 8·23	2·61 8·24
6·8 68·	2·61 8·25	2·61 8·25	2·61 8·26	2·61 8·26	2·62 8·27	2·62 8·28	2·62 8·28	2·62 8·29	2·62 8·29	2·62 8·30
6·9 69·	2·63 8·31	2·63 8·31	2·63 8·32	2·63 8·32	2·63 8·33	2·64 8·34	2·64 8·34	2·64 8·35	2·64 8·35	2·64 8·36
7·0 70·	2·65 8·37	2·65 8·37	2·65 8·38	2·65 8·38	2·65 8·39	2·66 8·40	2·66 8·40	2·66 8·41	2·66 8·41	2·66 8·42
7·1 71·	2·66 8·43	2·67 8·43	2·67 8·44	2·67 8·44	2·67 8·45	2·67 8·46	2·68 8·46	2·68 8·47	2·68 8·47	2·68 8·48
7·2 72·	2·68 8·49	2·69 8·49	2·69 8·50	2·69 8·50	2·69 8·51	2·69 8·51	2·69 8·52	2·70 8·53	2·70 8·53	2·70 8·54
7·3 73·	2·70 8·54	2·70 8·55	2·71 8·56	2·71 8·56	2·71 8·57	2·71 8·57	2·71 8·58	2·71 8·58	2·72 8·59	2·72 8·60
7·4 74·	2·72 8·60	2·72 8·61	2·72 8·61	2·73 8·62	2·73 8·63	2·73 8·63	2·73 8·64	2·73 8·64	2·73 8·65	2·74 8·65
7·5 75·	2·74 8·66	2·74 8·67	2·74 8·67	2·74 8·68	2·75 8·68	2·75 8·69	2·75 8·69	2·75 8·70	2·75 8·71	2·76 8·71
7·6 76·	2·76 8·72	2·76 8·72	2·76 8·73	2·76 8·73	2·76 8·74	2·77 8·75	2·77 8·75	2·77 8·76	2·77 8·76	2·77 8·77

Numbers at top of columns are additional decimal place numbers.
Thus $\sqrt{7·40} = 2·72$; $\sqrt{7·43} = 2·73$; $\sqrt{74·0} = 8·60$; $\sqrt{74·3} = 8·62$
$\sqrt{\frac{1}{10}}$ smaller numbers $= \frac{1}{10}$ larger answer. E.g. $\sqrt{0·74} = 0·860$
$\sqrt{10 \times \text{larger numbers}} = 10 \times$ smaller answer. E.g. $\sqrt{740} = 27·2$
$\sqrt{100 \times \text{larger numbers}} = 10 \times$ larger answer. E.g. $\sqrt{7400} = 86·0$

SQUARE ROOTS – *continued*

	0	1	2	3	4	5	6	7	8	9
7·7	2·77	2·78	2·78	2·78	2·78	2·78	2·79	2·79	2·79	2·79
77·	8·77	8·78	8·79	8·79	8·80	8·80	8·81	8·81	8·82	8·83
7·8	2·79	2·79	2·80	2·80	2·80	2·80	2·80	2·81	2·81	2·81
78·	8·83	8·84	8·84	8·85	8·85	8·86	8·87	8·87	8·88	8·88
7·9	2·81	2·81	2·81	2·82	2·82	2·82	2·82	2·82	2·82	2·83
79·	8·89	8·89	8·90	8·91	8·91	8·92	8·92	8·93	8·93	8·94
8·0	2·83	2·83	2·83	2·83	2·84	2·84	2·84	2·84	2·84	2·84
80·	8·94	8·95	8·96	8·96	8·97	8·97	8·98	8·98	8·99	8·99
8·1	2·85	2·85	2·85	2·85	2·85	2·85	2·86	2·86	2·86	2·86
81·	9·00	9·01	9·01	9·02	9·02	9·03	9·03	9·04	9·04	9·05
8·2	2·86	2·87	2·87	2·87	2·87	2·87	2·87	2·88	2·88	2·88
82·	9·06	9·06	9·07	9·07	9·08	9·08	9·09	9·09	9·10	9·10
8·3	2·88	2·88	2·88	2·89	2·89	2·89	2·89	2·89	2·89	2·90
83·	9·11	9·12	9·12	9·13	9·13	9·14	9·14	9·15	9·15	9·16
8·4	2·90	2·90	2·90	2·90	2·91	2·91	2·91	2·91	2·91	2·91
84·	9·17	9·17	9·18	9·18	9·19	9·19	9·20	9·20	9·21	9·21
8·5	2·92	2·92	2·92	2·92	2·92	2·92	2·93	2·93	2·93	2·93
85·	9·22	9·22	9·23	9·24	9·24	9·25	9·25	9·26	9·26	9·27
8·6	2·93	2·93	2·94	2·94	2·94	2·94	2·94	2·94	2·95	2·95
86·	9·27	9·28	9·28	9·29	9·30	9·30	9·31	9·31	9·32	9·32
8·7	2·95	2·95	2·95	2·95	2·96	2·96	2·96	2·96	2·96	2·96
87·	9·33	9·33	9·34	9·34	9·35	9·35	9·36	9·36	9·37	9·38
8 8	2·97	2·97	2·97	2·97	2·97	2·97	2·98	2·98	2·98	2·98
88·	9·38	9·39	9·39	9·40	9·40	9·41	9·41	9·42	9·42	9·43
8·9	2·98	2·98	2·99	2·99	2·99	2·99	2·99	3·00	3·00	3·00
89·	9·43	9·44	9·44	9·45	9·46	9·46	9·47	9·47	9·48	9·48
9·0	3·00	3·00	3·00	3·00	3·01	3·01	3·01	3·01	3·01	3·01
90·	9·49	9·49	9·50	9·50	9·51	9·51	9·52	9·52	9·53	9·53
9·1	3·02	3·02	3·02	3·02	3·02	3·02	3·03	3·03	3·03	3·03
91·	9·54	9·54	9·55	9·56	9·56	9·57	9·57	9·58	9·58	9·59
9·2	3·03	3·03	3·04	3·04	3·04	3·04	3·04	3·04	3·05	3·05
92·	9·59	9·60	9·60	9·61	9·61	9·62	9·62	9·63	9·63	9·64
9·3	3·05	3·05	3·05	3·05	3·06	3·06	3·06	3·06	3·06	3·06
93·	9·64	9·65	9·65	9·66	9·66	9·67	9·67	9·68	9·69	9·69
9·4	3·07	3·07	3·07	3·07	3·07	3·07	3·08	3·08	3·08	3·08
94·	9·70	9·70	9·71	9·71	9·72	9·72	9·73	9·73	9·74	9·74
9·5	3·08	3·08	3·09	3·09	3·09	3·09	3·09	3·09	3·10	3·10
95·	9·75	9·75	9·76	9·76	9·77	9·77	9·78	9·78	9·79	9·79
9·6	3·10	3·10	3·10	3·10	3·10	3·11	3·11	3·11	3·11	3·11
96·	9·80	9·80	9·81	9·81	9·82	9·82	9·83	9·83	9·84	9·84
9·7	3·11	3·12	3·12	3·12	3·12	3·12	3·12	3·13	3·13	3·13
97·	9·85	9·85	9·86	9·86	9·87	9·87	9·88	9·88	9·89	9·89
9·8	3·13	3·13	3·13	3·14	3·14	3·14	3·14	3·14	3·14	3·14
98·	9·90	9·90	9·91	9·91	9·92	9·92	9·93	9·93	9·94	9·94
9·9	3·15	3·15	3·15	3·15	3·15	3·15	3·16	3·16	3·16	3·16
99·	9·95	9·95	9·96	9·95	9·97	9·97	9·98	9·98	9·99	9·99

These Tables have been adapted from R. A. Fisher and F. Yates, *Statistical Tables for Biological, Agricultural and Medical Research*, Table 28 (Oliver & Boyd, 1963), by permission of the authors and publishers.

GUIDE TO SIGNIFICANCE TESTS

For Measurements and Ranks

1 Sample and an Average of a Large Group

S.D. of the large group is –
(*a*) known: zM Test; p. 152.
(*b*) not known, and the sample consists of 3 or more measurements: 'Student's' t Test, p. 160.

2 Sets of Measurements

(*a*) Unmatched –
(i) simple samples: Wilcoxon's Sum of Ranks Test, p. 166.
(ii) stratified samples: Wilcoxon's Stratified Test, p. 190.

(*b*) Matched –
(i) for difference between measurements of 1 characteristic: Wilcoxon's Signed Ranks Test, p. 179.
(ii) for correlation between 2 characteristics: Spearman's Correlation Test, p. 199.

3 or More Sets of Measurements

(*a*) Unmatched: Kruskal and Wallis' Test, p. 212.
(*b*) Matched: Friedman's Test, p. 222.

For Isolated Occurrences

1 Sample and an Average

Number in sample is –
(*a*) up to 40: Poisson's Test, p. 230.
(*b*) over 40: zI Test, p. 245.

2 Samples

Expected numbers are –
(*a*) the same in both samples: 50% Probability Test, p. 254.
(*b*) different in each sample, and observed no. in smaller sample is –
(i) up to 4: Binomial Test, p. 237.
(ii) 5 to 20 –
If $x > E$: Binomial Test, p. 237.
If $x < E$: zI Test, p. 245.
(iii) over 20: zI Test, p. 245.

3 or More Samples: χ^2 Test, p. 269.

For Binomial Observations (i.e. divided into 2 classes)

1 Sample and an Average

(a) Average is 50%: 50% Probability Test, p. 254.

(b) Av. of smaller class is less than 10%, and no. in smaller class is –

(i) up to 40: Poisson's Test, p. 230.

(ii) over 40: zI Test, p. 245.

(c) Av. of smaller class is 10% to 49%, and no. in smaller class is –

(i) up to 4: Binomial Test, p. 237.

(ii) 5 to 20 –

If $x > E$: Binomial Test, p. 237.

If $x < E$: zI Test, p. 245.

(iii) over 20: zI Test, p. 245.

2 Sets of Observations

(a) Unmatched, and number in combined samples is –

(i) up to 50: Fisher's Test, p. 292.

(ii) over 50: Yates' χ^2 Test, p. 285.

(b) Matched –

(i) for diff. between observations: 50% Probability Test, p. 254.

(ii) for association between 2 qualities, and sample size is –

Up to 50: Fisher's Test, p. 292.

Over 50: Yates' χ^2 Test, p. 285.

3 or More Sets of Observations

(a) Unmatched: χ^2 Test, p. 269.

(b) Matched: Cochran's Test, p. 318.

For Observations Divided into 3 or More Classes

1 Sample and an Average: χ^2 Test, p. 269.

2 Sets of Observations

(a) Unmatched: χ^2 Test, p. 269.

(b) Matched –

(i) for diff. between observations: 50% Probability Test, p. 254.

(ii) for association between 2 qualities: χ^2 Test, p. 269.

3 or More Sets of Observations

Unmatched: χ^2 Test, p. 269.

Testing for Randomness or Clustering

Runs Test, p. 322.